ALSO BY GARRETT HONGO

Poetry

Coral Road
The River of Heaven
Yellow Light

Nonfiction

The Mirror Diary: Selected Essays
Volcano: A Memoir of Hawai'i

Anthologies

The Open Boat: Poetry from Asian America
Under Western Eyes: Personal Essays from Asian America

As Editor

Songs My Mother Taught Me: Stories, Plays, and Memoir
by Wakako Yamauchi

The Perfect Sound

THE
PERFECT
SOUND

A Memoir in Stereo

Garrett Hongo

PANTHEON BOOKS
New York

All rights reserved. Published in the United States by Pantheon Books, a division of Penguin Random House LLC, New York, and distributed in Canada by Penguin Random House Canada Limited, Toronto.

Pantheon Books and colophon are registered trademarks of Penguin Random House LLC.

Some of the material in this book originally appeared, in very different form, in *The Georgia Review*, the *Los Angeles Review of Books*, *The Massachusetts Review*, *Plume*, *SoundStage! Global*, *SoundStage! Traveler*, and *SoundStage! Ultra*.

Grateful acknowledgment is made to the following for permission to reprint previously published material:
The Estate of Cid Corman: Poem #4 by Cid Corman, originally appeared as part of "thirty-one poems & an interview by Gregory Dunne: A Special AP2 Supplement" from *The American Poetry Review* (July/August 2000). Reprinted by permission of Bob Arnold, Literary Executor for the Estate of Cid Corman.
Farrar, Straus and Giroux: Excerpts from "Dog Day Vespers" from *Oblivion Banjo: The Poetry of Charles Wright* by Charles Wright. Copyright © 2019 by Charles Wright. Reprinted by permission of Farrar, Straus and Giroux. All Rights Reserved. Poem originally published in the volume *The Southern Cross*.
University of Pittsburgh Press: Excerpt from "The Idea of Ancestry" from *The Essential Etheridge Knight* by Etheridge Knight, copyright © 1986 by Etheridge Knight. Reprinted by permission of University of Pittsburgh Press.

Library of Congress Cataloging-in-Publication Data
Name: Hongo, Garrett Kaoru, [date] author.
Title: The perfect sound: a memoir in stereo / Garrett Hongo.
Description: First edition. New York: Pantheon Books, 2022
Identifiers: LCCN 2021027399 (print) | LCCN 2021027400 (ebook) |
ISBN 9780375425066 (hardcover) | ISBN 9780593316429 (ebook)
Subjects: LCSH: Hongo, Garrett Kaoru, [date]. Poets, American—20th century—Biography. High-fidelity sound systems. Japanese Americans—Biography.
Classification: LCC PS3558.O48 Z46 2022 (print) | LCC PS3558.O48 (ebook) |
DDC 814/.54 [B]—dc23
LC record available at https://lccn.loc.gov/2021027399
LC ebook record available at https://lccn.loc.gov/2021027400

www.pantheonbooks.com

Jacket design by John Gall

Printed in the United States of America
First Edition
2 4 6 8 9 7 5 3 1

To the memory of Albert Kazuyoshi Hongo

Ke ʻala o ka hala
Hala o mapuana

Music, when soft voices die,
Vibrates in the memory . . .

PERCY BYSSHE SHELLEY

Contents

Preludio   1

Part One   The Perfect Sound   11

Part Two   I Started Out on Stereo   55

Part Three Tubeworld, 1   127

Part Four   Tubeworld, 2   163

Part Five   It's My Life   209

Part Six   Wandering Rocks, 1   255

Part Seven The First Amplifiers   283

Part Eight   Talking Heads and Singing Platters   345

Part Nine   Wandering Rocks, 2   401

Part Ten   Among the Bohemians   433

Outro   495

Acknowledgments   519

Notes on Sources   521

Preludio

From here in my stereo room, in my day basement ten steps down from my entry hallway, much can seem perfect if I close my eyes and just listen. I've all my gear arranged in front of me, across the immaculate, midnight-blue Chinese carpet and against the acoustic-paneled wall opposite where I sit in my leather club chair—an acquisition from Pottery Barn during my middle-aged bachelor days. So much born of savings and sacrifice, but I hardly care, since the sound here is so gorgeous it lifts me out of things into a pure fabric of wonderment, adrift amidst all the sublime welter of notes. I start with piano music in the morning, Mozart or Beethoven concertos performed by the likes of Alfred Brendel or Claudio Arrau, their right-hand runs so liquidinous across the keyboard it's as though a clear water of crystalline singing were running over a streambed of orchestral accompaniment. I glance at the gleaming enameling on my speakers, brows-high, piano-black towers that mirror the inset ceiling lights when I switch them on, and I begin to want to flutter my hands like seabirds barely aloft over a dance line of shorebreak waves and indulge myself in this rapturous sequestration with music all around me.

During the days, mostly all my own because of my university teaching job (I teach classes once or twice a week), I switch among genres all the time, playing fifties combo jazz when I tire of classical, classic blues-rock when I want to stir things up (Cream, Santana, or the Allman Brothers), or even an opera if I have two hours together before an errand or a duty calls me away. I don't like interrupting the drama of an opera, its outsize grandiosity commanding a kind of narrowed attention, otherwise the spell of its fiction would collapse from the recognition of its absurdity. You live in its concocted, sauced-up emotions and situations as much as in its music, so you can't just break a spell once it's been cast. You have to give in to its supreme fiction, mad scenes, murders, and suicides,

its love-on-a-tubercular-shoestring plot, or else it dies as egregious caricature compared to the sedate, suburban lives most of us lead.

My equipment has been painstakingly selected and then frequently changed over the years as I've evolved in the audio hobby. Fifteen years ago, I started out simply wanting a good CD player, then was captivated once I heard what it could do, and off I went down the rabbit hole into this new universe, just listening to music all the time, every day, all the day long. For two or three years, I spent way too much time browsing audio mags and the Internet, lusting after new and better equipment, auditioning and acquiring it, and then quickly moving on to something else. Now my gear list is a little exotic, and, I'm willing to say, it has gotten completely ridiculous. I've a German-made turntable, weighing forty-five pounds and sporting two tonearms for which I have numerous phono cartridges, both mono and stereo. I've a Japanese SACD player (with the appropriate brand name "Esoteric") that weighs thirty-one pounds, has a sculpted frontal look, can play SACDs and also upsample Red Book CDs (a mere 16/44.1 kHz in resolution) through five additional PCM settings on up to DSD64. Got that? Through its USB inputs, it can even accept and convert digital files stored on an external hard drive and sent through a computer. It's three years old, perhaps twice a lifetime in digital electronics, but it's also very good, and I like it for its warmth and smoothness, especially with orchestral violins and operatic voices, which I've found the most challenging sounds for a digital source to reproduce well. Oh, I've an up-to-date Wi-Fi streaming device as well, Chinese-made in Beijing (bless the New World Order), that can capture digital signals transmitted wirelessly. The device has access to a half a universe's worth of files stored in the cloud somewhere by Apple, Tidal, Qobuz, or Spot-the-Fly. So you can all relax. Despite my fondness for vinyl, I've not been left behind as things have shifted over to computer- and network-sourced audio. My tubed preamp and tubed stereo power amp, made by a sedulous boutique designer in Osaka, a quiet though intense man who named his company, quite cleverly, after the millennium-old pronunciation for the ideograms of his sixteenth-century name, morphing the familiar into the recondite, transforming *Yamada* into *Zanden*. His gear has vac-

uum tubes all over it. The myth is that sound gets increasingly better as technologies improve, and maybe so, but a lot of us in this hobby just don't buy it. We've gone backward to old audio tubes no longer manufactured (we find them at estate sales and through commercial dealers who dig them up in military surplus dumps, warehouses in Eastern Europe, and from old collectors) and hand-wound electrical transformers, heavily lacquered wood-bodied phono cartridges, mono LPs from the fifties and sixties, and even to 78s from earlier than that. In a lovely, modern home amidst deciduous woods on Long Island near Dix Hills, where John Coltrane lived, I once heard brand-new, two-way speakers fashioned after old Western Electric movie-theater speakers from the 1930s. But my own speakers, I have to confess, derive from more recent German technologies and are fairly complicated affairs. I won't go into it now, but take my word, there's some meticulous engineering involved. They're each as big as a human being, weighing 265 pounds apiece. And from them I get a sound that, as Sam Cooke used to say, sends me.

My listening room takes up just over half of a twenty-eight-foot-by-twelve-foot space in the finished basement in an attractive three-bedroom, two-and-a-half bath trilevel, built in 1979, the deed told me—the first house on my particular street in the South Hills of Eugene, before the subdivision of McMansions on the top of the hill at the edge of town, before the other nondescript, thin-walled dwellings filled in all the cul-de-sacs and lots between us super-peons and the sub-one-percenters above. When the house was shown to me—by my real estate agent, a handsome and enthusiastic fifty-something fellow from Ashland (or Baja Oregon, as we call it)—my future stereo space was a TV room with a sofa and chairs and walls painted the color of blighted temple moss. But I immediately puffed my chest out and thought, *Here is my dedicated listening room, finally,* in my mind throwing up the IKEA cabinets for LPs that would divide it from the other end of the room, where I'd install my study and bookcases for my poetry library (Loeb classics, Everyman's editions of the Renaissance and Metaphysical poets, and Atheneum and Knopf contemporaries).

I am a poet. It's odd to say it, but that's what I've done for most of my adult life—study poetry, teach it at universities, and write it.

It's more than unusual, but it was such a passion of mine since my undergraduate days, when poets galore would visit my college, read from their thin but elegantly printed books, and captivate all of us with lyrics about a bucolic childhood on an Irish farm, working as a punch-press operator in a Detroit automobile factory, or sitting zazen in a Japanese Buddhist monastery. They gave such meaning to humble things and experiences, such homespun familiarity to the most arcane of esoteric wisdom literatures. I even remember a man in a medical gown reading from a hospital bed that stagehands wheeled into the pit of our amphitheater classroom, where I'd listened to lectures on Shakespeare's comedies. The poet said he was a pacifist and read aloud, paging through a loose-leaf manuscript and a leather-bound notebook, poems against the war in Vietnam, his voice quavering at times with weakness or passion. In each case, no matter who the poet was or what they wrote about, I heard an unmistakable ring of eloquence and sincerity in their words, a syntactical music that seemed to empower their willingness to speak their minds, a kind of bravery, and I wanted to be that. It was a feeling from within my chest, stirrings like a forest of seedlings suddenly sprouting through the litter of withered leaves and veining grasses on the broken asphalt that was my unformed identity, battered from the start in poor, urban public schools.

Though born in Hawai'i, I spent my school years in Los Angeles in a suburb called Gardena, bordered on one side by Torrance, where mostly whites lived, and on the other side by Compton, where mostly Blacks lived. Gardena, originally flower and strawberry fields before World War II, was full of Japanese Americans who settled there after they were let out of the internment camps, needing an enclave of their own. My family moved there in 1963—the year the Dodgers beat the Yankees four straight in the World Series, the year John F. Kennedy was assassinated, and just before the Beatles came to America. I took a gamut of classes at first—wood shop and metal shop and American history and Boys Choir, which I loved the most—until it was discovered I tested well and I got put into Advanced Placement courses along with all the other Japanese American kids. In those other classes, I mixed with everybody—Blacks, Mexican Americans (they didn't yet call themselves Chica-

nos, which at that time, like the even slangier *cholos* in gangsta-nese, was a term of some derogation), and poor whites whom we referred to as "Paddies," which I learned later in college historically referred to Irish immigrants. In high school, my friends were mostly Japanese, though, and I'd walk through the morning fog to the bus stop with a small gang of boys (to this day we're still in touch, albeit on Facebook), peppering each other with taunts and teases about girls, sports figures, and who knew the most about the latest hot rods and funny cars at Lions Drag Strip in Long Beach.

Did I say I loved Boys Choir? That's not quite true. I liked it—singing soprano and tenor parts to the dreary hymns and modern madrigals—but what I really loved was the doo-wop we'd sing out in the stuccoed breezeway before class started those mornings when a cold air would creep out of the ground along Artesia Boulevard and fan a fog over our town that was once a plain of wetlands. Choir was first period, the inaugural class of the day, and a few of us would gather around our vocal guru, one Gerald Hudson from Compton, who sang in his Mount Moriah Baptist Church choir and taught us those lessons before the academic day began. I think it started innocently enough, Gerald showing a few of us all the parts to "Swing Low, Sweet Chariot." Then, once we all got it down, he'd solo like a flying fish stitching through the face of our choral wave, syncopating guttural and plosive notes, hopping on his heels, driving us to pick up the pace, accelerando, until we could shade and scat and tailgate along with him just by a flicker in his voice giving us the cue. Pretty soon, we were doing four-part harmonies, two Japanese boys, a freckled white guy, and Gerald, half a head taller than the rest of us seventh graders, crooning out R&B hits like "Tonight, Tonight" by the Mello-Kings and "In the Still of the Night" by the Five Satins. But just as we were getting good, the whole choir crowding around us as we swung our arms, sassily snapped our fingers, and warbled before the bell rang to start school, our teacher, a tallish Mrs. Something, with graying hair she wore in a beehive and reading glasses she dandled from a chain around her wrinkled and powdered neck, busted us for "unauthorized commotion outside of class" and we were disbanded, then kicked entirely out of choir. I think, after a time, Gerald was reinstated, as he was such a grand

singer and could perform all the parts from bass to soprano. But the rest of us had to transfer into print shop or first-period gym. I felt it a tragedy at the time, my musical education abruptly terminated out of a mechanical prudishness. Mrs. Something said we were "hurting our voices" and "distracting others," but I think she felt her authority challenged and, culturally conservative, disapproved of what she thought of as "Negro music." Like I say to my university students, it was the Dark Ages when I was twelve.

Black music was all around us regardless of our own colorlessness, we who were groomed to perform an absence of ethnicity in our personae rather than suffer the curses of immigrant foreignness and wartime suspicion that all but damned our Japanese American parents and grandparents. Despite all efforts to assimilate to some ideal of the conservative mainstream, we grew out of a maelstrom of colliding histories—Blacks moving out of the South to work the oil and aeronautical industries that sprang up in and around L.A. during the war, whites from the Dust Bowl and postwar degree programs on the GI Bill, and we Japanese fresh out of the internment camps in the American desert and decommissioned sugarcane plantations of Hawai'i. What brought us together was dance music blaring from Sony transistor radios that dangled off the handlebars of our Schwinn Sting-Rays and ten-speeds, booming from the PA system in our junior high gym, the Japanese girls teaching us the steps they learned from the Black girls in cheerleading and on the drill team. By the time high school rolled around, we were dressing and stepping like our Black brothers and sisters too, buying our clothes from downtown merchants in touch with the latest from South Chicago, Detroit, and Harlem. I listened faithfully to KGFJ (a soul station on the AM radio) and bought Motown and Stax-Volt 45s in a record store run by a crew-cut Nisei (second-generation Japanese American) man whose business was in a mini-mall full of shops with names like Yuki's Hair Salon, Hikkari Travel, and Tosh & Jim's Take-Out. When we graduated, shuffling along in our rented caps and gowns in the slow diploma line up to the "Mohicans" stage at Gardena High (our mascot name egregiously incorrect), our parents may have allowed themselves a tear as Sir Edward Elgar's "Pomp and Circumstance" droned from an organ wheezing

from the pit, but it was Aretha's "Chain of Fools" that we hummed under our breath, ironic and mocking and assured of ourselves as Americans at last.

Four years later, I graduated, barely cum laude, in English and Asian Studies from a good private college in a cluster of colleges just outside of Los Angeles, then spent a year in Japan living in a Buddhist temple and traveling around with a backpack and a pretty girlfriend, an architecture student at Berkeley who could read all the street and rail signs and write them down for me in an angular, left-handed *kanji* as though drawn with a ballpoint pen by Thelonious Monk. I was the one who was fluent in the spoken language, having quickly acquired it over my senior year and the subsequent year in the temple. But I was never very capable with ideograms, and graduate school in Japanese literature proved too much of a challenge, perhaps even more so than the icicles and snowdrifts of Ann Arbor, where I studied for a year before dropping out after a summer of intensive language study in the morning and playing center field for our department softball team in the afternoon. What I remember most were the fly balls that I chased, hit to straightaway center, and the thunderheads in the distance forming in a *Soul Train* line scudding east from Lake Michigan after three-thirty every day. I'd turn my back to home plate, dash toward the darkening, bruise-colored skies as fast as I could, and arrive under the ball just as it descended and then settled with an almost pneumatic *thuck* into my glove. By five o'clock it would be raining and we'd call the game, whatever the score. And I remember winning the university poetry prize judged by the American Zen Buddhist poet I'd read in college. Pegged to the annual earnings from a large endowment, it was enough to buy a used compact car, a taxi-yellow Toyota that I drove out of town and across the Great Plains, Rockies, and desert back to L.A. In those days before cassette tape players were de rigueur in cars, I listened to Top 40s and country-and-western stations all along the way. It was perhaps the nadir of my musical experience.

And what was the apex? Well, I could say that I look forward to it every day, that I expect to have it the next time I play music on my stereo, when I slip an LP out of its paper sleeve, clean and black with its grooves shining, drop it onto the platter of my turn-

table, position the cartridge stylus above the lead-in groove just inside the edge of the record, then gently depress the cueing lever with the pad of my forefinger so that the tonearm and cartridge will start to glide slowly down. It could be the *Allegro ma non troppo* of Beethoven's Symphony no. 6 in F Major, op. 68, nicknamed the *Pastoral,* its sumptuous and sprightly strings bringing forth the illusion of a bucolic excursion through the countryside of early-nineteenth-century Germany. Or it could be Duane Allman's screaming slide guitar intro to the rollicking "Statesboro Blues" from the live Capricorn recording *At Fillmore East* (yes, I still own my 1971 original). But, much as I love these, especially the Bruno Walter interpretation of the Beethoven, when he conducted the Columbia Symphony Orchestra (I have both Odyssey and Columbia pressings), it would not be quite the truth. Instead, I will have to cite something I heard live with my own ears, felt in my own body, in Europe, in Italy, over fifteen years ago, when I was lucky enough to occupy a mid-hall orchestra seat before an evening performance of Giacomo Puccini's *La Bohème* at the Theater of La Scala in Milan in midsummer.

Part One

The
Perfect Sound

The Perfect Sound

I had been a casual lover of music for the longest time during my adult life when a lucky accident happened. Some fifteen years ago, as I said, the CD changer in my stereo micro-system suddenly broke down, and I wrote a friend of mine, Peter Morrison, a former surfer turned English professor in Southern California whom I knew to be an audiophile, asking for advice about getting a new "CD changer" to replace it. I'd just come back from Italy, where I heard opera performances at La Scala that were life-changing, they were so grand and beautiful. The magnificent first act of Puccini's *La Bohème,* about struggling artists in nineteenth-century Paris, had completely changed my attitudes about poetry and music. I suddenly could not do without more of this remarkable art form in my life—to me, a compelling blend of grand music, Romantic poetry, and high melodrama. But living in a small town in Oregon, I'd no regular access to the grand halls of the American metropolises, let alone La Scala.

So, I resolved to build my small collection of CDs to include as much from Italian opera as I could and listen to them every day, storing up capacities for romance, melodrama, and music myself. Here my broken changer thwarted me. Supposedly capable of loading three CDs and then playing them back in order, it rattled and hummed and then stopped working completely one day, clattering to a halt, eating CDs of the Puccini, Verdi, and Rossini operas I'd loaded into it.

To replace it, once I extracted my recordings, I was thinking of a two-hundred- or three-hundred-dollar carousel-type unit, maybe a Sony, like one I'd seen at another friend's place in Santa Fe. It could take up to five CDs in a revolving tray that spun out of a sleek black casing to load, then slowly retracted like a sleepy

dragon before playback. But Peter's answer was for me to consider a "single-disc CD *player*," which was only the preface to a long email essay about audio components and audio systems in general that, after my incessant questioning, he subsequently sent me. This, along with my newfound passion for opera, opened up an entirely new world of music and sound reproduction I hadn't been aware existed.

What started out as merely a wish to hear *La Bohème* and *La Traviata* playing daily on a stereo in my living room has morphed into another passion—not only for music and its great archive, but also for audio equipment and what each piece of it could do to bring me closer to what I thought I wanted to hear. I say "what I thought I wanted to hear," as this has changed over my time in this pursuit, years long now. As my listening acquired more focus and expertise, my ear successively asking first for more romance and warmth, then refinement and tonal purity, then things audiophiles call *soundstaging, air,* and *bloom,* I'd identify different components I thought would help me in these pursuits, rapidly changing equipment throughout my first year in the quest, thirsty to hear more and more nuances, reach grander heights, and have more powerful and inspiring musical experiences through an obsessive refinement of my audio gear. I'd want a system capable of reproducing string and timpani attacks in Beethoven's *Eroica* symphony, then one of clarity and timbral richness in the forty-plus choral voices of *Spem in Alium* by Thomas Tallis, and then another of great dynamic range and separation of voices (both operatic and choral) and instruments in Mozart's Mass in C Minor. These are not easy tasks for any one system. Ranging across musical types and performances becomes one of the challenges in the hobby, inspiring a seemingly unending quest for the right sound—the "perfect sound."

The pastime took on a life of its own, giving me more to know and tweak as I tried to dial in the sound I was after—not a stable thing, as I said, but one that shifted as I learned that many kinds of sonic representation were considered more "colorations" or even caricatures of sound than actually "realistic." What I found was that each type of system had its own vocabulary through which it constructed music from these shorter pieces of sonic communica-

tion and that its acoustic representation of a recording could, in fact, influence and even school the listener's ear to accept its particular style of representation as opposed to "getting it right," which is to say, matching the original master tape's capture of the live or the studio performance. During that first year, it turned out that, unlike most audiophile purists who'd spent years in the pursuit, what I particularly wanted was a sound more like vintage hi-fi, rich and involving, better-than-life for its romance and sonic colorations—like old postcards of Hawai'i with lush paints and exaggerations of splendor. My favorite sound at the time was "old-time," my choice in equipment more on the retro side of things. For instance, after an in-home audition (graciously granted by the retailer), I rejected one very highly regarded CD player for my system because, although it got a lot right, performing through the range of music I listened to in a completely balanced fashion, not overemphasizing any one set of frequencies, not playing orchestra poorly while privileging choral, it also dampened the gorgeousness of midrange bloom, that lushness I'd coveted. Its evenhanded neutrality worked at cross-purposes to the sound I was initially pursuing, to the sound that had indeed initiated me into the love of audio.

As my obsession grew, I discovered there was something else behind it that was driving me—a search to reconnect with my father, who passed away over thirty-five years ago, when he was fifty-eight and I just thirty-two. I got involved with using vacuum tubes for stereo accidentally, when, sometime after I acquired a CD player (I ended up buying the very one that my surfer-professor friend happened to be getting rid of), I saw a photo of a new, Dutch-designed, Chinese-built amplifier that was visually crafted along retro lines. It reminded me of my father's equipment and, willfully, I wanted that amp to be the one that suited my system. In the end, I bought the amp and rebuilt my system around it, learning how to shift the character of the equipment's sound by changing its tubes, getting closer and closer to an early 1960s style of fidelity, which then let go a flood of memories of my father building his own equipment, testing it, swapping tubes, asking me to "listen" for him, tell him what the music sounded like.

Back in the early sixties, it had been my father's hobby, and he'd

built amplifiers of his own, filling our Los Angeles suburban living room nearly every night with big band sounds and Hawaiian hotel music. But there was a terrible irony to his listening—he was rapidly losing his hearing as an accumulation of lifetime misfortunes started to take away the easy and fluid perception of sound from his life. A case of scarlet fever in infancy had damaged his inner ears to the point that, as a soldier, gunfire didn't bother him—a trait that allowed him to place himself behind an abandoned German machine gun and fire it at counter-attacking Wehrmacht infantrymen during World War II, as well as to work a jackhammer building runways for military jets after the war. These experiences had contributed to further hearing loss, which he must have recognized.

Near the end of my first year in this quest for the perfect sound, I listened a lot in the mornings, shifting away some from Romantic Italian operas to Renaissance choral music with its polyphonic vocal lines and profusion of harmonics. This is extremely difficult music for audio systems, exposing limits in their upper frequency extension and lack of transparency, often veiling and smearing the choiring voices into a tangled hash of treble sounds and tweeter scratchings. I was listening to Allegri's *Miserere,* then the masses of William Byrd, the thirty-six-part canon *Deo gratias* of Ockeghem— music written out of devotion and for religious ceremonies. And my hodgepodge system's sound was like that of an apse in Assisi. One morning, as I had the Huelgas Ensemble's performance of Desprez's twenty-four-voice psalm *Qui habitat* on the stereo, the Montana writer William Kittredge called. When I answered, he heard the music surrounding our telephone voices and exclaimed, "Heavenly choirs!—Hongo, are you dead? Or am I?"

It *is* a touch of heaven, this music, and that an audio system can catch a lot of it amazes and fascinates me. Friends and family call it an obsession almost like gambling, and it is, draining a bank account, dominating a day's thoughts, and lingering into the night sometimes. I think of husbands who built bomb shelters in their backyards and basements in the sixties, Dostoevsky running through thousands of roubles in a day full of casino losses at roulette. As a gambler himself (mainly on horses), my father would always say, "Bet your winnings," so I follow his advice. When I get

an honorarium for a poem or a royalty check for a book, I go online and find more tubes on eBay or a piece of used gear on AudiogoN, an audiophile auction site, that might get me closer to what I'm after—fleeting clouds of music with a leading edge of silver from a sun I can't see.

In the end, the main story lies in my own passionate quest, evolving from an audio ephebe into something else—not an expert, but someone who can mix a batch of disparate machines and equipment, tweaking things with cables and different audio tubes, taking a pair of American Sylvania "Bad Boy" 6SN7-GTAs from 1952 and combining them with an English military Brimar CV1988 pair, placing them in the input and driver sections of a KT88 stereo amplifier so that they together might form an electronic synergy that will play arias from Puccini and Donizetti that can make a grown man weep.

My First Audiophile Experience

My astonishment with audio and what it could do began after I received that single-disc CD player from my friend Peter in Southern California. He was upgrading to something called a universal player and sold me his old "Red Book" player (one that played only regular CDs and not SACDs, DVDs, etc.) for about a hundred dollars more than I'd budgeted for that carousel Sony I'd had in mind. His unit was not quite ten years old, made by an audio specialty company, now defunct, called California Audio Labs, and it was built like a tank. It had originally sold for something like $1,200—an impossible amount to me then. It arrived in a huge double cardboard box, floated on Styrofoam inserts, and bagged in tightly, intricately wrapped plastic. Peter had kept all the original packaging and, in reverse, had meticulously replicated all the moves the company used when they'd first shipped him the player.

Called a CL-15, it had the same footprint as a home theater DVD player, but it was at least double the height at almost six inches tall. It was also as heavy as a carton of wine or bourbon bottles—thirty-five pounds and a far cry from the trifling thing that sat within the top tier of my old micro-system stereo. "Industrial grade" is what

my friend called it, big and painted a flat black with an illuminated digital panel that tracked operation and progress in little balloons and stripes of black lettering in an amber-lit window. Its tray, operable from push buttons or the heavy remote (reminiscent of a Blackberry in size and style), opened and closed as though on Teflon rails, smooth and deliberate, formidable rather than flimsy. My friend told me its parts were all of metal, not plastic like my old player, and might never wear out. He'd included a handwritten instruction sheet on how to hook the player to my existing system. I followed it and off I went.

Once I got a CD into it, I was in another world. It was *Blues and Ballads* by Duke Ellington and His Orchestra and began with "In a Sentimental Mood," that classic tune with a melody like a Harlem flâneur sashaying down his boulevard in the early morning hours after a night on the town. I pushed the ON button on the Cal player and its LED screen lit up. The amber-lit, black numeral readout across it said "0" on the left, then "0:00" on the right. I pushed the switch for loading the tray, and it rolled out in a quiet, finely machined movement. My old changer had clanked, clattered, and whirred whenever it stuck out the tongue of its tray, but the Cal player's tray moved smooth and stealthy as a panther on glass. The digital readout ticked off its progress as it spun—*01.00:01, 01.00:02, 01.00:03 . . .*

After the first, sudden chord struck by Ellington on his piano, the music rose and fell like a fleet of gorgeous fountains in timed choreography, its sprays and spindrifts of notes curling through the space in front of my speakers, painting on its furling canvas a near believable illusion of the entire band. I thought I could see Ellington's piano. The stand-up bass seemed to come up from out of the floor, its peg digging through my carpet, the sensuous, f-holed, lacquered, and wooden body of it reverberating and making my tiny bookshelf speakers disappear. Ben Webster's big tenor saxophone took up center position and I could feel the intimate breath and flexed embouchure he used blowing through the mouthpiece and wet reed into the routes and valved curvatures of his instrument. I felt closer to the music than I'd been before, involved in its articulate movements from not only one isolate note to the next,

but dwelling within the filigrees of its shifting shapes, in its arrows of arrival and fleeting tails of achingly regrettable departure. When Webster held back on a note, releasing it in a halting syncopation and shaping a kind of fatness and swell to its hesitant birth, I held back a shade of my own breath as well and then let it out in a measured relief as though I were following him in an esoteric, even tantric dance. As the music lifted me, I became its acolyte, attending to all of its muscularity and finesse, skipping along with the buoyant galley of my heart.

And it went like that through the rest of the recording, tune after tune, a miniature of the Ellington orchestra performing before me, vivid in the air.

"That's called *imaging*," Peter said, bemused, when I phoned him, right after the first track had played. "You're in for a big ride."

Though not quite holographic, it was as though a scaled-down version of Ellington's wonderful swing orchestra had taken up residence along the short wall of my living room, pulled out its sheet music, and started blowing its collective ass off. It was like his orchestra was in miniature, spatially layered in front of the little birch-laminated British bookshelf speakers I had, a tromp l'oeil.

Yet, for all of this, it was not the music's realism, per se, that I think inspired and charmed me. It was its irradiated facsimile of the real that was so captivating. It was a loveliness once removed, like between the dead and the living, a distance barely more than a heartbeat and a breath, as a poet once said. Let me explain.

The Japanese have a traditional theater that brings together sophisticated puppetry, narrative chanting, and *shamisen* (a three-stringed, banjo-like instrument) accompaniment into what's called *bunraku*. Before the modern period, Japan's best plays—love suicides and samurai revenge tragedies—were written for the puppet theater. I saw a performance of it at the Music Center in Los Angeles when I was in college. Our teacher, a passionate translator of *bunraku* scripts, told us to look at the puppeteers, often aged men dressed in black robes, two on each puppet (one to handle the head, body, and right arm; another for the legs and left arm), and how they danced nimbly around the wooden figure, animating it, shaping its movements to mime the human, working its cutout mouth-and-chin

piece timed to the dialogue spoken by the chanter at the side of the stage. Then he said to wait as the story took charge of things, the desperation of the lovers (a poor apprentice to a soy sauce maker and a woman who worked as a domestic) driving the plot forward as they made their pact of precipitous suicide rather than suffer the rest of their lives being apart. "Don't you forget the handlers?" he said. "Doesn't the fact it's all an illusion dissipate? Are you not charmed by these imitations"—*Mini-Me* had not yet entered the language— "as though they'd a life of their own?"

In *bunraku,* what's forgotten is the real and what's present in its place is this beguiling representation, itself alive with dance-like movements, a dramatic plot and heartfelt speeches, and a surge of emotions when tragedy strikes. You give yourself over to the trick of these vivid substitutions for the real. You invest belief in them, laughing and crying, feeling hopeful in their love or plots of revenge and then wretched with them when they despair of deliverance. It is the *as if* of all representational art—we give up our cynical hearts to let them beat (as one) with those onstage or on TV or in the movies so that we might live, albeit until curtain or commercial interruption, as lustily as the characters before us. *Alas, poor Hasselhoff, I knew him well . . .*

I think audio struck me that way, its *as if* and mimesis adding powerful elements of charm and artificiality as much as any mirror of the real thing. I'm not saying it was supra-real, or that the sound was colored and sauced-up to make it more appealing. Not at all. What I mean is that stereo sound can be so compelling it displaces the real and captures your soul even with its *irreality*, its difference from the real, plus the added value of its gestures toward the real. Audio is like a costume romance, the fakery part of its gorgeousness, its nearness to and distance from the real thing almost coequal aspects of its attraction.

Opera Punks My Stereo

To my disappointment, for all the salutary effects the audiophile CD player brought to my humble system, I could not get it to

sound good with opera singers. Eager to hear arias the way I'd heard Ellington, I tried one CD after another—Renée Fleming aria compilations, then highlights from *La Traviata* and *La Bohème*. The music would sail along for a while, wondrously woody bassoons and piquant oboes and piccolos sending out their pleasing notes like chirping birds in a rainforest. The violins would sound lavish and sweet too. Then a Fleming or a Bergonzi would crank up their voice, declaiming in dramatic Italian some kind of hurt, wrong, or frustrated love. Was it George Bernard Shaw who said that the essence of opera is a tenor and soprano being prevented in love by a baritone? Well, my tenor and soprano were prevented in love by a *stereo*.

I listened to the great voices thinning and crackling as they rose up the scale to the grand, shimmering top notes of their arias. It was almost as if I were listening to an FM station as the signal shifted away suddenly and slipped out of range, the purely beautiful voices collapsing into a shuddering scratchiness and the sharp, glassy fibers of static. I'd play Luciano Pavarotti singing "Che gelida manina" from *La Bohème* or Joan Sutherland's "E strano" from *La Traviata* and, just as they came to the fullness of emotion and the extreme dramatic pitch of the music, my system sound would shrivel like the inner flower of a tidal anemone, as though a mischievous boy had just urinated on its pristine infield of vivid, tempera-colored flesh.

As astonishing as my sound was on jazz classics, it was atrocious on opera standards. And it didn't matter which recording I tried. Whether they were digital transfers of live performances from the fifties or the latest studio recordings by the likes of Russian soprano Anna Netrebko, they all ended up sounding, during the best showcase passages of operatic art, like a cat coughing up a hairball. *What is the problem?* I wondered. What made opera so difficult for an audio system to render? My car stereo system couldn't handle the music either, Rolando Villazón's burnished tenor voice shattering into static when I tried to play his *Italian Opera Arias* as I drove through town.

I called around, first to the repair guy at the electronics store that couldn't fix my changer (its plastic parts had worn out), then to

a place named the Stereo Workshop out on West Eleventh near the long strip of miscellaneous mini-malls, side streets of light industry and warehouses. I called my SoCal audiophile friend Peter too. And they all said the same thing: "Your amp's probably clipping."

They might just as well have said, "*Gnomes* go bowling in there and that's what you hear in your speakers. They like to mess *witchu*."

What was *clipping*? Along with all these new machines (amps, CD players, preamps, etc.), this new foray into the world of stereo plunged me into churning waves of new, confusing terminology. Some of the terms had to do with electricity, its behavior and characteristics. These were vaguely familiar from my science classes in junior high and high school and not that difficult, taken one by one, to understand. *Amperes, volts, resistance, capacitance, ohms, current,* and even *watts.* All categories of electrical measure, relating to its flow, resistance to flow, and storage—these were no problem. Then there were terms like *frequency response, megahertz, bass, midband,* and *treble* that related to the audio band. Easy enough to understand for a guy with a GED and a little help from *Audio for Dummies.* But there was also a little storehouse of vocabulary that related specifically to how electronics measured and acted during *operation*—that was all new to me. I'd studied literature, poetics, traditional poetic meter and form, philosophy, and critical theory, but here was a language of description specifically tied to electrical performance in audio gear.

Peter, in his characteristically meticulous way, explained that *clipping* meant that the power the system needed to reproduce the sound of an opera singer's voice exceeded the output of my amp at its peak ability. It came down to a simple matter of watts—a question of electronic power and the likely limited output of my micro-system's amplifier, only a desktop unit after all and never designed for the kind of high-end audio gymnastics I was now requiring it to perform. He told me to look for its "power rating," perhaps noted on the back of the unit. Sure enough, there it was in raised plastic lettering—"25 Watts." That was the output power rating at peak—enough to drive the little speakers on most music, particularly at moderate volume, but not enough to handle the dynamic peaks of the opera singers, whose voices demanded larger amounts of power,

particularly at volume levels close to live performances, which is what I wanted and why I had goosed the volume of the little desktop amplifier way, way up.

The little thing had simply run out of juice. It was like a mouse trying to make love to a giraffe. The speakers were underpowered for the demands of the signal being fed to them, resulting in fairly severe distortion and making them sound like they were breaking up. That was the static I heard that crushed grand, operatic voices until they sounded like wicked witches shrinking and smoking, dissolving into black, ashen piles evacuated of their bodies.

Or, as Peter explained, an opera singer was like a long train of heavy freight, and it took more than a little engine to haul it.

I'm gonna need a bigger amp, I thought.

A Visit to Wheaton's

Except for the pearl of a university where I teach, Eugene, Oregon, is a nondescript sprawl of kindly suburbs (on the map, subdivisions of fanlike poker hands stippled with single-family dwellings) surrounding a few aging central neighborhoods and a downtown eking out its revival in the form of coffee shops, wine bars, and a former movie palace repurposed for hosting a quirky program of rock concerts produced by the fun-loving descendants of local-hero novelist and former Merry Prankster Ken Kesey. If you are too young to know who the Merry Pranksters were, let's say they were pioneers in the be-in, tie-dye, hippie-ganja-hair, and patchouli-oil culture of the sixties, who sponsored a series of what another writer has called "Electric Kool-Aid Acid Tests" all around the country, from their home base near San Francisco to the Midwest and beyond, wreaking hilarity and incinerating synapses wherever they roamed. A scattered coven of their like—our current anarchists and hemp-loving counterculture revivalists—still live amongst us mild suburbanites, raging against the machine on occasion, spreading their rainbow-colored wings at an annual event known as the Slug festival, wherein grown, full-sized men tie on imitation Venetian carnival masks, cross-dress in tights and tutus, throw on long, pea-

cock trains of crinoline and taffeta, and roller-skate through a main street cordoned off for the express purpose of protecting their narrow band of hippie nostalgia for a day.

The population is around 170,000 here, so it's a small city, actually, though I find hardly anything here I could comfortably call "urban." There's a quaint combination headshop and shoe store run by a South Asian entrepreneur (he sold Doc Martens to my younger son), a newly launched flotilla of micro-breweries on the Berzerkeley side of town known as "the Whiteaker" (where all the hippies, anarchists, and junior faculty live), and even a couple of sushi bars my non-Japanese friends insist are, paradoxically, both inventive and "authentic." But I really wouldn't know, because I've made it a policy never to patronize a sushi restaurant that isn't on the coast or in a city of at least half a million. The chanciest I've allowed myself were places in Laguna Beach and Carmel, both boutique and wealthy towns I figured snobbish enough to get themselves the real thing both in fish and *itamae* (chef). In Eugene, the best fare to be had, *sasuga-ni* (contrary to expectation), is Tuscan, at an inn and ristorante owned by a former soccer star from Florence. I eat there every chance I get, on a teaching day springing for a lunch of Dungeness crab ravioli and a glass of Pinot Grigio, or, when it's cold and I've a sniffle, a bowl of bread-thickened *ribollita*.

Across from a wine shop near the public library downtown, we used to have, up until a year or two ago, an establishment I'll call Wheaton's Home Entertainment, our only non-big-box retail store specializing in new stereo and A/V equipment. I found it "looking through the Yellow Pages" (a thing that, fifteen years ago when I began all this, still existed). I called them up, talked to a good-natured man about what they carried (all the brand names meant nothing to me at the time), and was invited down. I think I went right away, anxious to know what I might find as a potential upgrade to the kit amp that went with the defunct micro-system changer. I'd called a couple of the electronics stores in the malls, thinking I might find an integrated amplifier I could match with the audiophile CD player I'd just gotten, but all they could offer were other micro-systems like the one that had just broken. I'd have to buy all

three things together—speakers, amp, and CD changer—in a new all-in-one deal for about five hundred dollars. So off to Wheaton's I went, my first venture into the world of retail hi-fi.

Wheaton's was a spacious place, of a kind hardly to be found anywhere anymore, with four separate rooms all dedicated to stereo and A/V gear. Its main showroom featured three or four "stations," each with its own little sofa, club chair, or love seat, and an assembled system before it curated by the staff. Potted dwarf palms and little end-tables replete with squadrons of remote controls at the ready were strategically placed everywhere. One of the owners, a slightly built, brown-haired man with a thick mustache, showed me around, explaining things. When I told him what I was looking for and my price range, he steered me over to one of the smaller rooms, which was jam-packed with electronics, amps, and CD players in silver and black, along with about a half-dozen pairs of speakers lined up against the far wall, all with attractive wood finishes. He called the ones on stands "monitors" and the tall integral units "floor-standers." The monitors were "two-way," which meant they had a midrange driver and a tweeter in each box, while the floor-standers were "three-way," meaning they had three drivers—a woofer for the bass, midrange driver for the mids, and a tweeter for the highs. A driver was the actual speaker module, comprising the cone, surround suspension, magnet, and basket structure behind it all. What we colloquially call a speaker is actually made up of the cabinet and the drivers, damping material, and electronic innards stuffed inside.

The owner (I'll call him Jason) showed me a stepped suite of CD players while we waited for the preamp and amp to warm up some. He'd flipped them all on when we walked in, but he wanted me to get acquainted with the CD players first. One was made in England and had a brushed aluminum faceplate with a line of function control buttons spread horizontally across it. When I pressed the OPEN/EJECT button, the motorized tray slid open silently and then retracted with a soft, attractive *clunk* when I pressed it again. It was $699. Another player was made in France, and it featured a faceplate with handsomely beveled edges that made it look vaguely stealth. Its tray operated in complete silence. It cost over a thousand

dollars. Finally, Jason showed me an American-made player that had no tray at all but sucked CDs in with a soft *snick* through a horizontal slot near the bottom of its faceplate. When I pushed the EJECT button on its remote, a substantial, weighty thing of the same type of anodized metal as its black chassis (it had sharp edges, and the heft of an ingot of steel), the CD slipped salaciously out of its slot and hung out like the gladdened half-circle of a silver tongue in front of me. This would cost me a few thousand. Each one was a deep rectangular box about twice as wide as my micro-system tower, but flatter, with a height of only about four to six inches. Why they varied so widely in cost, I'd no idea, but, ignorantly, I asked if any of them had plastic parts.

Jason looked at me with surprise and said that only the trays were plastic in the two players that had them, but that all other parts were metal.

"It's not going to break or wear out inside, if that's what you mean," he said. "Everything here is not only warrantied for at least three years, they're also fixable if a part does happen to break down. We send all of our warranty and repair work to a tech here in town."

Reassured, I let him move along to the demonstration. He pulled a CD out from a short stack of crystal cases lying in front of one of the CD players and popped it in the English one first, then the French, and finally the American one. It was a Bob Seger album, as I recall, his song "Like a Rock" the demo track Jason selected for me to hear. With each player, the song sounded pretty much identical to me, just the blue-collar anthem that it was, which had at some point become a musical shill for Chevy trucks. Though it was certainly clearer and more spacious than a car stereo, I still thought the overall character of sound emanating from the two sets of speakers was more alike than different. Sure, one did sound "pushier" than another, the guitar slightly louder, drums snappier, the bass with more thump. Seger's voice would be grittier via one player (like a small jet engine rather than a power sander), and then boomier (as in a movie theater) from another. The costly American unit seemed to throw ghostly sonic images of Seger and his band-mates in a space that floated in front of and between the speakers.

But, at that stage of my audio education, these distinctions seemed hardly worth the hundreds, even thousands of dollars of difference in cost, when matched with the humble gear I owned.

We turned our attention to the amplifiers then. First, there was the matching stereo amp and preamp from the same English company that made the CD player I'd just heard. With all-English electronics, the sound was very finely grained on Seger's rock tune, punchy and full of drive that suited the music. This combo seemed fine to me, but Jason moved on to show me what a French amplifier could do, this one an integrated unit, matched to its own player that I'd just heard. This sound was more laid-back, "polite" being the audiophile cliché I'd learn later, but also more supple and kind of sinuous. There seemed to be no grain in the treble range and, oddly, it robbed Seger's voice of some of its characteristic grit, cutting back on its cheesiness and emphasizing instead more delicate shifts of tone and timbre. I said so and Jason complimented me, saying this combination might better suit my listening preferences "for classical." Finally, we moved on to try the American combo, three black boxes with silver controls, Jason dialing each one deftly, impressively using the meaty side of his hand rather than his fingers, as though brushing the hollows of a lover's body (or carefully calibrating turns on the knurled control to a spinning lathe). Again, here was the most impressive suite of electronics, the system putting up a lovely stage of sound, so deep and wide you could step into it, the images of instruments and singer lithe and morphing before us, their sounds clear and delineated. But the cost was more than I paid for my Toyota sedan.

I went away thinking that the relatively affordable English amp and preamp would be plenty fine, but told Jason I wanted to mull it over. I got up from the cushy club chair (upholstered in a freckled, Mediterranean orange) in Wheaton's listening room and asked him to write down all the prices and model numbers on a card for me. To this day, I can still see the deft dollar signs he wrote in blue ballpoint ink on the back of his business card. Though these choices weren't quite the price of an automobile, they each represented more money than I'd spent on anything that wasn't a plane ticket to Italy or a washer-and-dryer combo from Sears.

Electrocompaniet ECI-3

It took me not quite a month to figure out what I should do next. Since I couldn't play my beloved opera CDs, I simply put on what sounded best in my makeshift system—combo fifties jazz, Ellington's orchestra, probably some rock 'n' roll from the late sixties and seventies. My CD collection then wasn't the massive thing it became later—it was maybe just over a hundred titles, their crystal cases neatly lined up, spines facing outward from a cheap, oakwood case I'd bought on Amazon. I had the usual standards—Miles, Monk, Mingus, and Coltrane in jazz; Clapton, Dylan, and the Stones in rock; a smattering of chamber music and violin concertos with orchestra. Outliers were things like Austin rocker Jimmy LaFave, some Hawaiian groups, and my tiny collection of Renaissance choral music. I played some of these, idling in my listening while I researched and plotted my next move.

For advice, I'd talked to two composer friends with recording and playback studios in their homes, and they'd introduced me to a bounty of potentials that only managed to confuse me. They spoke of tube versus solid-state amps, hybrid amps, balanced versus single-ended operation, acoustic treatments for my living room, flat frequency response versus a heightened midrange in speakers, and a holy host of audio arcana that was more than I could hold in my head. I told myself what I needed was simply a system that sounded good and could play opera. As it was turning out, this was a taller order than could be realized without more time and pain (both acoustic and financial).

I remember maniacally Googling for e-zine reviews on every amplifier and set of speakers that struck my fancy and falling under the spell of sumptuously produced online ads showing off their product's curves in sensuous photographs and sexy snatches of prose calculated to stir you up like a Vegas stripper. Some of the websites featured scenic photography and recorded music—simple things like a forest and lake with a sound loop of piano and maracas, maybe a glockenspiel and chain of reindeer bells all chiming as you paged through the products line, the speakers in richly veneered wood cabinets or glittering amplifiers placed on tidy stands at one end of

a spacious, immaculately furnished living room. It was more than a bit like shopping for a new car when you finished browsing all the dealer ads and studied equipment specifications, noting the highway mileage, admiring the fuel efficiency, but really being seduced, in the presentation of these rational facts, by the photographs of snazzy interiors and curvaceous cutouts around the vehicle's headlights, by the tactile feeling of the beveled edges and cleverly ovoid shapes of the cars (or electronics) themselves.

It wasn't that I hadn't liked the gear I was shown at Wheaton's. I did. But it seemed a limited selection, given all that was out there, and I wanted to be sure I'd surveyed the entire field of possibilities, done my due diligence, as I did every time I approached a major purchase. It was a habit I'd gotten into with one of my first adult decisions—choosing colleges when I was in high school. I started by going to the counseling library and browsing the reference volumes—thick encyclopedias of reports, ratings, and reviews of just about every institution of higher learning in the United States—then ordering catalogues from ones I'd identified as likely candidates. I hit upon Occidental College, UC Santa Cruz, and Reed early on. Thick as trade paperbacks, their catalogues became dog-eared and full of check marks as though I were a boy dithering over the statistics of his baseball heroes in *The Sporting News*. It was no different with audio, except that now everything was online. I sorted through the available literature, sensitized myself to an evolving list of evaluative categories, identified what among the formidable multitudes might be suitable, and also algorithmed in a good measure of potential romance before I identified my targets.

Cost was not left out of my calculations, so I determined that I required an integrated amp—one that combined the preamp functions with the final stages of audio amplification just as the amp in my little micro-system had (saving the expense of buying separates). And it had to provide a decent level of output too, so it wouldn't start clipping just as an opera singer got into the high notes of an aria. *More watts,* I thought.

I'd augmented my e-zine studies by checking out primers and general guides about audio from the public library. The various guides to home stereo, the audio rags I found in the magazine

stacks, and the audio websites I found online all made up a new bible and Apocrypha of references. I was no longer reading the sonnets of Sir Thomas Wyatt or annotating my copy of Shakespeare's *Venus and Adonis,* but studying pages of *Stereophile, The Absolute Sound, 6moons.com, Hi-Fi+, SoundStage!,* and *Dagogo.* Like a teenage girl fantasizing her wardrobe, launching herself into a more glamorous world, I identified several types of amplifiers, both tubed and solid state, and created a short list of amp choices: a stylish Italian hybrid amp with triode tube input and a MOSFET (transistor) output stage; the Wheaton's-approved British solid-state amplifier with its hedgehog-like appearance half full of heat-sink fins on its face; a Chinese-made tube amp with its double phalanx of old-style vacuum valves; and, finally, a Norwegian-made unit called the Electrocompaniet ECI-3 with its aluminum casing painted a liquid midnight, a clear and thick acrylic faceplate, an array of gold push-buttons, and blue LED lights beaming like cat's-eyes in the dark.

I'm not sure why I turned away from the other products, but, from what I'd read, the Electrocompaniet had all the qualities I wanted—stable performance through high frequencies (for those mighty, Caruso-like top notes or La Divina roulades), midrange warmth (for orchestral strings), and enough power to handle the most difficult and complex music. It put out over 70 watts—I mistakenly thought this was about three times what my micro-system amp did (in reality, because of the way watts are calculated, 70 watts was only about one and a half times as much power as my 25-watt micro-amp) and assumed that should be enough. I read scores of reviews and highlighted those that mentioned classical music, particularly opera, attending to their descriptions of a tenor or soprano performing the splendors of an aria. Comments like these were scarce, but a few reviewers would make them, so I grew to recognize their names and looked to them for guidance. One of them mentioned the Electrocompaniet.

On AudiogoN, the website that advertised used and demo audio gear, I found a floor model for sale by a dealer in Chicago. You could go to the site, and on any given day, there would likely be listed for sale a used or demo unit fairly close to what you were looking for. The procedure of purchase was this: I called the num-

ber given, haggled a bit with the man who answered (he turned out to be the owner of a hi-fi store), agreed on a price, sent him my money via PayPal (cost + insured shipping via UPS + 3 percent fee), and, a week later, the "lightly used" unit was in my system. It was a lovely black-and-gold thing, shining under the track lights suspended from the ceiling and looking perfectly matched to my friend's old CD player. It had four gold button controls arranged in a diamond on its Lucite faceplate—a configuration that reminded me of my father's 1959 Plymouth Fury. The Fury had a three-speed Hydroglide and a push-button gearshift on the dash, and it too was glossy black like my new amp, its late-fifties super-fins a touch of pure ghetto gaud. The exotic Electrocompaniet had the same air of swankiness and swagger my father had gone for in his car. Despite all my careful research in making the choice, I think I bought it mainly because it reminded me of him.

Shrine

Looking at the gleaming Electrocompaniet amp every day, its quartet of round, gold push buttons arranged in a diamond pattern, its overall snazzy gestalt, I couldn't help but think of my father and his polished black Fury, gleaming in that gauzy sunlight of late afternoons in L.A. My father had style. The amp that helped me remember him took up a spot on the middle shelf of my stereo console—a mahogany piece I'd bought from Pottery Barn. Above it on the top shelf was the formidable CD player, making a very nice presentation together—all black. Never one for meticulous, attractively arranged interiors, I suddenly got the urge to compose a coherent look.

I brought out some of my father's old things that I had stored in a box in my study—his Sony transistor radio in its black leather case, the raffia gambling hat, the framed photograph we used at his memorial service, and a Sony Watchman he'd take to Dodger games so he could also monitor the stretch runs of the races at Santa Anita and Hollywood Park when they were shown as clips on the nightly reports. I was aware that this was unusual, but I didn't care. These were what I had left of the man, such a gentle and supportive

presence in my early life, so I built a shrine out of them and the stereo. The last thing I added was a necklace, a lei really, of tiny florets like cowrie shells that I draped over his framed photograph. It was my modest tribute.

For a while, I'd considered whether my father's more expert enthusiasm for stereo emerged out of "the Playboy philosophy," that early 1960s notion of excess and a canned sophistication fostered by Hugh Hefner's "The Playboy Advisor" column in the early days of his magazine. It declared that a man's stereo was as important as his car, his drink of choice, and his lifestyle. To be a man at all meant that you had to have a stereo. Audio gear was part of the image— like Hef's smoking jacket, his ascot, and his fish-horn-shaped pipe. For leisure, you slipped into something more comfortable, mixed a martini, and turned on the hi-fi, spinning a vinyl album on the turntable. A man's stereo was a sign of taste and success—a two-channel phallic symbol contained within a lift-top mahogany console. It was an extension from the fairly constrained area of the three-bedroom tract home into an illusory space of aspirational leisure and luxury.

But the Playboy image didn't square at all with who my father was in his increasingly silent, work-filled life. I dismissed it. My father scoffed at following trends or allowing his sexuality to be defined by anything as highly capitalized as a slick, soft-porn mag. No, his stereo, his own hand-built collection of components, had something more to do with the DIY craze—the do-it-yourself hobbyist culture that caught fire among GIs returning from war, inspired by magazines like *Popular Mechanics* and *Audio Engineering.* My father was one of *those* guys, short on cash and long on the ethos of can-do he picked up during the war. He'd found his taste for big band music while in the service, too, ignited by attendance at stateside USOs and then perfected by hanging out, postwar, at Gibson's Bar in Honolulu, where Trummy Young, once Louis Armstrong's trombonist and then a longtime bandleader at various Waikīkī hotel lounges, played swing tunes with the local players from the Royal Hawaiian Band on weekend nights. His jam was a plush, sashaying kind of music that went with billowing trousers and pleated skirts, an ex-soldier's slicked-back hair, and epaulet-shouldered dresses on

the squealing, gleeful Japanese girls executing the Lindy Hop at the segregated dances.

But, more than either of those hyper-masculine models, his lust for sound, I think, had to do with the fact that, year by year, he was losing his hearing—and fast. He wanted to listen to his music one last time before he couldn't hear at all. That's why he built his home stereos and that's why he built from kits and why he designed his own final amplifier, before the world turned off its sound and all he could hear was the faint, diastolic *whissh-hope, whissh-hope* of his own heartbeat.

Further Dissatisfactions with Sound

As for myself, I was driven by a thirst for a certain quality of sound closest to what I remembered hearing at La Scala. I wanted to be moved by the passion as much as the musicality of opera, but passion was something that could not be experienced in reproduced sound unless you could escape from the technology and feel as one with the music itself somehow. There is that definite charm of artificiality I spoke about, that allure of the marvelous illusion of a living, vibrant, and chimeric thing being cast before you in your living room like a beguilement put upon Jason as he voyages through Aegean isles. But that too is swept away once you connect with the music itself. It is like the tilted realism of the stage set, the apt costuming of the characters (full of period detail), the sumptuous replica of life, an artificial truth that startles and then captures you as the curtain lifts and the intricate pattern of lights guides your eyes to the dramatic focus of things. You're swept up, of course, but your heart waits for all these to fall away as the music takes you up in its phantasms and splendors. And then it does: the poet Rodolfo, in *La Bohème,* grasping seamstress Mimì's cold hands as they first meet in the Parisian garret and rousing himself into extravagantly lyric declarations, his voice moving swiftly from recitative into the beautiful aria I'd come to treasure. The tawdriness and poverty of life seem simply the murk from which the blushing petals of a lotus rise then, and all our speech, however poetic, merely the prelude to human song.

But it seemed that each component in the chain of audio equipment was taking its turn getting in my way. First it was the CD changer, then the amplifier, and finally it would be my speakers. Outfitted now with the audiophile-grade CD player and amplifier, I thought the humble Mission bookshelf speakers that came with the desktop kit of electronics inadequate and decided to replace them. They seemed too small for the amplitude and grandeur of the music I wanted to hear—not only opera, but orchestral music as well now, symphonies and concertos. The symphonic soundstage felt shrunken to me compared to what I enjoyed hearing; the Ellington orchestra, for example, seemed absurdly miniaturized for the full sweep of the musicians who would be arranged across a stage. And the instrumental voices of the classical orchestras got concentrated into a kind of oval swarm at times. There wasn't the differentiation that I marveled at when I listened to Ellington's music. I couldn't "see" the instruments alive before me. Instead, the music roiled like a storm of commingled strings, woodwinds, and brass. My audiophile friend Peter explained that hardly any audio system could create imaging in a soundstage out of the multitude of instruments that an orchestra had, but that, during certain passages and interludes, the timpani, say, might emerge and be "seen" standing out from the rest. What I heard was that the music seemed to break apart and collapse like a wave falling onto itself, its action more a foaming than a gorgeous curling, the music a froth of confusion. And finally, the sound I got was still spitty and metallic during an opera's most spectacular moments. If I played my highlights CD of *La Bohème,* at the height of Rodolfo's first aria, just at the cabaletta with its gleaming and burnished top notes, my stereo, even with its latest upgrades, would squawk and shriek like a battered tomcat crawling back from a night of prowling.

The Craftsmanship of Sound

I fooled around for the next three months or so, shopping for and trying out two more sets of speakers, neither of which sounded good with the Norwegian amp or the CD player built like a tank. My sound was still unlovely in the passages of music that were, in

life if not in audio, meant to be the loveliest. I wanted that moment back at La Scala when the arias had grabbed me by the spine and lifted me from my seat, making my lungs shudder, and caving me into an emotion as though I'd been struck by the universe with a pure bolt of energy and insight. I was not getting this with audio so far. I bought a pair of tower speakers from Kentucky that were not good for what I wanted. I bought a pair of bookshelf speakers from New York that weren't right either. Still, I kept them, but I searched the Internet for speaker reviews that mentioned opera, for a reviewer who mentioned classical or operatic music, for ads from manufacturers who quoted from operatic literature, who gave their speakers names like Callas, Caruso, or Pavarotti.

I found a company that claimed their designs were modeled on musical instruments, shaped to suggest the bodies of lutes and cellos. It was an Italian company named Sonus faber, famed for its lushness of sound, with tweeters and midrange drivers supposedly built to handle the vocal dynamics of opera and the richly complex tones of a section of violins without the harshness I'd heard in other speakers. These Italian speakers had lovely veneers and, with their curvy side panels and raked-back tilt to ensure proper time-arrival of the various frequencies, they looked like instruments left onstage in their cradles, waiting for their masters to return and play them. I found an online dealer in Southern California who sold them and, letting go of my wallet, promptly bought a pair of floor-standing, three-way speakers made in Vicenza, Italy.

Although I'd been cautioned by my expert friends not to expect too much "right out of the box," as the speakers needed breaking in (the rubber surrounds of the drivers wanted softening and plain electricity, erratically flowing as ice-melt spring water, needed to find a consistent riverine course over the new wiring inside each speaker), I put the svelte Italian towers on their metal-frame stands right away, arranged them in my living room, hooked them up to my system, then jumped right in and dropped CD after CD into the tray of my player, anxious to hear some semblance of a real human voice emanate from my stereo at long last.

I did. I heard French soprano Véronique Gens singing an authoritative "Come Scoglio" from Mozart's *Così Fan Tutte,* and the

Concerto Köln orchestra accompanying her sounded even better—richer, fuller than the singer's voice—making for a lovely contrast of timbres, the one singular and stirring, the other sumptuous and complex. A choral work by Palestrina sounded spookily beautiful, the polyphonic vocal lines each distinguishable and lifting eerily in air as though a cathedral's small apse rose above me. I played the toughest and most dynamic operatic voice I could think of, one that had destroyed my system each time before—the Maltese tenor Joseph Calleja—and the blessed Italian speakers had him fine. There was no hashiness, no glare that I could tell, and only a shade of brightness that I attributed to the speakers' lack of break-in. I listened to Jascha Heifetz playing the Brahms violin concerto, and the power and sonority of his instrument were fabulous via my new combination of the Norwegian amp and Italian speakers. There was only a shade of sharpness in the attack strokes Heifetz made on his violin, an expressive bite that commanded attention rather than lulling you off into somnolence. It was thrilling. And on an old RCA recording of the Brahms Double Concerto, the orchestral basses and Russian master Gregor Piatigorsky's cello were grave and magnificent. Each section sounded very distinct from the others now—strings, woodwinds, brass, and soloists. I thought this remarkable and could not then imagine a grander sound. Finally, I played a jazz quartet, John Coltrane's *Ballads* album, and his tenor sax sound was round, warm, and fat, with McCoy Tyner's accompaniment ringing and chiming on the piano and a pleasant, almost tactile metallic shimmer emanating from the cymbal work of Elvin Jones during the crescendos. I wrote to all my friends *in gloria, in exuberant excelsis.* I finally had my shit together.

Electrified by Music

For many weeks, I spent every successive day electrified by music. At first, I wanted to hear everything I used to listen to in the past—digital versions of recordings I'd owned on vinyl during my college days, jazz acts I'd caught in Seattle, L.A., and New York over the years. I wanted to hear symphonies, chamber ensembles, and lots of opera of course. I'd jump into my Toyota Camry, make the short

drive over to the shopping mall across the Willamette River on the other side of town, and spend hours ransacking the CD offerings of our local Barnes & Noble. Or I'd dash over to the other shopping mall—the one with a taco joint, a Trader Joe's, and a Starbucks (I called it "the portal to California")—and browse like a moose in river weeds over the CD offerings at Borders (still a going concern then). I'd spend sixty dollars a pop on a small stack of old classical recordings—Isaac Stern performing the Wieniawski/Bruch/Tchaikovsky violin concertos, George Szell conducting the Cleveland Orchestra in Beethoven's *Eroica* symphony, Carlo Bergonzi and Joan Sutherland singing highlights from *La Traviata*. Quickly, in a matter of weeks, I accomplished the near complete duplication on CDs of my old vinyl collection of jazz records—classic LPs by Miles Davis, John Coltrane, Charles Mingus, Thelonious Monk, and Sonny Rollins. I'd play them and feel not only reconfirmed in my love of this music, but also as if small measures of the past were being returned to me. I was reconnecting with scattered remnants of myself I'd abandoned all those years ago, and they were mostly all still there, dormant within me and revived by elapsing strains of a once familiar music.

Nostalgia was certainly part of this pleasure, that effect of an oldies station on the car radio blaring out "Sugar, Sugar" from 1971 and your recollecting the silly kick line of stoned and sangria-ed classmates at a dorm party once upon. But that wasn't all of it. I'd play "'Round Midnight" by the Miles Davis Quintet and remember the first time I'd heard it in my room in college, the clunky Zenith turntable dropping the LP onto the spinning platter, Davis's Harmon-muted trumpet laying down the melody in a slow and smoky opening solo followed by Coltrane's swaggering, muscular one on tenor sax. I listened to this music as though it were a poem I could read that, within its orderly lines of verse, possessed the track of my own developing consciousness at the time, a musical portrait of the artist as a young man, a space of youthful mind I could invoke and conjure as though it were a childhood landscape infused with ecstatic memories and self-blandishments. For it wasn't just the music I sought, but what was within it, something that the music meant to me at specific times in my life. And that I could

return to it like some primal scene in the eternal sequence of notes in these jazz melodies meant I could re-occupy the ghost of myself as though on a visitation, hover over the young ephebe of a dumb life and slip him twenty bucks for admission and a couple of drinks at the Lighthouse in Hermosa Beach, give him some talisman of a fuller consideration than I was then able to produce. "Emotion recollected in tranquility," as the Romantic poet William Words-worth once said, and now music was my tranquil hermitage and an infinite mirroring, a reflection of the past upon the present and a simultaneous reflection upon the present by the past as they receded into each other in a long tunnel of infinitude. I looked upon this old music of mine as artifacts of a prior self and habitation, and in playing it I inquired into the vale of my own making, how I pro-nounced approval in sensibility and taste over all my failed enter-prises (a two-act play planned to be three, competency in Classical Chinese, dating a girl with a Brazilian wax), how a certain tune, heard more clearly this time, might've been the requisite catch in the throat that would have moved me to change and enact a finer tone of consciousness.

But I could play anything on my stereo finally—opera, choral music from the Renaissance, jazz, solo piano, and classic rock from the sixties. There was an aural solidity and presence to jazz record-ings that made it seem the players were in the room with me. There was a fine liquidity and delicacy to operatic voices as well as their awesome power. I played music all the time, from when I awoke to just before turning in for the night. I was a junkie for recorded music and I needed my fix every day. And when I ran out of the familiar, I sought the new; I prowled CD World out on West Eleventh Street, the Musique Gourmet downtown on East Eighth, as well as Ama-zon, MovieMars, and every other Internet avenue. I wanted to hear everything—piano trios, chamber ensembles, ancient and baroque music, the Orchestra of the Age of Enlightenment, Bach cantatas, masses and requiems and motets. I reveled in uplifted voices, bands of angels singing.

In choral music, specifically of the Renaissance, what I loved were all those voices choiring together to create a kind of hush-ing cloud of polyphony. Listening, I could imagine wings of notes

feathering the air, then soaring upward together as though flying through a sun break. My new foray into stereo was beginning to awaken many memories of music like this—disparate, intense, and isolated in that I'd not pursued them at the time I first heard them. But now I felt free to pursue these subtle suggestions, little jabs of ephemeral pleasure recalled so many years later. I started collecting recordings of compositions by Josquin Desprez, William Byrd, and Thomas Tallis—music written out of devotion and for ceremonies of worship. Their effect was like stepping into a chapel centuries old, then gazing upward at the magnificent painting covering its domed ceiling. Grand earthly and patriarchal figures sat in stately repose nearest the walls, while above them, in ascending rings of clouds, angels and cherubim floated in the soaring vault of sky.

I'd spend a morning listening for the Evangelist parts in Bach's *St. Matthew's Passion,* afternoons with Beethoven's Fifth on repeat play, and evenings with the jazz saxophones of Stan Getz and Gerry Mulligan playing live at Carnegie Hall. I wasn't making a study of things. I wanted to build a stronger feeling for each kind of music rather than expand any intellectual grasp. I was living with all this music that had entered my life again. I loved how it enveloped me, how it could bathe me in tides of a symphony like the Hawaiian nightclub music my father once played on his stereo back in Gardena, a liquidinous aural pool that rippled with a breathing sensuousness.

Bereft

My father had passed away unexpectedly from a heart attack in 1984 at only fifty-eight years of age, a time when he was looking forward, I'd assumed, to retirement. I was just starting out as a university-employed poet, having quit my Ph.D. program at UC Irvine to take a one-year job at that same school. The appointment had been engineered by my own mentor, Charles Wright, an eminent poet who was leaving to take a job in Virginia. The academic rule is you can't serve as a full-time member of the faculty while simultaneously being enrolled in a degree program at the same school. So I resigned from the Ph.D. program just after

passing my exams and getting qualified to write a dissertation. The chair of the English department, a slightly rotund man who wore pastel-colored turtlenecks under his revolving repertory of Harris tweed and gaudy sports coats, was absolutely shocked. He expected (and likely hoped) I'd turn the job offer down. But I took the one-year gig as a way out of the desiccating experience of spending a projected two years writing a dissertation on Edmund Husserl and the phenomenology of mind in the poetry of William Wordsworth. I lusted after the decidedly unacademic activity of writing poems, and I wanted, as they say in Zen, "to eat what was in my bowl."

What was in my bowl that winter became my grief over my father's sudden death. My parents lived in Gardena, and Irvine, where I taught in Orange County, was only an hour away by car. I took a week off from teaching, made all the arrangements, and gave the eulogy at the funeral to an audience almost as stunned as I was that he was gone. It tore at me inside that I couldn't mourn for him publicly or in private concert with my family because I was in charge of holding everything together—notifying everyone, negotiating with the mortuary, arranging the program, keeping track of everyone's travel, even cooking the food (a miserable roast of prime rib complemented with trays of catered *makizushi*) at the wake. Everyone else had cried and sobbed and wailed at the funeral in the chapel and then did it all over again a few times more at the graveside where an old bandmate of my brother's played a mournful "Amazing Grace" on his tenor saxophone. I never did weep. I stood apart and watched my brother, cousins, and my father's friends shudder with anguish as they held on to each other and took turns in a halting processional tossing flowers, leis, and handfuls of earth into my father's grave. As his oldest son, I thought it was my job not to give in.

Albert Kazuyoshi Hongo, born in Honolulu and raised in McCully, a district of flower shops, chop suey houses, shoeshine stands, and tiny clapboard shacks behind the main streets, got his training at L.A. Trade-Tech after the war and made his living as an electronics technician at Learjet in Santa Monica, working on helicopter control panels. But when I was a kid, he'd repaired old black-and-white cathode-ray TV sets in his spare time and worked on

tuners, radios, and stereo amplifiers, too. I remember him sitting on the floor of our apartment, holding my mother's vanity mirror behind the opened back of a TV set he'd propped on a dining chair and periodically twisting its vertical and horizontal controls on the front. After his burial, I cleaned out his closet and, in the garage, found most all his electronics gear—a rusting oscilloscope on a tilt-up stand, a multimeter big as an old Macintosh computer, small and intricate tools in various shoeboxes, and old vacuum tubes in rusting Hills Brothers and Yuban coffee cans. There were his manuals from L.A. Trade-Tech (so many, they took up over a yard-long shelf of a metal rack) and big, Coke-bottle-shaped tubes swaddled in paper towels. I think my brother and I packed most of these in boxes, along with other household things, and stored them against a far wall of the garage. We gave his clothes away to a thrift store. We dumped most all his manuals (though I kept two for myself). I remember one of my uncles asking for my father's golf clubs, a cousin wanting fishing gear—small giveaways of grief like that. My mother wanted it all out of the house if not disposed of and I tried to oblige, thinking myself dutiful. With my brother, I filled a big rental truck with things from his bedroom and storage shed, but I cannot remember where we went with them. I kept only a few mementos for myself—as I said, his gambling hat, his Sony radio in its black case of thick and perforated leather, and a small, slotted lure box filled with a few old vacuum tubes nested in yellowing, crackling paper. I didn't even know why I kept them, but I did.

His prize possession was long gone. It was that jet-black '59 Plymouth Fury, the first car he'd bought brand-new. It had twin tail fins each as tall as an orca's dorsal. When I was a kid, we lived in an apartment complex on South Kingsley Drive in Central L.A.— where Koreatown is now. And I remember him carefully washing his automobile down, hosing its slick, black body with a fine spray at first, then scrubbing it, and then hosing that off before chamois-drying the whole gleaming thing meticulously, even to folding and re-folding the chamois and re-rinsing it in a small tinned bucket of steadily browning water, wringing it dry, using it again. He'd bought it from a dealership, not from a classified ad or used car lot, and that mattered to him, so he kept the car clean and glossy at all

times. He eschewed Turtle Wax, but found something European, with more natural polymers. And he wiped it on easily, in slow and small circular caresses. Our apartment was on the second floor, in front, overlooking the small, severely crowned street, and I watched him all the time. I was six, seven, and eight, and I watched him.

The evening he bought his Plymouth, he took me along, probably because I was out of school and needed minding. I remember waiting (for hours and hours it seemed) in our old car—an Edsel, I think, green and bulbous, shaped like a cartoon drawing of a cartoon character's car—while he negotiated with the salesmen inside the dealership. I could see him inside its office through the big picture window—my father dressed in khaki pants and a brown, Polynesian print aloha shirt, the car salesmen in ties, white shirtsleeves, and hiked-up gabardine trousers. While my father sat there, scribbling on a notepad from time to time, the salesmen seemed to come at him in shifts. In the end, he sauntered back with a big grin on his face, a toothpick shoved in the corner of his smile, and helped me out of the Edsel. "No more diss wan arrszz," he said, using the Hawaiian pidgin that he spoke, and walked me over to the Fury, a killer whale of a car if there ever was one.

My father was from the "old Hawai'i," from before jet air travel brought the world so swiftly to the islands; in his day, tourists came by Pan Am prop planes and the SS *Lurline,* a Matson Line cruise ship that docked in Honolulu Harbor by the Aloha Tower off the Farrington Highway. He grew up half an orphan there, abandoned by both mother and father, raised by an elderly woman who'd been their housekeeper. Fukamachi was her own family name, though my father kept his. He told me he shined shoes on the corner at McCully and King Street so he could make change "for go ball game." He'd seen Hugh McElhenny of the 49ers run back a kick for a touchdown at the NFL All-Star game in old Honolulu Stadium. He'd seen Joe DiMaggio play center field in an exhibition game there. He'd seen Japanese planes come in low over the town on their approach to Pearl Harbor on December 7, 1941.

He was fifteen then, a sophomore at McKinley High, tailback for their varsity football team. He told me he and "Bighead" Dan Inouye, their center, ran up to watch the attack from the Makiki

hills above the McCully district, but that Inouye didn't stay. Instead, the older boy, a high school senior and future U.S. senator, rode his bicycle down to the harbor to see if he could help whoever was injured. My father didn't have a bicycle, so he just stayed on the hill and watched all the bombs going off, ships smoking, and planes like dragonflies zipping and diving over the fleet. He told me about it one summer afternoon in the backyard when I was home from college. I never knew what to do with that story. But I still can see him smiling wanly and waving his hand, a kind of stiff-arm with the palm thrust forward and tilted up, waving it in a semicircle like shaking a finger.

"Was kid still yet. No can do nutting."

He was kind of a genius at making a quiet statement.

Two years later, when he was seventeen, my father enlisted. He joined the army, shipped out one morning on the SS *Lurline,* the old luxury liner converted into a troop ship and bound for Oakland. His older sister, my aunt Grace Goya of Hilo on the Big Island, gave me a photograph taken that morning. It's a fading black-and-white, and shows Mrs. Fukamachi and my grandfather with my father, his shoulders buried in leis, looking like he could be one of my own sons at his high school graduation—lean and unvisited by the calamities yet to come. Once he reached landfall in Oakland, the

army put him on a train for Fort Bragg, North Carolina, where he joined other boys from Hawai'i. They were to be trained as replacements for the 100th Infantry Battalion, called One-Puka-Puka, a battalion made of nearly all Japanese American soldiers, who'd suffered big losses in Africa, Sicily, and at Monte Cassino in Italy. He never talked about this, but, from reading books and combat narratives kept by the various platoons, I reconstructed a brief of his subsequent travel. It's uncertain (his service records burned in a fire at the Saint Louis archives some time ago), but my father may have linked up with his battalion at the beachhead landing at Anzio on the western coast of Italy, where the Allies got pinned down for weeks. The group had been combined with the 442nd Regimental Combat Team, as that larger force now included Japanese American draftees and volunteers from the wartime internment camps, as well as what was left of the original Japanese American troops from Hawai'i. After hard losses on the beach at Anzio, where the Allies were caught waiting for weeks, there was finally a breakout and then a swift push toward Rome, to Lanuvio along the Appian Way, next north along the coast of Tuscany to Livorno and the harbor there, then to Pisa and its airport. I interviewed vets, listened to oral history tapes, and watched video interviews archived in museums by veterans' memorial associations, and in 2007, I made a trip over the line of march that the 442nd made in the spring and summer of 1944, walking some of that same ground. I went with my older son, Alexander Kazuyoshi, then twenty-two and himself named after my father, who died when I was thirty-two, and whom I've never stopped mourning.

First Moon with Vacuum Tubes

Just before I bought the Electrocompaniet integrated amp, I'd been looking at online images of amplifiers that used vacuum tubes—those pre-1960s glass bottles that glowed in the dark from the electricity running through them. There was a romance there that I couldn't quite put my finger on, a kind of mystery wrapped up in this older technology from the earlier part of the twentieth century. Sound coming from transistors, those tiny, fingertip-sized devices

inside amps that pushed vacuum tubes into obsolescence, somehow didn't have the same mystique. Transistors were hidden inside the casework for one thing. For another, they were small ceramic tiles with circuitry printed inside and their workings were invisible. Not so with tube amps. Not only did their tubes have intricate filaments that glowed, but they came in varied alphanumeric designations and exotic shapes—EL34s like oversized test tubes, KT88s like broad-chested vitamin bottles, GZ34s like old Coke bottles. There was an aura of the laboratory about them, and they conjured thrilling days when a man might capture lightning in a Mason jar and draw it down through an arcing Jacob's Ladder (an impossibility, I know) to feed some mysterious engine.

Along with images of the Electrocompaniet and other solid-state amps, I printed five-by-seven pictures of tube amps as well, shuffling them like a deck of cards, fanning them out on my desk, trying to divine my own druthers via whatever pull any of them had for me.

Even as I was elated over the sound I was getting with the new combinations in my fairly modest stereo rig, I still wondered about tube sound and what it might give me. I'd been reading how tubes were supposed to be "more liquid" and organic than transistors, if less resolving and punchy. That seemed exactly what I wanted—something that could render the micro-details of performance in the human voice and shape its movement not so much as a stop-and-start thing, but as a continuous flow, the heads and tails of notes sliding into each other like small waves in a softly pitching sea. The glass bottles such amps depended on seemed to call out to me, stirring memories of my father's own tube equipment.

Browsing one of the audio magazines, I came across a photo of a new, Dutch-designed amplifier (albeit built in China) that was visually much like the tube amps of the early sixties. It featured a row of four tall output tubes jutting above its chassis that were stamped with the logo of the manufacturer—PrimaLuna—and the designation "EL34." These were tubes twice as large in circumference as the aluminum cylinders some cigars came in, and about as tall as a Romeo y Julieta #2 Corona. The amp had two pairs of signal tubes as well (for preamplification), smaller than the output tubes by

more than half, the size and circumference of a cigar butt, arranged in a neat phalanx before the row of EL34s. Behind them was a large metal case, running nearly the full width of the amp, that housed the mysterious power and output transformers: wire-wound metal-core inductors electrically strapped to storage capacitors inside the amp and to the output tubes on top of the amp's chassis. The output transformers channeled the electrical energy stored in the power supply that drove the speakers. In front, on the handsome faceplate, were a push-button power switch and the two large knobs that governed volume and source selection—whether one had a CD or tape player, phono, or some auxiliary source like a radio receiver. One could hook them all to the amp in a single glorious, multi-source audio system.

Subconsciously, I wanted that PrimaLuna to be my new amp. I thought about it for a couple of weeks, reading reviews, then I realized it would be silly not to buy it. I put in a call to the dealer in Southern California who had sold me the Italian speakers. Kevin Deal, owner of Upscale Audio, was the U.S. importer, and he assured me the amp would be a good match for my speakers, and offered a guarantee. Unasked, he offered me a discount as well. I didn't need the amp—the Electrocompaniet was doing splendidly—but I wanted those tubes. It cost less than a thousand dollars and I'd just gotten a royalty check for about that much. I went all-in.

The PrimaLuna arrived and I placed it in the system, supplanting the Electrocompaniet I'd acquired only a couple of months before. Its sound was even more gorgeous, and it sent me to an even higher order of deliverance among the acoustic angels coasting on invisible clouds across my living room.

Amps from Kits

One day, as I was listening to a captivating bloom of choral voices from the stereo, the gloriously tubed PrimaLuna amp producing a bounty of harmonic richness, a flood of memories came back to me. They were all about my father when he built his stereo.

During the early sixties, at our small, three-bedroom tract home in Gardena, my father spent his evenings fixing things—

mostly the malfunctioning cathode-ray tube TV sets of relatives and neighbors—but one day he got the notion to build himself something.

He decided to build a stereo.

A Heathkit came first—a box with all the parts for a wonderfully intricate amplifier with tubes inside a bronze-colored birdcage of protective metal. In the kit were gray metal bits, a sheet metal chassis, brown Bakelite tube sockets, different sizes of vacuum tubes, brown and green and yellow resistors (some striped), white- and black-jacketed wires, a silver coil of solder, a black blob that was what he called a "varistor," black plastic control knobs, bronze rods, and two heavy and black-painted blocks of metal that were the output and power transformer. There were raw aluminum chassis parts, phalanxes of carbon film resistors, and electrolytic capacitors lined up in rows like army men. He had Phillips and flathead screwdrivers, needle-nose and bent-nose pliers galore, wire cutters, and strippers with their calibrated jaw-holes, but his most important tool was a big black Weller soldering gun. It had two penlights, like missiles at the ready, on either side of its metal snout. It had a heavy, outboard control box he could dial up or down, according to needs for high or low heat. At a touch it melted a coil of solder and made it run like silver snot down the fine, copper wires. When that was done, he'd place its tip on a wet sponge in a coffee cup saucer by his knee.

He assembled all the parts guided by a booklet of instructions. The Heathkit motto was "We won't let you fail," and help was supposedly just a phone call away—only my father couldn't call, because, from long travail, a voice on a phone line was something he just couldn't hear. On his own, he laid out the schematic drawing, marked each component part with masking tape and a number correspondent to the instructions, and, solder-joint by solder-joint, put the thing together over the course of a few nights. He'd fire the gun, wet the tip with a bead of solder, then carefully bend over the evolving inner structure of the piece, adding a resistor here, coupling a cap there, completing circuits with mercurial runs of gleaming metal that hardened into small, silver pearls to knit that amp together. Puffs of gray-blue smoke came up from the Weller

gun as he worked, and he'd blow against the tip like it was a pistol each time he was done.

Soon he made another—a Dynaco Stereo 70, something that looked not at all like the caged bird the Heathkit was. The Dynaco had its tall bottle tubes (EL34s), black-belled transformers, and blue chimneys of capacitors all sticking up from the nickel-plate chassis of the amp, making it look like the above-decks of a steamship. I found out later that the amp did come with a cover—its own brown cage. But my father must have dispensed with it, preferring, as I did, to see the glowing tubes at night like a small tea tray full of tiny hurricane lamps or a washtub filled with seawater and floating candles to light an evening's meal at the beach.

This was before the Beatles came to New York and were on *Ed Sullivan,* when the *Perry Como Show* was the top program, when singers wore penny loafers and cashmere sweaters in pastel colors, and the music on the radio was still fifties pop—hair-sprayed crooners like Paul Anka, Bobby Darin, and, worst of all, Fabian. But my father, a man of his own generation, a soldier during World War II, listened to none of these. He liked forties big band music— Tommy Dorsey, Harry James, and Glenn Miller, his favorite. He liked what he called "Hawaiian music" too—Arthur Lyman on the vibes, Alfred Apaka on the 'ukelele, and the Sons of Hawaii, a group of virtuoso musicians who sang in Hawaiian. Before he built his stereo, our record player had been of the kind we had in school—a single-unit affair with turntable, amp, and speakers all in one heavy box, and the sound it made was like that of a PA system, hollow and scratchy. It shouted like noise from the inside of a tin can when you turned the volume up. And my father always had to turn the volume up because of his bad hearing.

We'd have to speak up so my father could hear. Our TV was loud, we sometimes had to shout to him as though the dishwasher were on when it wasn't, and we always had to speak in pidgin, that mellifluous and gloriously abridged creole of English that evolved in the islands among Asian and Portuguese workers and their Scots and American bosses in the sugarcane and pineapple fields. Pidgin was always easier for my father to understand.

"I going mah-kett!" I'd say, when I ran an errand to the grocery store. Or, "You come eat—*kau-kau* ready," when it was dinnertime.

Bit by bit, we added hand motions, making the gesture of scooping with chopsticks as a call for mealtimes, a shaka sign to the ear for a phone call that was waiting, hands-on-a-steering-wheel for taking a drive or needing him to drive us somewhere.

The stories varied about why he was this way. My mother said that he'd just never protected his ears, working on the tarmac without aviation muffs at the Kaneohe Marine Corps Base as jets landed and took off, without earplugs operating a jackhammer when he did roadway construction in Honolulu. Much later, after he'd died, one of his older sisters told me he'd had a fever when he was a child and that it seemed to damage his hearing.

"It was always faint," the decorous woman said, "and we had to raise our voices like we were on the playground even though we were just sitting at the dinner table."

He'd tried hearing aids, but gave up on them as they only made what was muffled much louder, not clearer.

But he could hear certain voices, certain frequencies, very well. He'd gravitated to a Sony transistor radio the size of a thick paperback book. It was a birthday gift from the family and had a tiny earphone that attached to a jack he could plug into its top, and he'd listen to Harry Henson, the baritone announcer for the track at Hollywood Park, calling out the day's results over the radio, almost every afternoon. He could hear Vin Scully, the great voice of the L.A. Dodgers, clear as a baritone bell. And he could hear vibraphones and guitars and trombones.

While he built his hi-fi system, it was my job, as an eleven-year-old, to sit in the "sweet spot" of listening between the two channels of his stereo and report exactly what it was I was hearing—a shimmer of the brass chorus behind a Harry James trumpet solo, a peak of over-driven treble, a muffled splash of a drummer's cymbal strike—and to describe it to him. He would then both note it in a small book he kept and make corrections by tweaking his system—re-soldering wires, swapping out vacuum tubes, and repositioning speakers. He'd say "capacitor!" and change out a tiny red

oblong of plastic that looked like a shiny piece of cherry bubble gum stuck on the circuit board he'd just pulled from the case of an amplifier or preamp. Or he'd utter some spell of numbers and letters—"Six-DeeJay-Eight!"—pulling a pair of RCA tubes out of their glass sockets at the top of an amp, swapping them for two Dutch Amperex tubes he at first just gently shook in his hand like a pair of dice, then blew on for luck, then shook again. When he rolled them out on the carpeted floor, completing the mime from his days as a GI, he called them "Bugle Boys" and told me they were "top secret." He grinned at his ex-serviceman's joke, picked them up, plugged them into the amp, and, like magic, the stereo would sound completely different after he flipped everything on—sharper or warmer, richer or full of clearer high notes. And every night this vivid display of do-it-yourself I witnessed was then packed up and put away like a magical troop of toys well cared for, stored in tool chests, shoeboxes, and Tupperware tubs, and then neatly removed from sight. It was a breathtaking ritual, and I looked forward to it like a puppet show I got to see enacted every night.

For the longest time, I didn't know what happened with his kit amps. They were gone before I knew it and my father was then at work on something he'd dreamed up on his own. More than anything else, he must have wanted to understand the theory and applied technology behind these audio circuits; the Heatkit and Dynaco electronics, great as they sounded, disappeared after only a few months. My father then built yet another amplifier, an integrated one from plans and schematics he must've gotten from a hobby magazine, or maybe it was based on a circuit of his own invention. He had to have designed the chassis, gotten it cut and bent into shape, assembled all the parts himself, tested them against specs, and then put them all together in a metal box he could run a cable through and plug into the wall. Night after night, I remember him stripping down to long pants and a sleeveless cotton undershirt—the electronics were hot, the test gear was hot, and the season must've been summer or fall (always hot in L.A.)—spreading out sheets of the *Los Angeles Times* or the *Racing Form* over the carpeted floor, and meticulously laying out his gear. His project went from a small village of electrical parts arranged over old newspapers

to a tidy, gleaming metallic thing the size of a samurai doll in a glass case, a compact miracle with the power to make a vibrant lagoon of music appear in our living room.

From these memories invoked by my acquisition of that Prima-Luna amp, I realized that my father had built his stereo system as a way to take in the last of whatever he could hear—the horn choruses of "Moonlight Serenade" or a vibraphone's gorgeous shimmer—sounds he'd loved as a seventeen-year-old soldier during the war. That he could not precisely hear them anymore was something I didn't fully appreciate until late in that first year of my own immersion in audio. And when it hit me, I wept. A warrior, he'd come back to practice peace. And, through his pure love of sound, he'd imparted a legacy.

Bugle Boys

It was the summer of 1963. I could hear TV noise drifting in from the family room—a game show with its odd cycles of a moderator's maundering voice to the jittery speech of contestants and enhanced laughter from the audience—while I sat cross-legged on the living room carpet, my back against the sofa. My father had drawn the drapes closed and switched on a small desk lamp over his stereo. He was standing across the room, fiddling with his amplifier, lifting it up from the table it was on like it was the lid to a wooden box, shining a penlight into its underbelly, reaching out with his index finger and jiggling a wire. A look of puzzlement crossed his face, gleaming in incandescent light. He set the amp back down then, softly, and started checking how firmly its tubes were seated in the amp's sockets. The tubes were squat Coke-shaped bottles, tall and slender bottles, and blunt little glass rockets with tips like nipples on them. Once he was done, he flicked the lamp off, flipped a switch on his amp, and fired it all up, turning to grin at me, a toothpick dangling from one corner of his lips. I saw a momentary burst of light from all the tubes, then, in that darkened room, the filaments inside each one started to glow red like embers inside a half-dozen Halloween jars set out on the lawn.

He'd assembled a marvelous system—an Empire 398 turntable

with a gold, brushed aluminum finish to its plinth and platter; the tube amplifier he'd made himself; bookshelf speakers he'd bought at Fedco (a local discount department store); and a huge table-and-cabinets he'd made woodworking in the garage. My mother called it "the Green Monster"—it was painted a mint green and was so big and deep, it held all of my father's records, some of his tools, and a bottle of Four Roses tucked inside it too. Our dowdy living room (with crocheted doily antimacassars draped across the arms and back cushions of the sofa) got a new and glamorous look just from having all this gear and new furniture imported into it.

He could not hear all the little details of sound he asked me about. He wanted to know what Sylvania tubes sounded like as opposed to GEs, and he swapped them in and out over and over again, asking me questions, writing notes in a shirt-pocket notebook as I answered, fishing out more tubes to try. Sometimes, he'd pick out a pair and drop them into the pocket of his aloha shirt, and they'd sway inside the blousy fabric and click and clatter against each other like dice as he worked. There were tubes marked in all sorts of brands—labels from England and other parts of Europe like Mullard, Siemens, and Telefunken—but my favorites were the little ones called "Bugle Boys" for the cartoons of anthropomorphic tubes blowing bugles stamped on their glass. My father said these came from Holland and were especially "sound good kine." Even with his faint hearing, he could appreciate their warmth and richness.

"Whatchu hearing?" he'd ask, looking up from over the spinning record, adjusting the amplifier's volume. "Too loud?—Good?—'Nuff?"—his other questions as he calibrated the system's level of sound.

What did I say, just turned twelve and full of happiness during those sessions? I can't recall one thing, except that swing tunes like "Shake Down the Stars" and "American Patrol" commingled and interfused with the lavish sounds of Arthur Lyman's vibes on "Ilikai" and "Taboo," all taking me up in their bounty like rolling waves inside a lagoon as the tide rose and surged in over the reef. It felt like warm, sticky waters swirling around me, amber-lit, vibraphones and trombones and shimmering brass floating like boated

candles in an acoustic bath. How could I have explained to him what I was hearing? How would I have translated gorgeousness into pidgin—our sweet and common language from the islands?

I t'ink diss wan izz good!

I remember my father grinning as he spun his LPs, pleased with me, and my feeling I could do something for him. It was just one summer of our lives, but night after night it went like that, my father changing tubes like dice in a game of craps, shooting for winners, the music like tropical waters coiling against our ears.

Part Two

I Started Out
on Stereo

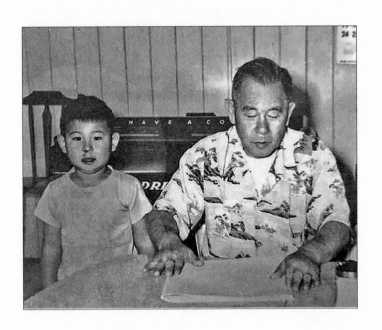

Beginnings

My aural life began among soft constancies of wind furrowing through the canefields and the chatter of worn coral and lava stones scumbling on the beaches along the North Shore of Oʻahu. These wind and water sounds were my first companions and the piping of seabirds over the sands and marshes my second. Then I heard blasts from the whistle at the sugar mill, signaling start and stop times, lunch breaks, and time for the cane to be bundled and carted and the trucks to haul everything back along the red dirt roads. The grinding of gears was next, the grunts of machinery and the susurrant surflike sounds of wet tires rolling along Kamehameha Highway on rainy days. On Sundays I heard the flat bang of a Buddhist gong at the Betsu-in, the priest knocking his delicate hammer (it looked like a dandelion on a long stem) against the fish-mouth of his wooden bell lying on a red satin pillow on the dais in front of him. He chanted, a deep warbling at first, then a series of cascading moans and mutterings that seemed to rise from within his belly and bowels, rhythmic and mesmerizing, signing the air with rapture and piety, a scroll of gutturals and nasal drones that transformed both innocence and guilt into penitence, the profane into the sacralized, if only for the moments their sound made of a boundaried territory.

I remember my first intimate friend was a tiny portable record player—some squarish, powder-blue thing with a built-in loud-speaker and tonearm the shape and weight of a Popsicle. I spun a banana-yellow 45 on it, a recording of "Mr. Tap Toe"—a rhythmic rattling laid over a catchy jingle that I played over and over again in the yard, the driveway, and while I sat on the smooth stump of a fallen eucalyptus tree that was my mount. I shouted, "Hi-Ho, Silver!" like the Lone Ranger, the vivid overture from Rossini's

Guglielmo Tell trumpeting in my memory as my horse galloped under me in my imagination. It was my first acquaintanceship with orchestral music, though, of course, I never realized this until only a few years ago. It was also my introduction to fanfares, opera, and bel canto, though how could I have known that then, a lifetime before I took in Joyce DiDonato in a tenderly sung *La Donna del Lago* at the Met? Music was a Hawaiian band at a family *lū'au*—a bass, *'uke,* and guitar trio singing lovely harmonies around a falsetto lead that made the grown-ups applaud, laugh, and cry as they passed plates of *lau-lau,* coconut cake, sushi, and sweet fish wrapped in *ti* leaves to each other around a circle of belonging and welcome. Music was stately Japanese *enka* and rousing *min'yo* folk songs sung by the neighborhood seamstress or local tofu-maker at the Bon-Odori, our festival for the dead, their anguished warblings, shouts, and *kobushi* melismas backed by a band of cane workers and field bosses who played saxophones, clarinets, hand drums, *shamisen,* and guitar while the community of dancers in summer dress circled around them counterclockwise, the sleeves of their lightweight gowns and *happi* coats fluttering like kites in the evening trades. Music was "The Stars and Stripes Forever" played by the Filipino marching band practicing with the drill team in full regalia on the wide lawn beside the Catholic church on the edge of town.

When I heard "Melancholy Serenade," the violin-drenched theme song for *The Jackie Gleason Show* on our green-screen, cathode-ray, black-and-white television, it was to me a heaven of sound, luxurious in its pace, rich in brass and horn harmonies, and a little miracle of time in how it ordered its romantic articulations in measured and progressive sections that enhanced its tiny drama of revelation. It would repeat itself and rise up in a flourish of strings and brass together in a grand crescendo. And then the hilarious shows would ensue . . . Crazy Guggenheim and Joe the Bartender with his pomaded hair. I think I heard Julie London sing "Cry Me a River" on the show once, likely lip-synching, but would I have known that? She was tall, thin, and elegant, with a coiffure of short curls, the fine ridges of her collarbone and a hint of cleavage showing over a strapless, sequined sheath dress. Her voice was breathy, then throaty and dulcet with a huge dollop of reverb echoing in the

mix, a laconic stand-up bass and deliberate electric guitar her only accompaniment.

Shrines were the Shingon-shu temple, a movie theater, and those countertop jukeboxes I found whenever the family would drive out to Haleʻiwa, where there was a restaurant that had them placed at every third chair along its counter. I was five and rushed to them immediately and begged a nickel or a dime so that I could play a tune of my own choosing. One of the Sun singles by Elvis Presley might be among them—"That's All Right," "Blue Moon," or "Don't Be Cruel"—but I think I always preferred novelty tunes like "Sixteen Tons," a country ballad by Tennessee Ernie Ford; "The Great Pretender" by the Platters; or the folky "The Wayward Wind" by Gogi Grant. But the grandest shrine of all was the glorious, push-button-operated Seeburg jukebox my grandparents installed in their own establishment on Kamehameha Highway in Hauʻula that they called the Crescent Café. You could see the 45s lined up like upright piano keys inside the bulbous glass cabinet. You could see the little robot shuttle work deliberately back and forth as it found each selection. A single tune cost a nickel and you got six plays (and choices) for a quarter. My younger brother, three years old, was so obsessed with Elvis ever since we saw him on *The Ed Sullivan Show,* he would torture me by picking six straight plays of "Hound Dog" whenever he got the chance. But even then I understood curatorial selection and taste and carefully chose a sequence of songs, never the exact same playlist twice. I played one or two American hits, but mixed in among them would be "Hiʻilawe," a slack-key guitar song to please a Hawaiian policeman who'd come in for lunch every day, a Western swing tune to please the servicemen who would also come in (I think this is how I first came to know "San Antonio Rose" by Bob Wills and His Texas Playboys), a bluesy and quaverously sung Japanese *enka* for my grandparents, and a big band tune by the Glenn Miller Orchestra that I thought my father would like. If you dropped in for a hamburger or *saimin* noodle soup lunch, you'd sit on wooden stools at the counter or around square, green Formica tables and hear a medley of 45s chosen by a five-year-old MC who had already understood music was chop suey, just like American culture itself.

Taboo

In November 1959, Kīlau'ea-Iki on the island of Hawai'i (where I was born) had a spectacular eruption that sent a fountain of glowing lava spuming from the south wall of the summit crater. The fountains grew from fifty to a hundred feet high along a fissure almost a thousand yards long. Lava cascaded down on a forest of 'ō'hia trees on steep slopes above the caldera, and the fountaining heights rose to over two hundred feet at times, building up a huge ashfall and scattering shards, bits, and strands of pumice over the landscape and upon the heads of locals and tourists who'd started to gather to witness the spectacle. My cousin Sandra, a pre-teen at the time, told me she'd hold a fold of newspaper above her head to block the cinders of hot pumice, stand there, and continue to marvel. The eruption created a huge lava lake, full of floating black islands of hardened rock, sucked down by periodic whirlpools of drainbacks sinking into the vent of the volcano, and also a huge mound of built-up tephra eventually named Pu'u Pua'i, which still stands there today. But it was the magnificent fountaining episodes that seized the world's attention, photographers from *Life* and *Time* capturing the events in memorable color photographs: magentaed, stories-high chanterelles of molten rock, demonstrating the unique sublimity of Hawai'i's active volcano.

And so, for the 1961 mono pressing of Arthur Lyman's *Yellow Bird,* an LP of vibraphone music dubbed *exotica* in its brief late-seventies revival and *New Age* on the iTunes of today, its issuing company, the plainly labeled Hi Fi Records, chose a particularly gaudy photo of one of Kīlau'ea-Iki's fountaining episodes for the cover shot on its outer sleeve. The image features a three-armed, pink, Balrog-shaped fountain rising from the black pit of the crater, itself backgrounded by a wash of purple—the walls of its sheer, clifflike slopes—and an amateurish fringe of the skeletal trunks of charred, juvenile 'ō'hia trees in the foreground. Overlaying it all in floating block letters are the title and Lyman's full name in woodsy, Frontierland yellow and fuchsia, suggesting an aura of Polynesian primitivism, I would guess.

This was one of my father's favorite records and, along with

his big band music, I heard it over and over again so many times, I think its character and style penetrated into my nascent musical soul, taking up permanent residence and becoming a point of reference for all the listening, naïve or tutored, that would follow. It blended into an interior imaginary landscape and signified, to my alienated and diasporic consciousness, things Hawaiian, faraway, and paternal, a sound that authorized my own childlike nostalgia for my island beginnings and for my father's sponsorship, aesthetic and existential. But Lyman's exotica style was in no way "authentic." It blended cool jazz of the period with the clichés of Chinese gongs, birdcalls, and other theatrical noises to evoke a steamy pastiche of music from an imaginary tropical jungle.

Lyman's band was a quartet, consisting of himself on marimba, xylophone, vibraphone (his main instrument), bongos, conch shell, and various other exotic wind and percussion instruments; a pianist who also played celeste and glockenspiel; an upright bassist who handled birdcalls and 'ukulele; and a percussionist who doubled on marimba, xylophone, or bass when called for. Lyman also handled birdcalls—his signature feature, supposedly improvised one drunken night in a performance with a prior band—and, together with his bassist, they hooted, chirped, and squawked up a storm at times, simulating the Foley tracks of cacophonous bird noises in a Tarzan movie, much to the delight of his faithful Waikīkī club audiences of tourists and his fans in the land of hi-fi. Besides "Taboo" (its melody accentuated by a scratchy guiro and ambient birdcalls) and the popular calypso tune "Yellow Bird," his playlists included titles like "Waikiki Serenade," "Bamboo Tamboo," "The Love Song of Kalua," "China Clipper," and "Adventures in Paradise." Blessedly, the band also recorded versions of Ellington's "Caravan" and Ravel's *Boléro*, featuring a melody played sequentially on the celeste, flute, vibes, marimba, and doubled vibes and piano. Lyman performed a chiming, easy-listening kind of instrumental music mostly, the sound of his lavish vibraphone with a magical depth that rippled sinuously from my father's small bookshelf speakers, seeming to echolocate in sensuous waves of notes sent vibrating directly into the interior cavities of my torso and sending the blood flow in my body into a kind of rhythmic pitch like a small sea affected by

a monophonic moon. At times, it was as though I were the jouncing strings of a pedal steel guitar and my father's stereo the tremolo
pedal stretching and bending the gently strummed chords within
me in small, waving swells of grandeur and bemusement. Its effect
verged on the addictive, especially as rendered via the magisterial
capabilities of my father's Dynaco tube amp and preamp, blessed
with their own sensual inner auras.

Kainoa

After we moved from Hawai'i to Los Angeles in 1957 (I was six), we
moved again to the San Fernando Valley for a few years, and then
finally, in 1963, to Gardena, where we again, as in Hawai'i, could
live mostly among Japanese Americans. It was there that my father
started up his audio hobby. Though it was he who mostly played
the stereo, my mother sometimes did too. She favored a postwar
Japanese album she picked up from an electronics store in Little
Tōkyō near City Hall in downtown L.A., where she stayed on in
her civil service job and worked all day. This was *Sayonara Farewell
Tokyo: Souvenir Songs of Japan,* recorded on a ten-inch "Unbreakable Vinylite" that was a vivid, transparent red. The artists were
the Club Nisei Orchestra, a Japanese American band in Honolulu
that covered arrangements they learned from Japanese swing 78s
brought home by GIs on their way back stateside after serving in
the Occupation forces. On the 49th State record label (the company
had jumped the gun on the exact order of Hawaiian statehood), its
cover featured a color photo of a Japanese woman dressed in a pink
kimono and holding an old-fashioned bamboo umbrella, also pink,
a dark-green-and-black background (was it a lotus pond?) tastefully
blurred behind her. The music was an amazing blend of American swing band style applied to *ondo* (traditional Japanese festival
dance melodies) and *enka* (sentimental narrative ballads, sort of like
Japanese country and western). I recognized many of the melodies
and lyrics from summer Bon-Odori festivals I'd gone to as a child
in Hawai'i and the jukebox records my grandparents had in their
diner. They signified to me the spirit of another time and the place
we had come from.

The album had titles like "Tokyo Serenade" (pronounced *seh-rah-nah-deh* by the contralto singer), "Tokyo Boogie Woogie," and "Japanese Rhumba," so you can imagine the cultural crossing, the incredible mash-up all this was. I suppose if you weren't from the islands or Japanese American, you might react to all of this as some sort of parodic pre–*Blade Runner* musical Walpurgisnacht, a farcical brew of East and West, and just a bad and racist Mel Brooks joke. And you might be right—there is a klezmer-like blend of brass, woodwinds, and novelty percussion, a mimicry of American hip culture pulled through the ragged strainer of severe ethnicity, something unmistakably Other about its sound, yet also an affectionate, if off-kilter incorporation of American swing arrangements. But to my ears, it sounded just like home—the Crescent Café my grandparents ran in Hauʻula, the festival band up on the *yagura* tower at Kahuku Betsu-in's Bon-Odori, dancers in colorful summer robes and *happi* coats cycling counterclockwise around it, penitents earning alms for the betterment of karma for their ancestral dead. Sleigh bells would shake, temple blocks would tock, and a quaverous female voice, backed by a flute, would tell the tortured story of a young woman's sacrifice, spurning her lover's offer of elopement out of loyalty to family, damning herself to a life of loneliness and servitude, her passionate heart, once inflamed, turning cold in the falling snow that surrounds her.

I'd two favorites among all of these—the famous 1939 Japanese hit "Shina no Yoru" ("China Nights") and "Ginza Kan-Kan Musume" ("Ginza Can-Can Girl"). The first tells the story of a soldier, stationed far from home in Manchuria during the Japanese invasion of China, full of longing, imagining returning to *the green, green grass of home,* sung in a lugubrious, pentatonic melody. He imagines stepping on a junk that will sail him back to his love, who waits for him as the war draws on. The melody is catching, with unusual intervals, captivating syncopations, and critical performance ornaments that made the tune an earworm even for American GIs who heard it in bars all over occupied Japan. Affectionately, albeit casually colonizing and sexist, they gave it the English title "She Ain't Got No Yo-Yo." Navajo code talkers brought the tune back to their drum circles. And third-generation Japanese Ameri-

cans, knowing nothing of the meaning of the lyrics, would sing it at family parties and make their parents cry. But "Kan-Kan Musume" is up-tempo—a raucous, jump tune with a sinuous beat and lyrics that tell the story of a sassy dancer in Tōkyō's famous entertainment district, not unlike a song Liza Minnelli might sing in *Cabaret*. The Nisei band really swings on this number, horns hitting hard, the musicians echoing the lead singer in the chorus verse: *Kore ga / Ginza no / kan-kan musume* . . . "This is none other / Than the Ginza / Can-Can girl . . ." The song has panache, and I can remember a cousin singing its lyrics in jubilation, whirling his hands, making up a slick dance move, sliding backward on his heels in an ur-moonwalk across the packed dirt under a jacaranda tree as he celebrated victory in a game of marbles. *Kono ko sugoi,* he sang, "this boy is great," gender-bending the lyrics, stretching out, getting his schwerve on, Buddhahead flamboyance let loose through his triumphant play and the irrepressible insouciance of this show-tune straight outta Ginza-chō.

It wasn't only this hybrid Japanese American music that my mother played. She liked Hawaiian music too—mostly the songs of Marlene Sai, a woman of *kānaka maoli* (Native Hawaiian) and Chinese descent who was a kind of Sarah Vaughan of the islands. Sai had a smooth alto voice with a dulcet vibrato perfect for the Hawaiian and English-language songs written in praise of romantic love and island things like sand and seashore, fragrant flowers with Hawaiian names, and other features of the landscape. Unlike Lyman, Sai performed mainly songs written by local artists and had prominent local musicians in her band. Her songs ranged from love ballads to jaunty numbers written to *hula* rhythms derived from the Spanish-influenced songs of Hawaiian *paniolo* (cowboys). She began her recording career in 1959 with a 45 rpm recording of "Kainoa," a *hapa-haole* song (with lyrics that are partly in English, partly in Hawaiian) by a friend of her uncle, the popular Hawaiian singer and *'ukulele* player Andy Cummings. It was a song written by a man promising love for his wife beyond his death, and it had particular appeal to the local audience. It became an immediate hit in the islands and Sai, then only eighteen, moved from a casual gig in a club on the windward side of Oʻahu (with the not-yet famous

Don Ho as her organ accompanist and boss) to Duke Kahanamoku's supper club in Waikīkī. Before the year was out, she recorded a full-length 33⅓ LP of mostly Hawaiian songs. Lehua, her local record label, entitled the album *Kainoa* after the big hit. My mother played it as she sat on the living room couch doing up her hair in curlers, paying bills, and eating pistachio nuts. She never sang along, but it was about the only time she was quiet and not snapping at us or my father. I imagine now it brought her mind back to her own child-hood and teenage years growing up in Lāʻie, the sugar plantation village out on the windward shore of Oʻahu. I knew from photos that she'd practiced *hula* herself and was a high school cheerleader. I saw her smiling moon face lying prone behind a row of pom-poms or kneeling beside an array of *ipu,* the large, polished gourds used as percussive and rhythmic instruments to accompany the *hula.* It saddened me that she was so cheerful in the photos, while at home with us she was always gruff and dissatisfied, chronically sharp with orders and complaints.

But when she listened to Marlene Sai on the stereo, our cranky house would still itself and be at peace while the gentle music, set to hypnotic *ʻauana hula* rhythms and commemorating flower leis, comparing a lover's face to a flower in the fields of Paoakalani and praising the children of Hawaiʻi, flooded through the rooms, vibrant and comforting at once. One evening, for no particular rea-son, my mother told me about the hand gestures of the *hula,* how a palm held upward with the fingers cupped and then deftly revolved and released could mean a flower, or a kiss if it touched the lips; that rain was a sinuous wave of fingers on both hands arcing diago-

nally in front of her body. She said you had to breathe slowly and "watch yourself from inside" so that you could control each loving motion, natural to your own body swayed to music. That was the only time she ever spoke that way. It was 1964 and she was alone nights, my father working the swing shift, my brother and I about to go to bed, the scent of her cold cream dominating the living room air but for the sinuous music of Marlene Sai curling around us.

Over fifty years later, during her last illness, while she lay in a hospital bed in a care facility in Carson, California, not far from her Gardena home, I sat by her and operated a cheap cassette player I placed by her bedside. I'd found scores of tapes in shoeboxes in a family room cabinet, and among them were the albums *Kainoa* and *Club Nisei* (the latter the same recording as on the original ten-inch red vinyl, plus bonus tracks, but now renamed, stripped of the dated title that had appealed to the foreign and the exotic). I played both of them—and others by the Sons of Hawaii and Japanese *enka* singers— during the last, diminishing weeks my mother's life. It calmed her and her wrinkled face, furrowed lifelong by scowls of an inner anger and disappointment, smoothed and freshened despite the noisy hall- ways and ghastly fluorescent lighting of the place. She spoke of *home* and wanting to go back there, to sit under a plumeria tree, look at the pretty white-and-yellow flowers and smell their scent.

Cottonfields

I wish I could say that great music touched me early in my life, but that is just not so. I think a chance encounter at a department store led to my own first passions in listening. It would have been during the ritual of late-summer shopping with my mother for new school clothes when I was twelve or thirteen. She'd have taken my brother and me to the May Company or Sears in one of the shopping centers with those huge parking lots along Hawthorne Boulevard in Tor- rance. We happened to pass by a section of electronics that had been playing a new record by the Brothers Four, New Christy Minstrels, or Kingston Trio. What struck me was the purity and clarity of their voices, vocal images alive in the enspirited air like Christo draperies

hung from the branches of a mesquite tree in Tucson or Santa Fe. This was during the second wave of the folk revival, commercial labels signing groups more scrubbed, polished, and youthful than the politically progressive Weavers or rootsy Carter Family. The crew-cut, coat-and-tie-wearing Limeliters had established themselves on the national club and college circuits and shown the music industry the new, successful blueprint for "folk singers." I think my mother, uncharacteristically, recognized my rapture and wanted to give it validation. She opened her purse, got her checkbook out, and paid the store clerk for a single LP. It was *The Kingston Trio* on Capitol Records.

So, with allowance money, I began acquiring more LPs of all these groups, nagging my parents to take me to Wallichs Music City near the May Company on Hawthorne and Artesia Boulevard, flipping through the browser bins, carefully selecting the three records at a time I was allowed to listen to, and lining up for my turn in one of the half-dozen semi-isolated listening booths provided for auditioning. Wallichs was a chain business with stores all over the Los Angeles basin. The first one, started in 1940 by Glenn Wallichs, an owner of Capitol Records, was located in Hollywood on Sunset and Vine and became the place to go for vinyl (45s and LPs), instruments, phonographic equipment, sheet music, and concert tickets. My local store opened in 1963 and it became a kind of mecca for me, a place where I could spend an hour at a time sampling dozens of LPs, ranging through what was in the Folk section, and also trying out the recordings of acts I'd seen on *The Ed Sullivan Show.* Though my parents favored mainstream artists like Frank Sinatra, Steve Lawrence and Eydie Gormé, or the execrable Tijuana Brass, the free listening sessions at Wallichs introduced me to my own, independent alternatives—Johnny Cash, Nat King Cole, and the Beatles. I bought Cash's *Ring of Fire,* Cole's *Ramblin' Rose,* and *Meet the Beatles!* after playing battered and groove-scoured demo copies at Wallichs.

An attendant would be watching you from a dais perched slightly above floor level, and when a booth became vacant, he'd nod to whoever was at the front of the line to pass through a ratch-

eting turnstile to an aisle leading to a bank of six or seven glassed-in booths, each equipped with a glass-paneled door, two or three stools (so friends could join you), and built-in counters with a clunky phonograph and a stereo set of four-inch scratchy speakers (mounted under slatted grills) embedded in them. The phonograph player had adjustable speeds for 45 rpm and 33⅓ rpm records, a 33⅓ spindle with a retractable and circular pop-up around it to accommodate 45s, and a hideously heavy tonearm made of cheap metal alloy that wore down the grooves of the demo records in almost no time. As you listened, you had to filter out the harsh scouring sounds, pops and ticks, and general scratchiness of most every session in order to make your judgment. Sonics were impossible to assess, and in any case, more refined hi-fi issues like compression, resolution, and separation of images were hardly in my consciousness. I was for melodies, pleasing tones to the voices, and maybe the clarity of guitar playing. I noted lovely string-picking, the pulse from a bass, and that was about it. Yet the entire experience seemed luxurious in the extreme to me, a child of working-class parents whose only leisure hours of the week were spent in front of either the television or the home stereo. At Wallichs I sampled, tried to make some kind of pre-adolescent discernment of taste, and usually, paying $4.99 full retail, took only one choice record, wrapped in its crisp cellophane, home with me.

At home, the stereo was usually free (my father tended to play it only on weekend evenings), and I'd burst through the door and go straight to the Empire turntable. I'd lift off its square, protective bag of pellucid plastic, unwrap the new album, slide the fresh black vinyl from out of its cardboard and paper sleeves, and fit it onto the spindle. I was careful to flip the proper switches in sequence, preamp first, then amp, and then there would be that soft, mid-bass *whoosh* that told me the speakers had activated, catching the amp's current in their drivers. I'd cue the tonearm up by the finger-lift and lower the stylus carefully into the lead-in groove of the album. There was that pleasing swirl of sound from the stylus searching to track within the groove, then a soft tick that was barely audible, and the music started, clean and clear, palpable in the air—a twang from a guitar, then the rhythmic strumming followed by voices

raised in a simple melody, a sweet chorus of backup singers warbling in harmony, and I was in a different world. In that room cluttered with bowling and Ping-Pong trophies my father had won in lunch-hour tournaments with his co-workers, a set of *Funk & Wagnalls* encyclopedias, and a small shelf of miscellaneous volumes from the Reader's Digest Book Club no one ever read, what dominated was my father's stereo, sitting on "the Monster"—the huge cabinet table he'd built. Inside it lived our record collections—my small but growing one, my father's big band and Hawaiian LPs, and my mother's Marlene Sai, Andy Williams, and Sarah Vaughan. A few years later, my brother's blues albums by Albert King, B. B. King, and Peter Green's Fleetwood Mac joined them. Other than the loving meals my grandmother prepared and the few summer fishing trips the family took, what emanated from these recordings was more the life of my heart than almost anything else during those years.

But listening to my pre-adolescent collection of folk music albums today inspires mild embarrassment: their earnest presentations, button-down and funk-cleansed voices, cloying lyrics and corny sentiments all combine to indict the zeitgeist of the entire era as, if not delusional—ignoring civil rights, class conflicts, post-immigrant and ethnic identities, and the emergence of feminist issues—then certainly naïve. The music that compelled me was faux folk, after all, bearing little resemblance to the ethnographic field recordings of John and Alan Lomax or the great music from early 78s with original artists like Memphis Minnie, Buell Kazee, or the Elders McIntorsh & Edwards' Sanctified Singers (collected by Harry Smith in the *Anthology of American Folk Music*). The style of the New Christy Minstrels, for example, was a somewhat saccharine blend of different commercial approaches to group singing, taking the hard-strumming "folk" guitar style and combining it with blueprints derived from the chorus of Broadway musicals, the Mitch Miller Sing Along Chorus (a relentlessly cheery television group founded by the stiff-armed, Van Dyke–bearded producer and A&R man at Columbia Records), and the Norman Luboff Choir (which backed pop singers like Doris Day, Harry Belafonte, and Bing Crosby). Yet among its playlists would be African American

spirituals, songs from American gospel, and tunes by blues great Huddie Ledbetter and modern composers Woody Guthrie, Pete Seeger, and Bob Dylan.

A song I'd play again and again, picking up the needle from the groove after it ended and placing it carefully back in the spacing before the track, was, in fact, Ledbetter's "Cotton Fields," recorded by the New Christy Minstrels in 1965. Ledbetter, known as "Lead Belly," had gained fame doing radio shows in the forties and recordings for RCA and the Library of Congress. Earlier in his life, he'd been an inmate in Angola Prison Farm in Louisiana and at Imperial State Prison Farm in Texas, convicted of various crimes, including killing a cousin in a fight over a woman. The legend is he got pardoned after John Lomax made a plea to the governor for his release written on the back of the sleeve of a 78 recording of "Good Night, Irene," Lead Belly's signature song. But it was his strong tenor voice, huge catalogue of compositions, and engaging personality that made him popular. To accompany himself, he played a loosely tuned twelve-string guitar, strummed emphatically, and he sang to rhythms reminiscent of railroad work songs sung by Black gandy dancers as they laid track.

In the Minstrels' version of "Cotton Fields," a soft-voiced tenor began the tune, slightly syncopating, dipping down to the baritone range at one point. Then a full complement of singers, perhaps over a dozen, joined in, humming and cooing, their immaculate voices hanging like a softly shifting curtain spanning the living room. The clear voices lifted in separate sections—men on the left, women right, and the tenor soloist foregrounding them at the center-fill, as though the choir were arranged on miniature risers in front of me. It was a smallish, technologically accomplished miracle, a stereo-created illusion that beguiled me out of the boredom that was my life. An affordable passion. An alto soloist would pipe prettily from the center, displacing the tenor, and the full ensemble would join in again, hootenanny style: *It was down in Louisiana / Just about a mile from Texarkana* . . . A banjo got picked, a sparkling tambourine chattered on the beat, a drum kit kicked in and what had been a controlled and carefully groomed sound suddenly got raucous and revival-like. The studio-recorded performance closed with

the tenor soloist again and then a backing chorus of plush baritone
voices humming in a way that set a new pulse streaming through
my blood. It had taken its beat from an old work song, infused a
shouting in my veins rising from a field holler half a century away,
Lead Belly's lyrics still reaching me even through the scrubbed
voices and perfect diction of perky white singers who had suffered
neither incarceration nor scars of the kind Huddie Ledbetter bore
throughout the rest of his life.

A Guitar at Wallichs

I picked up the golden Spanish guitar with its rococo ring of deco-
ration around its sound hole. I felt its amazing lightness and fragility, its delicate girth under my arm and on my hip. I tried fretting
it and then strummed, producing a tangled *whang* of sound that
buzzed and sent a briar patch of notes into the air. I re-fretted and
strummed again. Another tangled sound whanged out. I felt embarrassment and dismay. My mother stared blankly at me, clutching her
purse.

The salesman gently reached for the instrument, saying "May
I?" I handed it to him and he grabbed it by the neck, deftly swivel-
ing it around so he could play and began strumming, very lightly,
producing chords of mellow sweetness that lingered in the air. Then
he started frailing, his fingers spraying out in a sequential fountaining that made a bolder and more bracing sound, exciting the
soundboard of the guitar, making the notes ring and vibrate. When
he plucked it, picking at single strings with the curled fingers of
his right hand, the brilliant notes he made morphed from a biting
attack to a vibrato that he emphasized and sustained with a controlled shake and precise pressure of the fingers of his left hand on
the dark wood of the fretboard. He frailed a chord again, emphatically, damped the strings with the heel of his right hand so they
silenced. He twirled the neck of the guitar in his left, deftly spinning and reversing the instrument around, and handed it to me.

"Eventually, if you play long enough," he said, "the guitar
becomes your friend. You respond to it as much as it responds to
you—believe me."

I was twelve and I'd convinced my mother to take me to Wallichs Music City in nearby Torrance. I'd become captivated with guitar music, my ear kidnapped by the clanging solos of Al Caiola on theme songs for movies and television shows like the westerns *Bonanza* and *The Magnificent Seven* and *The Guns of Navarone,* a war pic starring Gregory Peck (who'd been a favorite since I was nine). And I'd no idea of the differences between electric (what Caiola played) and acoustic (Spanish or steel-stringed). I just wanted to get a guitar and learn how to play it.

My mother opened her purse, fished out a pocketbook thick with checks and cash, and bought the demoed guitar for me. We brought it home, and, for a few weeks, I picked and strummed at it before I gave up. I'd gotten a *How-to-Play* book from the public library, an instructional LP with stepped exercises and lessons, and tracks to play along with, filling in your own solos. But I couldn't do the first thing right—I couldn't *tune* it. I'd hear the pitches and loosen the keys on the headstock of the instrument, tensioning them up slowly, trying to approach the note, and move above it and not quite be able to tell. Then I'd loosen the key again, twisting it, dropping the pitch of the string, then slowly twist it back up by the uncalibrated integers of my reckoning and move past the pitch again, over-tensioning the string so much it would be stretched past being able to hold any note at all. I think I was tone-deaf.

Slow-Dancing with 45s

At the age of fourteen, during my last year in junior high school in Gardena, I began to realize that if I were ever to get a date, I'd have to revolutionize my musical points of reference. The faux folk music and light pop I'd once cherished had to be exchanged for something completely different. I was among mostly Japanese American kids who took their cue regarding dance music not so much from the new rock 'n' roll coming out of England—the Beatles, Rolling Stones, and other mopheads—or even the catchy tunes from East Coast rockers like Dion & the Belmonts, but from love ballads sung by Black artists from all over—what the guys I'd wait with at the bus stop near the McDonnell Douglas railroad tracks on

186th Street and Normandie Avenue called "slow-dance" music. These Japanese teens styled themselves after Chicano low riders, wearing their hair in versions of a Duck's Ass, slicked at the temples with Tres Flores pomade, and tucking a stylist's thin-handled comb into their back pockets. They wore pressed khaki trousers, polished wingtip shoes with steel taps on their toes and heels, dingy white tees, and Pendleton shirts. The music that blared from their pocket transistors was invariably from singing trios and quartets like Little Anthony & the Imperials, the Mello-Kings, Penguins, Chantels, Duprees, Miracles, and Impressions. It was a doo-wop treasure trove of teen heartache, harmonized paeans to blue-balls frustration, and earfuls of echoing reverb, all to a slow and steady, two-note bass beat targeted for dancing up tight with a girl, who, for me, remained only a distant, idealized, and exquisite corpse of imagination only.

I started borrowing 45s from the guys who'd lend them to me, once from our stoic and taciturn ringleader, who worked on cars already (his father was co-owner of a garage and gas station) and built motor scooters on weekends. The most erotically advanced among us (it was intimated he'd *done it* with so-and-so), he kept his records in a dedicated case upholstered in black plastic with white, sequined stars dotting the outer surface. The case had file separators for each disc, not alphabetized but arranged according to a loose narrative of crushing, hookup, devotion, and breakup, escalating on a scale of sexual intensity as well, mildest in front to most intense at the back. I think "You're a Thousand Miles Away" and "Maybe" came first, then "Tonight, Tonight," followed by "For Your Precious Love" and "Chapel of Dreams," and then, inevitably, "Bad Girl" and "You Cheated." Eventually, I began buying my own collection of these vinyl mantras, dropping by the local music store on Redondo Beach Boulevard.

Designed to replace the fragile 78 shellac records of the previous era, the seven-inch 45 rpm record was a format developed by RCA Corporation and released in 1949. 45s were at first only monoaural rather than stereo and there were tracks on both sides—a featured track (with its best song) as well as what was called the "B-side," where a song considered a lesser hit probability was pressed. "Hound

Dog," Elvis Presley's breakout hit on the RCA label and a song considered to have launched rock 'n' roll nationwide, had on its B-side the tune "Don't Be Cruel." Cute, cheap (selling for less than a dollar), and very portable—and thus collectible and easily tradable—the new format had immediate appeal for teens and for parents buying music for their children. And they were terrific for jukeboxes in diners, bowling alleys, and malt shops, where you could choose a single play for a dime and three plays for a quarter. Records at first came in standard black vinyl for most pop, green for country music, and yellow for children's songs. The category of "race records" (what RCA called "blues and rhythm," which also included gospel) were on vinyl RCA called "cerise," a variation of orange; classical came on red; and other instrumental and international music was on blue. The most popular classical music came on a special midnight-blue vinyl. Eventually, RCA dropped the color-coding as too expensive, and other labels developed their own list of 45s, mostly on black vinyl. My slow-dance records were issued by companies named Roulette, Tamla, Standard, Dootone, Paramount, Rama, Chess, Arctic, Brunswick, and Herald. And the bigger corporations developed their own catalogues, too, Columbia Records eventually outproducing its rival RCA. Each had their own distinctive color scheme for their labels, and many made use of trapezoidal patterns or irregular bars of silver against the background of a primary color, but a few, like End Records, had inspired logos, like two sausage hounds traipsing across the eye-catching blue-and-yellow label.

The store I found, called Kay's Music, was in a tidy, half-block-long strip mall and next to a shop where my mother got her hair done. I probably just wandered in there one Saturday afternoon when my mother made me go with her. It was one of those standard storefront shops with an all-glass front, a glass door, and an inner accordion security gate that the owner stretched out when he closed up. Kay (short for Kazuo, I was told) was a Japanese American man in his forties, his crew-cut hair was cigarette-ash white, and he spoke in short, gentle bursts, a bit taciturn but kind in his manner. I remember he dressed like a pharmacist—black oxfords, dark slacks, and short-sleeved white shirts, always crisply pressed. He was used

to dealing with all ages, pre-teens and teenagers included. He had stacks and bins of sheet music, books on various instruments, and musical instruments for sale. In back were rehearsal booths where lessons took place. Behind the long counter were all the records, mostly 45s, which he kept in cubbies that stretched along almost the whole length of the store on one long wall. It was amazing, and I don't think there was ever any record I asked for that he didn't have in stock. Kay could fit more than a hundred single records in one foot of bookshelf space. He might have stocked over two thousand 45s and another five hundred 33⅓ LPs in that little store. He had scores of records I'd not heard of yet, and once in a while he'd suggest something he thought I'd like and spin it for me. He wouldn't let you handle the records yourself. You had to make a request or approve of his suggestion and then he'd turn around and go hunting for it, pushing a wooden ladder anchored on rollers at the top, then climbing nimbly up there, fetching the record, climbing back down. He went over to a turntable in a cubbyhole below the stacks of 45s, slid the record out of its crisp paper sleeve, then placed it on the player and lifted and lowered the tonearm, settling the stylus carefully in the lead-in groove. I think I first heard "Angel Baby" by Rosie & the Originals there. And dozens of other songs I'd never have discovered just listening to the radio or talking to friends. The man stocked surf music, Pat Boone and Paul Anka, Japanese *enka*, *shakuhachi*, and *sankyoku* music, Elvis and Carl Perkins, and Wilson Pickett too. But slow-dance sides outsold everything.

Though I'd have to ask my mother's permission, after dinner and homework I'd try to spend every weekday evening in the living room spinning one slow-dance 45 after another, popping up that brushed brass adapter on my father's Empire turntable, moving its rubber belt down from the 33⅓ pulley to the one for 45 rpm with a cat's-cradling touch of my two hands. It was as though I'd cocooned myself in a malt shop jukebox every night, soaking in melodies of teen angst and hopefulness, with their naïve and soaring harmonies, trying to imagine the body and personality of the opposite sex swaying and dipping under my rhythmic caress on the dance floor. The Penguins crooned "Earth Angel" and I fell into my first lessons on love and poetry, ardor and longing:

Earth angel, earth angel,
Will you be mine?

The next year, I was fifteen and threw a party, inviting my gang and all the girls I could think of who I thought would come—about a dozen. I think I even sent invitations I'd printed up in my shop class, using fancy card stock and type I set myself. My mother and her sister helped prepare the house and make the party

snacks—onion dip, chips, broiled chicken livers wrapped in bacon with water chestnuts, and two cases of Coke and orange soda. I put together the 45s in three stocky towers of gleaming black vinyl stacked around wooden spindles my grandfather had fashioned out of unpainted doweling and pine planks. My mother decorated the dowels with red and silver ribbon wound diagonally around the spindles to make them look like candy canes. Our whole living room was packed with Japanese kids in party clothes—the girls in sleeveless, monochrome sheath dresses and the guys in pre-disco flash with lots of cologne splashed on. The place reeked of Brut, English Leather, and Jade East.

I remember mostly that I got no action, except for a few dances with different girls who only tolerated my clumsy efforts at slow-dance suavity. Despite the scores of carefully selected 45s, choice sides like "Maybe" by the Chantels and "Yes, I'm Ready" by Barbara Mason among them, I got about as far as a Buddhist at a Baptist dance party. Pretty soon, I realized I'd do better serving snacks and popping sodas, dodging amidst the dancers as they hunched against each other in decorous embraces. I handed out Dixie cups filled

with a sprinkling of peanuts to the stags and wallflowers, trying to make time with Cheryl Morita or Janie Ebata as they glanced over my shoulder, their black eyelashes thick with mascara, their faces smelling of lightly scented powder. That was about as close as I got to "grinding" that night, though I admired others hard at work, shuddering to the slow beats slamming from my father's speakers through the floorboards, rising like Federico García Lorca's *duende* up through their heels to their teenage torsos freaked with *amor*.

But I'd been introduced to the game, and, slowly, I started doing a little better, going to carnivals and school dances, girls chatting with me more and more, even dancing, riding up with me in a Ferris wheel at night and not jumping out as it swung us to the stars. I stopped needing to talk about records and doo-wop singing groups and gradually coaxed out innocuous facts and admissions from the girls—the names of their pets, where they'd go on weekends with their friends, what desserts they liked when they went out. If it turned out that they liked records too, they were nowhere near as obsessive as I'd been. And nowhere near as policed in their tastes as I was by my gang of male friends. A lot of them liked novelty tunes like "Leader of the Pack" by the Shangri-Las, "These Boots Were Made for Walking" by Nancy Sinatra, and the Japanese hit "Sukiyaki" by Kyu Sakamoto. I hated those, but nodded and proposed casual meet-ups, then dates.

Early on, though, I made the faux pas of praising a white singing group to the gang at the bus stop. I'd taken a liking to a tune by the Capris, a quartet from back east somewhere that Kay had introduced me to and thought was pretty good. *There's a moon out tonight*, they crooned. *Let's go strollin'* . . .

The next day, though, talking to the guys waiting for the bus, I went on way too long about the song and got the silent treatment. On the bus, I asked one of the guys what was up, and he told me, "*Fool*, the Capris is a *white* group—they ain't no brothers, man." I was shocked. "But they sound so Black," I said. And my friend turned away and fell silent like the others.

I brooded on this for days thereafter, spinning the Capris'

single and trying to hear *whiteness* amidst their singing, hoping to spot a telltale giveaway they were doing the musical equivalent of blackface—late-nineteenth-century song-and-dance men, all white, who'd blacken their faces in shoe polish to both parody and co-opt a style of musical theater that had originated in African American culture. *Ooo-oh-oh-ooo,* crooned the Capris. And I listened carefully not only to their harmonies, which sounded soulful to me, but to their diction and enunciation, how they'd slur a word or sing a syllable in a particular way I could recognize as definitely *white* or *Black.* It's true their lead singer enunciated clearly, and I thought I detected an easy suburbanish style of pronunciation as he sang most of the lyrics. But his falsetto was the height of cool, sailing as high and as purely as anyone from the Penguins or Heartbeats, two bona-fide Black groups. I could hear no major difference across all these records I was learning to love and that were shaping my attitude toward my own fitfully emerging sexuality, even as I'd done nothing but moon for imaginary girls and dream of getting an embrace, a kiss.

What was race? I asked myself. Why was it critical in terms of my new listening, identity, and sexuality? And what were its signs in terms of musical styles? Was it fake for white kids to "sing Black," to imitate soul brothers? If that were the case, what about us Buddhaheads, Japanese kids loving and singing the same songs, dressing in styles we learned from our Black classmates from Compton—the pegger gabardine slacks, high-collar Kensington shirts, and pointed imitation Italian shoes we'd get at Flagg Brothers in downtown L.A.?

I gave up after a while, but I stopped publicly praising the whiteface Capris, not letting anyone know I was still listening to "There's a Moon Out Tonight," or imagining taking a girl out for a walk under its silver light. The song simply settled back in the same pile of 45s I'd stored in my father's living room cabinet. I just told myself I wouldn't loan it out, that I'd hide it inside the sleeve of some innocuous LP whenever anyone came over. Throughout that last year in junior high, "There's a Moon Out Tonight" was my backdoor 45.

R-E-S-P-E-C-T

I was a sophomore in high school, sitting in the bleachers alongside the 440 track at Gardena High, where I was among about fifty Japanese American kids watching the meet between us and Jordan High, a school situated in Watts, a predominantly Black district called "Central." At Gardena, a "Marine District" school, there were a thousand Japanese American kids attending alongside another thousand African American students, a thousand white kids, and a few hundred Mexican Americans. Although we were called the first "integrated school" in Los Angeles, the actuality was we were self-segregating, kids mostly sticking to their own ethnic groups and rarely crossing lines, particularly socially. If there was any coalition, it was rhetorical: on occasion we called ourselves "The Three Bees"—bloods, Buddhas, and beaners. If you looked up into the stands from the green field of grass or from the cinder-colored track circling it, you'd have seen rows of African American students sitting to the left of the band dressed in green-and-white uniforms, and a few rows of Japanese American kids sitting to the right and along a row or two in front as well, making an elbow of sorts with the band in its crook. Behind them all would've been white students in scattered small clumps going upward higher on the bleachers. They were the fewest in attendance. I don't recall that any Chicano schoolmates came to the meets.

Our team was composed mainly of African American sprinters, field athletes, and medium-distance runners. Two of our sprinters were wide receivers on the football team as well as co–city champions in track, running the 100-yard dash in just over 9.6 seconds. Another sprinter, Richard Imamura, our lone Japanese American, ran a consistent 9.7 and usually finished fourth or fifth in these head-to-head heats against other single schools. Our medium-distance runner was future Olympian Wayne Collett, who ran record speeds in the 440, shattering his own personal best just about every other race. He ran the 880 too, but not quite as well, often finishing second or third. Our miler was a slender, red-haired white kid, who always made it a race between himself and whoever else was

the fastest and most resolute among his opponents, never winning or losing by much, but creating a lot of drama at the stretch and lunge at the tape, bringing the crowd to its feet at the finish. We loved him. Finally, our 440-relay team was breathtaking—one of our champions would start, then the Buddhahead and Collett himself ran the curves, and our other champion anchored, making for terrific dramas just about every meet.

On this particular afternoon, likely a Friday, I sat alone, apart from the other Japanese kids, and alongside my Black schoolmates for some reason, enjoying the boisterous cheering and calls from the stands to our squad, encouraging them to run faster, jump farther, vault higher. There was a line of girls, not dressed alike, but cheering and applauding as one, chanting together and doing playful routines with their hands, leaning one way, then the other, standing up briefly, applauding, and then sitting back down, all done in unison, in various cheers they'd rehearsed together beforehand. They rocked the stands, the crowd around them picking up their spirit, yelling and applauding louder as the lineup of girls worked through their vivacious routines.

At one point, one of the girls pulled out a portable record player—the kind that opened up out of a compact carrying case and operated on D batteries. Its speakers were embedded under the plastic deck of the player, under slotted perforations alongside the metal platter that spun a 45 rpm record. What I heard next, during a lull in the competition, was something I've never forgotten. It was "Respect" by Aretha Franklin, that tune released by Jerry Wexler's Atlantic Records in 1967. The song had a thrust and torque like a hard shore-breaking wave spinning up from a steep bottom and churning with inevitable, renewable power, driving the chugging shoulders and spinning hands of the lineup into funky, symmetrical gestures simultaneous with the heavy beat. They danced while they sat, shouted *Ooo!* and *Justajustajusta justa li'l bit* together with the backup singers, sang *Sockittome sockittome sockittome* and bobbed their heads like rockin' on a bus in time with the bass, toms, and kick drum that locked them all together. Though I'd listened to plenty of R&B and danced to it too, I'd never heard anything quite like Aretha's "Respect" before, nor seen how a song could move

through a crowd like a spiraling anaconda, rocking everyone from their seats in a whirl of rhythm and soul hungry for its own spirit.

I didn't know one thing about the session men at Muscle Shoals or saxophonist King Curtis or Otis Redding's original with its band of legendary musicians or about Aretha herself. All I knew was that the tune rocked something inside me and that I wanted more of it. Its music was like a three-o'clock school bell, freeing my own pale, wretched, and hairless adolescent body that had been up to then deprived of the rhythms of history, untutored in the postures of regret and resistance, and empty of any pride in its own flesh, bones, and blood. Aretha's bluesy shouts implored my soul to strut and join a dance line. King Curtis's saxophone solo—braying, hiccupping, and whinnying through the bridge—thrust the rhythm of the melody down my spine and ripped it out again, jagged and bedizened with raw nerve endings. And the chorus of Aretha's sisters took away all shyness and reticence and jumped like salivating banshees onto my hips and sex, furled and unproven. Redding's lyrics testified to kisses that were "sweeter than honey" and, after a devilish rhetorical syncopation, rhymed it in the next line with "money" so that it shoved the point home. And Aretha shoved it good.

From the repeated figures of the horn chorus at the tune's intro and the funky flatted guitar riff at the lead-in, it was obvious the black-and-red-labeled Atlantic 45 single turning on a portable player in the grandstand was "something else." Despite the pain of the song's dramatic setting, what I heard was Aretha singing in jubilation, the same way my Black schoolmates talked and shouted. The tune had the rhythms of their speech and celebration embedded in its grooves. And I felt tutored by it all, by these fortunate elders of mine in American culture, what kānaka ʻōiwi (Native Hawaiians) called kupuna—those who came before and knew the ways of persistence and survival on a land particular to our peoples, this land called The Angels. The operatic drama of a woman pleading for "respect" from her man reached out to me as an immediate rhythmic punch and emphatic grind, its synched momentum of piano, tambourine, organ, and guitar strumming a repeated chord. I might have realized that its lyrics were from the perspective of a woman going through erotic trials, but love itself was still a myste-

rious thing to me then, its lore more story than experience. Dance and passion were what reached me, the spirited physicality of that line of girls at the track meet embedding itself like a creature burrowing through my body.

Baby Love

So, I found myself moving along fitfully, no steady easing into sexuality, but fraught with the problematic of being pubescent, male, Japanese American, and bathed in concepts of physical identity swarming around me from television and the movies that not only told me my body was all wrong, not only barred from heroism (James Bond and the cowboys in Westerns) but in fact made the target of comedy and ridicule. Here was the era of Mickey Rooney taping up his eyes and wearing a bucktooth prosthesis in his mouth, playing the role of the salacious and preposterously impotent Japanese landlord to Audrey Hepburn's fetching Holly Golightly in *Breakfast at Tiffany's.* Here was Marlon Brando as an obsequious Okinawan, slant-eyed in makeup, loose tongue slavering like a serpent, dressed in a short *hakama* made of rags, sucking up to the American military administrator played by Glenn Ford in *The Teahouse of the August Moon.* Here was Hop Sing on TV's *Bonanza,* the black tail of his queue set swinging as he stalked away, muttering in incomprehensible singsong Chinese complaints to himself that the Cartwrights, his virile boss ranchers, wouldn't eat his fried rice while it was hot. *Hah-hah-the-fuck-hah.* It pissed us all off and we tried to ignore these things, even as the Japanese girls all around us at school saw Asian women exoticized as sexy China dolls, compliant *geisha,* and demure but available sarong-wearing sirens in old moving pictures like *The Road to Hong Kong.*

And so we immersed ourselves in Black music—the rhythm and blues coming out of Detroit's Motown—Black bodies our models for dress, savoir faire, and how to step, sway, and shake to the cool jerk and the Slauson shuffle, Black voices our incorporeal enchantment, imagining our male selves the "breathtaking guy" in a song by the Supremes, imagining our female counterparts wearing the emerald chiffon dresses and white satin gloves that would stop us in

the name of love. In a culture where we Asians were never realistically represented, African American music gave us a presence, if only to ourselves, in a way that the absence of our bodies or ridicule of our stereotypical images on TV and in the movies did not.

Bouffant-coiffed Diana Ross cooed like a dove in lyrics about a burning *deep inside,* ending the verse from the Supremes' hit "Baby Love" with long, drawn-out *Ooo-ah* that came from her sweet, warbling throat. It was money and I chased it relentlessly every day in the halls of high school, out in the parking lot after football games, and at weekend dances till it hurt so bad. I asked girl after girl out, called them up to make time on the telephone, my mother hovering over my conversation, listening to every piece of persuasion I tried to encode as I talked on the wall-mounted kitchen phone. It developed my love of symbolism and metaphor, insinuation, and double entendre, perhaps my first practices in a language of spectacular deception and hidden meanings.

Cheryl Morita went out with me to a fancy date in Malibu where I took her for Virgin Marys and a plate of oysters on the half-shell. She wore a beehive, lip gloss, a white patent-leather raincoat, some kind of loose shift with big, Mod circles printed on it, and knee-high white André Courrèges boots that stomped pleasingly on the wooden floorboards of the seaside resto as she strolled in there with me. She looked great but we just couldn't get anything going, our talk stalling out over slurps of blue-gray oyster flesh, our mouths drying out from the big grains of salt sprinkled on the mouth-sized bulbous bodies that we swallowed. I think I took her out one more time—to the Hong Kong Bar in Century City to catch Cannonball Adderley and his quintet. Alas, what I remember most vividly from that night is the catchy tune "Mercy, Mercy, Mercy," written by Adderley's pianist Joe Zawinul, later of Weather Report fame. An R&B number with an easy, sashaying rhythm rather than the hard, bluesy bop the band was famous for, the song knocked me off my feet more than my date did, and, I'm sure, it did her too. We never saw each other again.

But my listening was changing. I remember Ken Kawada going off at me and a couple of other guys one afternoon while we were standing around our lockers, bullshitting each other, talking about

girls and the Temptations, Wicked Wilson Pickett, or something we'd just heard on KGFJ, "the soul of L.A.," the favorite R&B AM station back in the day.

"Why you always only gotta listen to *blood* music?" Kawada shrieked, using the ghetto insider's slang term for Black. "Why you always gotta put guys down for listening to the Beatles, Beach Boys, or the fucking Rolling Stones, huh? *Why?*"

He was high on something—yellow jackets or bennies—that was clear, his voice whining and impassioned, a tone all but forbidden in our style of teen cool and impassivity. But Kawada was right: *Why did we only listen to Black music?* I'd internalized the prohibition in pursuing a culture to belong to, one I could join in without being put down by its imperious needs to colonize and ridicule my body, which is what we all sensed was the deleterious side effect of participating in white culture.

But so-called white music was changing. It wasn't the pastel sweaters and aftershave stuff of Perry Como, Andy Williams, or Paul Anka anymore. The sixties not only brought us the British invasion bands, but counterculture pop and blues-influenced hard rock as well. The AM airplay of singing groups (backed by studio contract musicians) like the crossover 5th Dimension (our senior prom band) and the dreamy, turtlenecked Association gave way to songs by the new rockers who excelled in playing their own instruments—the scruffy-haired, country-influenced Buffalo Springfield; San Francisco acid rockers like the Grateful Dead and Jefferson Airplane; and the folky Greenwich Village duo Simon & Garfunkel. American mainstream music was diversifying, getting freaky, and derived as much from a mix of Chicago electric blues, Appalachian balladry, and other traditional music as it once may have from its commercial, culturally specious origins in Tin Pan Alley and Vegas lounge acts. And its lyrics *meant* something—dropping acid ("White Rabbit"), a police riot ("For What It's Worth"), the burden of white American guilt ("The Weight"). At an outdoor assembly, I'd heard classmates cover "Sunshine of Your Love," that breakthrough acid rock tune by Cream with Eric Clapton playing a driving ostinato on electric guitar throughout the piece. There was no doubt of its propulsive power, the astonishing virtuosity of the three British musicians who

made up the band, and my Gardena schoolmates did a creditable job in faithful mimicry, the bass line booming over the asphalt grounds and concrete breezeways of our campus. That the lead vocal was sung by my old doo-wop mentor Gerald Hudson made hearing it all the more remarkable, a Black kid imitating Jack Bruce's operatic singing style. Here was American Black blues culture, imported to England, now being brought back via the Black voice.

I can't say exactly how or when, but I started listening differently after that, pursuing the new music, tracking Cream, Buffalo Springfield, the Band (Bob Dylan's backing band when he went infamously "electric" on tour in 1965), and Jethro Tull and Fleetwood Mac (when both were blues bands). I stopped hanging out at Kay's and Wallichs Music City and found new record outlets away from Gardena and South Bay. I started flipping through record bins out in Hollywood, Westwood near UCLA, and along the Sunset Strip.

About the same time, I'd grown bored with most of my classmates and, reciprocally, they'd gotten completely sick of me. I was reading *The Atlantic* and *The Nation,* looking for explanations about our changing national culture, which was full of unrest. Vietnam was going on, there had been student protests at Berkeley and Columbia, and race riots had inflamed our inner cities across the country, starting with Watts just a few miles away from Gardena. I'd seen the smoke from my rooftop when I was thirteen. My classmates were still focused on dances and the football team, college entrance exams, and keeping their noses clean so they could move up into the economic and social niches their parents expected them to. Recently, a classmate told me that I'd yelled at a pretty Japanese girl for being petty and politically ignorant when she tried to solicit me for some kind of club raffle the day after Robert Kennedy was assassinated. I was also told I'd pounded a Japanese guy's head into a locker after he taunted me for singing "Cielito Lindo" in the hallways. I must have been imitating the *ay, ay, ay* yodeling of the Mexican singer who recorded it, intoning the polka-influenced verses about a lovely lady with dark eyes.

I could have been singing it because I'd gone to a Mexican dance over the weekend, dressing like a *pachuco,* slicking back my hair

with Tres Flores, wearing high-water gabardine slacks with billowing pantlegs, black high-top shoes, a tight-fitting maroon shirt, and a borrowed, broad-shouldered black jacket of a long and stylish cut. I was seeing Alina, a pretty Greek American girl who sat in front of me junior year in a creative writing class, and we couldn't hang out at the segregated dances for white kids or Japanese. We were known too well, and racial mixing was strictly taboo. Our friend, Felipe García, a gay Chicano who sat across from me in class, suggested we go with him to a Mexican dance in his neighborhood. He'd coach us on what to wear, how to dance, and we'd have a good time on his turf, more open and without the social segregations of white or Japanese culture that prohibited us from dating.

Alina was vivacious and a good talker, not to mention a beauty; I saw her as a teenage Lauren Bacall, full of quips and anecdotes about her life. She spoke easily about paperback novels she had read—books like Ken Kesey's *One Flew Over the Cuckoo's Nest,* Thomas Berger's *Little Big Man,* and Joseph Heller's *Catch-22.* She found them in her mother's beauty shop while sweeping up and closing the place. They were left by customers, culled by her from a ragtag library of Westerns, mysteries, and bodice rippers. We talked all the time because our teacher, who was retiring after that year (he'd announced it), was checked out, telling us we could do anything we wanted so long as we kept it down and didn't riot on him while he studied his chess manuals. And I got off on talk, was starved for it among the taciturn Japanese American kids I was usually in classes with. I took creative writing on a whim as an elective, coaxed by my retiring English teacher, who promised I could read any books I liked and practice my "weird poetry." I'd handed poems in to him as an exercise once in a previous class, and he recognized I might be a good recruit to fill up his enrollment. (I think that what I'd handed in was plagiarized from song lyrics by the Southern gospel quintet the Statler Brothers, who once sang backup for Johnny Cash—*Counting flowers on the wall / That don't bother me at all . . .*)

Alina spun around one day, spraying her long glossy, light caramel hair in a whirl in front of me, taking my breath away, and slapped a paperback book down on my desk.

"Here. Read this," she said, commanding me (*Obbedisco, mia*

donna!). "The guys in here are a lot like you. You're both McMurphy and the Chief rolled into one."

It was Kesey's novel set in a mental hospital, but it had not yet been made into the famous film starring Jack Nicholson. In it, McMurphy is a boisterous con trying to nettle the head nurse and rile up his fellow inmates, all overmedicated and cowed by their imprisonment and its degradations. One of his pals is a tall Native American called "Chief" who speaks to no one until one day muttering "Thanks" as McMurphy opens a pack of chewing gum. They bond over their favorite flavor (Juicy Fruit) and mutual disregard for obeying the rules. They play two-man basketball and beat the shit out of everyone and then lead a kind of prison break where they get all the inmates to jump the grounds and take a recreational fishing trip that they all love.

"One day you're outrageous and crazy and the next day you're silent as Chief. Which are you?"

It was a challenge and I accepted it, reading Kesey's book over the weekend, then talking to Alina all the week long thereafter, she handing me more and more novels to read as the term wore on, schooling me in doubled ways with increments of literature and the engagingly proffered romance I began to anticipate would be ours.

No One Told Me About Her

I got to hanging out at Alina's house after school, where there was inevitably a crowd of other boys and a few girls gathered around her portable record player, which she put out on the concrete stoop by her front door, swing-out stereo speakers blaring a stack of hit 45s. The guys were a mix, mostly white, with one Chicano and one guy with a striking face and a head of black curly hair. He was called Azzie (for Aztec), and turned out to be half Black, half Filipino. Why he was called Azzie can only be explained as ignorance and some kind of inverse respect, I suppose, the name an honor of ancestry, though not his. He'd arrive riding shotgun in a chopped Japanese compact with chrome wheels. The driver was the Chicano, a chunky guy called Bags who wore Pendletons and his pants always baggy. But the others were "Paddies," brusque white guys

from the football team who grilled dogs on a portable barbecue they'd set up in the front yard, bringing cans of soda and beer in a washtub filled with ice that they'd stowed in the living room. The ritual was you drank the beer indoors (so her neighbors wouldn't see) and brought the soft drinks out into the yard. I was certainly the odd man out there—the only Asian except for a big guy named Chow who played right tackle and had a buzz cut and a scowl that withered anyone. We'd been friendly in junior high, but he'd since grown to over six feet, muscular and tough as the trunk of an oak tree. He barely spoke to me but neither did most, except for Azzie and Bags and Alina and a Japanese girl from my class who was Alina's neighbor. She wasn't like our other Japanese classmates; she wore her hair long and unvarnished with Aqua Net (the popular hair spray of the day), and dressed in Mexican appliquéd peasant blouses and tight jeans made of green or russet-red velour. I walked behind her many times after school on my way to Alina's, eventually unnerving her so much, she invited me to walk with her along the way. We became friends too. Near dinnertime, the crowd would dwindle, the washtub would be gathered up, Azzie and Bags would climb into the tiny Toyota, lower their windows, and call out a high-pitched, laughing *grito* as they burned rubber and sped away, sashaying the spray-painted Corolla down the street. Alina would move her record player back inside, the washtub empty of drinks and put away, and invite me, last to leave, inside to "rap."

We sat on a pebble-grain fabric couch and she'd play records as she brought out a bowl of chips. She had an eclectic collection—45 singles of the Righteous Brothers, Beach Boys, and Beatles. I remember listening to Dusty Springfield too. But the records that stick out in memory were by the Zombies, a British rock group from the mid-sixties, from London but obviously influenced by the Mersey beat sound (the Beatles, Gerry and the Pacemakers, et al.). The Zombies had hits like "She's Not There" and "Tell Her No," love ballads of lament and rejection, in the clear, slightly nasal tenor voice of the lead singer, backed by tight vocal harmonies, heavy beats, a melodic bass, and busy keyboard work from the rest of the band. Alina had a lot of their records, not just their hits— singles that her older brother, a soldier stationed in England, had

brought back on trips home. And a lot of these were terrific covers of American blues and R&B—Chicago blues-master Muddy Waters's "Got My Mojo Working," "Goin' Out of My Head" by crooners Little Anthony and the Imperials, and "This Old Heart of Mine" by the Isley Brothers, a soul group highly favored among my old crowd of Japanese kids. She loved these and seemed completely innocent of issues like the race-crossing and cultural mimicry that these sides raised in my own mind, so I never went there when we talked and started making out. "Mojo Working" played on repeat in the background, the Zombies cranking in a jumbled, herky style like Manfred Mann and His Men, while we got busy.

Alina had a way of unhooking her bra under a knit blouse, reaching back with one arm, sliding the rig around, and, with a grin, pulling it out from under like a rubber chicken yanked from a top hat. It was thrilling. But though she let me fumble around in there, my hands groping and pinballing with her plush breasts, my arms hitching the lower hem of her blouse up to her heaving belly, she never took her top off all the way no matter how passionately she seemed to sigh and writhe under my caresses. It made me crazy and maybe made her crazy too, both of us fumbling to enact the script written from sixties fantasies about the opposite sex.

I'd worked up an especially obscene lather one afternoon, when Alina's father, home from work early, walked in on us. The music was turned up so loud, I guess we didn't hear his pickup pulling into the driveway or even his door key working the lock. When he came in, the whole house seemed to take in a breath while her father stood for a second at the entryway. He saw what we were doing, the Zombies still working their mojo, then snapped his head aside, looking past us, and walked straight through the living room toward the back of the house. I was frozen, immobilized in my ardor, abashed and still rigid under my school slacks. But Alina, calm and more experienced than me, sat up, straightened and smoothed her knit top, twisting her slim hips, then leaned down and put her unflustered face in front of mine, commanding me to *Wait here.* She didn't bother getting her bra back on—it had been thrown limply and now was spread out like a dead, beige-colored octopus on the floor. She walked over to the kitchen, bouncing and

shifting freely under her top. I heard the freezer door open, then a rattle and a crunch, a ceramic tinkling, and a tap opening and water running. She came back to me with a wide-mouth, rubberized bag that looked like an inflated shower cap (it was bulging with ice), and handed it to me, saying, "Here, put this on Mr. Johnson and you'll calm down." I did as I was told, icing myself. Her father called to me from the kitchen then, scraping a dining chair along the floor. He'd taken a seat at their Formica table. I hesitated a moment, looking down at myself, handed Alina the ice bag, and stood up. There was a cool dampness on the front of my pants. There was a dark, spreading circle straddling each side of my fly. Water had leaked from the faulty bag onto the fabric and, fool that I was, I hadn't noticed it for all the chill I'd been desperately applying to myself.

"Son? Are you coming?" her father asked.

I had no option but to walk in there, hands draped along my sides, wanting to cover up. Yet how could I? There was no decorous way and the little voice inside me kept yelling, *It's not what you think Not what you think Not what you think at all!* But I stood there, hangdog and silent by the framed entryway to the kitchen, waiting to face the music.

Her father was a huge man, obese and blubbery under his work clothes, a long-sleeved shirt over thick poplin slacks, and I imagined him, in God's rage, swallowing my whole body in the folds of his flesh. But he was acting oddly calm, even friendly. He offered me a beer, which I meekly declined, staring at the floor, not wanting to meet his eyes. He called for Alina to come help him with his "tools" and she scurried behind me into a bathroom down a hall next to the kitchen, and under the sound of the Zombies still working the mojo, I could hear her rattling a few bottles in a medicine cabinet. She brought back a package of gauze bandages and a hypodermic needle, placing them all on the dining table in front of her father, where there was already a small glass bottle with a clear fluid in it. He started taking his shirt off in front of me, asking at first if I minded, but I knew it was useless to protest. He lay back in the frail dining chair I thought would buckle under his weight, stretching his belly to its full expanse. Then he got it all going—popping a new needle for the hypodermic with his teeth, attaching it to the

nose end of the hypodermic, dipping the needle through the rubber cap on the bottle of insulin, then stabbing himself, the needle jerking into his roll of belly fat. He pushed the plunger quickly and extracted the needle, seemingly in one motion. This exposed a little bead of blood at the injection site. Alina tore open a paper package, plucked a square piece of gauze from it, and handed it to him so he could daub himself.

"You know why we live here in Gardena, son?" he said, speaking in a voice thickened with age and strain and without any hint of the rage I had expected. "Because we admire the Japanese American people."

It was the first time I'd ever heard anyone white use that particular locution when speaking about us, and I was astonished. Usually, we were referred to as *Japanese,* making no distinction between us and Japanese nationals, *Orientals,* or worse, as *Japs,* if the speaker carried a prejudice from World War II times. Our Black schoolmates called us *Buddhaheads,* picking up the term from our fathers who had served in the U.S. army. It came from Hawaiian Japanese and was originally pronounced *buta*-head, meaning "pig-headed," referring to immigrant field-workers so stupid and resolute that hard labor never daunted them.

"We moved here from Texas as soon as I was well enough," Alina's father said. "I was wounded in a battle in France and my whole battalion got trapped by the Germans up in this forest, cut off from the rest of the regiment. It was freezing and you couldn't tell if a crack was from a rifle or the limb of a tree snapping off somewhere. We were given up for lost by command and thought, *Well, this is it, boys. Better make your peace.* We were getting picked off, one by one, losing GIs all over the place, hunched down on hard ground we couldn't dig into. No foxholes. Machine guns pinned us down and sniper rounds would smack into your buddies left and right all along the line. Most was dead before they could say anything. I was so scared I pissed my pants—kinda like the cum you done right there."

He gestured at the spreading Sargasso on my fly.

"Well, son, guess what? First, a squad of Nisei boys come up, then another—just two or three guys each—and pretty soon what

was left of their whole battalion come to our position. That was the Four-Four-Two. You know who they are, don't you?"

Alina's father had been one of 275 soldiers of the First Battalion of the 36th Infantry Division from Texas, called the "Lost Battalion" for having been encircled by the German army in the Vosges Mountains near the border of Germany. On October 24, 1944, the all-Nisei 442nd Regimental Combat Team suffered over 800 casualties in a battle that rescued the 211 surviving soldiers from Texas.

Alina's father's eyes were warm, his voice so calming, but I could only answer stiffly. I said, "Thank you for telling me your story, sir," and made some kind of excuse that I had to go. I withdrew quickly, trying not to catch Alina's eye, bowing my head as I pulled open the front door, then pushed on the screen one, finally stumbling down the painted concrete steps of the house. As I walked away, my soggy fly feeling cool against my crotch in the light afternoon wind, a strange mix of things flooded through me—that frustration in my loins, the thrumming panic in my soul soothed at disaster averted, stray tears, and a kind of mixed grief and gratitude for the strange and intense recognition that Alina's father had bestowed upon me. It was all rivering through my body—a sticky flotsam of race, arousal, legacy, and music. No one ever had told me about it.

Songs to a Seagull

I took a black-and-white portrait of Alina in a shop class on photography, using a four-by-five camera with a bellows focus and a Fresnel screen that showed me an upside-down image of her as I hunched under the blackout cloth draped over my head and shoulders. She was a great subject, holding a steady pose with a cool, pensive expression on her face. I made several prints, one an eight-by-ten blowup I mounted for her mother to have, who'd been so excited when I gave Alina proofs to show her parents. Alina told me her mother praised the photo for the "mature" look on her face, which I captured using a soft spot and reflector umbrella. I'd studied the photos of portraitist Yousuf Karsh and advertising photographer Richard Avedon, but what most guided me was Karsh's remark that

one had to gaze into the "inner character" of one's subject in order to capture not only a memorable image, but an expression that revealed something of that person's inner life. The story was he'd gotten that famed portrait of Winston Churchill by yanking the stub of a cigar from the great man's drooping mouth and thereby unleashing Churchill's truculence and resolve, which then came to dominate the face Karsh captured on film. I felt Alina's character wasn't in any stiff grin that populated all the prom photos or the quick, easy one she might give off when greeting friends, but something I might see in her deep reflection as she thought about the poems she'd been reading to me as we drove around South Bay on our dates after school. Cruising Normandie Avenue to San Pedro or taking Crenshaw Boulevard down past Pacific Coast Highway to the sea, she'd incant the vowel-crowded strophes of Dylan Thomas or the esoteric pronouncements about life by Rainer Maria Rilke, a bona fide bohemian and possessed of the most exotic name I'd yet heard. I'd asked her to think of "soft roses you do not see" from one of Rilke's narrative poems, and when I snapped the shot, pushing the metal button on the plunge-trigger, I thought of the slightest shadow on her face as a "fold falling on soft brocade." The lighting I'd designed bloomed gently over the rich cascade of her long, straight hair that framed her oval face, painted a faint glister upon her lips and eyes, and illumined the soft feathers of her lashes into stark outlines, fine and sharp in focus.

The portrait put me in good stead with her mother, a Greek woman with a tight figure like Alina's and a welcoming, cheery personality that was a little unnerving to me, more accustomed to the stolid personae of Japanese American adults. She wore cat's-eye glasses and her tawny hair in a plain bob, despite her job as a hairdresser. I could divine that she and Alina had the same kind of gladness about them, but Alina's branched into an inner melancholy as well—one that I tried to let tutor me, as it served to tie so much of experience together in bonds more subtle and sublime than the conformist and regulatory culture of Gardena High School, the cheap, Vegas-inflected desires of our parents, or the instant frivolity practiced widely among our multi-ethnic peers. Alina's mom would pour me a grape soda in a daisy-speckled tumbler made for

iced tea and Alina would talk about it later as a column of "pure uprising" like a purple crocus in spring, interpreting the mundane with what she called the eyes of exaltation, a kind of double vision. And this was before acid, mescaline, and peyote. She'd gotten this ability simply from reading poetry and the novels of Hermann Hesse, a German who claimed that inchoate gods twisted within the blood of our bodies, ready to spring forth in moments of psychic breakdown or ecstasy. Alina said *Steppenwolf,* using the Germanic pronunciation of *schtepp* like a verbal needle, prodding me to move within my skin so the god might shift alive within me as I read the books I borrowed from her. She cast a spell over me, silken webs surrounding the trivial nubs of personae I'd been till then, cocooning me in silver shrouds of new and strange possibility. It all circled a damp core of budding sexuality that now had nowhere to unfold, given that her father had so nobly taken the fizz of our living room sessions away from us.

I proposed we go for drives after school and on weekends. I had my father's BMW 1800 at my disposal, as he worked nights and didn't need the car on weekends unless he went to the track. I'd pull into her driveway and she'd already be bounding off her front porch, her hair pulled back in a jouncing ponytail, and we'd be off, headed for a bowling alley and a motley group of friends, a record store out on Hawthorne Boulevard in Torrance, or down to the piers in Hermosa or Redondo. Eventually, we found a routine—we'd drive down Artesia Boulevard from Gardena to the beach cities, take Pacific Coast Highway to the cliffs between Torrance Beach and Redondo pier, park along the bluffs, and hike down through dunes covered in ice plants to the wide and deep beach below, where surfers would just be showering off from their afternoons of chasing waves. We'd sit on the edge of the concrete boardwalk or take a blanket, throw it down somewhere, and park ourselves so we could watch the combers curling in long, foamy spirals from way offshore. Speckles of light glinted in lacy trails as they crested and fell, and Alina would read something aloud from one of her books.

> . . . *draw near to Nature. Then try, like some first human being, to say what you see and experience and love and lose . . .*

seek those which your own everyday life offers you; describe your
sorrows and desires, passing thoughts and the belief in some sort
of beauty—describe all these with loving, quiet, humble sincerity,
and use, to express yourself, the things in your environment, the
images from your dreams, and the objects of your memory. If your
daily life seems poor, do not blame it; blame yourself, tell yourself
that you are not poet enough to call forth its riches; for to the
creator there is no poverty and no poor indifferent place.

(*Letters to a Young Poet,* Rainer Maria Rilke,
trans. by M. D. Herter Norton)

I can't say I felt the words she read to me with as much passion
or understanding as she had saying them aloud, but I could see that
what she read changed the expression on her face like an invisible
hand passing across the folds of a velvet curtain, that a bluster of
wind would take hold and buoy her voice upward into flights of
quiet emotion while I marveled silently next to her, admiring the
pearls of tears that winked alive in her eyes while she spoke to me
and beach sand kicked up around us in small gusts, lifting our hair,
speckling our skin.

It was like that for what seemed the longest time before we
began exploring each other again, my hands never at the ready
but stilled, kept away like floats drifting free of their net, without
the tracery of sea-soaked lace that held them to their purpose. But
Alina turned to me, recognizing my shallow despair, draping her
arms around my shoulders and hands around the back of my neck,
drawing me to the softest embrace and the briefest brush of a kiss.
I was encouraged, if still bewildered about what poetry had to do
with all this.

What helped was music again. Alina had moved along from
45s of white soul, British R&B, and the Zombies to something
new—an LP this time, an album by a songwriter who accompanied
herself on acoustic guitar, but completely unlike the cheesy folk
singers I'd listened to before. Her music was eccentric, ethereal in
a way I'd never imagined a song could be. My references in doo-
wop, faux folk, R&B, and white covers of the blues were all wrong
for this. The album was *Song to a Seagull* by Joni Mitchell, a record

Alina had spinning on her portable phonograph whenever I'd go over there. Mitchell's spooky voice warbled out her strange and melancholy lyrics that seemed like mini short stories of doomed love—one lived out in a foreign country of desperate passions and sorrow I'd only seen hinted at in TV soap operas. And her voice was so thin and quaverous, it seemed to string icy melismas that frosted the air across Alina's living room. Her guitar canticled like a tower of church bells ringing. Her voice hung in the air as though from a gothic trapeze of notes. I was mystified.

One day, I came upon Alina sitting on the floor in front of the record player, embracing her knees and rocking slowly on her haunches. She wept as Mitchell's voice rose above us. Alina was listening to the title track, a tune she'd played innumerable times before, but something different must have pierced her heart— Mitchell's singing could do that, a leaping note drive up through you like a sliver of ice, severing the knot of pent-up passions so its blood flowed. It could let a sorrow loose, a grieving that wasn't for anything in particular, but for the pain of life, gentle or acute by turns of Mitchell's phrasing, and Alina now wept to feel it immeasurably. Numb to its profound subtleties, I wanted to feel it too, though I didn't quite, inadequately prepped by my prior musical immersions in more conventional portraits of innocence and experience. Here was a song about a girl's dreams flying with a seagull out of reach that floated so far away she had to imagine its cry.

Rilke may have been the poet we read, but Mitchell was the poet we *felt*. Her images penetrated our bodies, prophesying the daggers of erotic grief that might later strike us down, if we were brave and lucky enough, if we could live as Rilke instructed and Mitchell lamented. I marveled at the plain things of life Mitchell transformed into shining talismans of seaworn glass that seemed wrenched from a nature that was better than the one we lived in, where beaches stared up past *dolphins playing in the sea* to skies emptied of pain. A relentless prophetess of erotic sorrows I could not pretend to understand, she seemed a dervish at the crossroads where I'd soon have to make a deal with love.

Eyes brimming, Alina leaped from the floor and gave me a giddy embrace, brushing away tears and putting on a smile, asking

if we would be going for a drive. I gazed down at the record spinning on the phonograph, trying to let the lyrics and melody wash
over the vague muddiness I felt inside, hoping it might impart the
purer order that Alina must've felt. Except for the sheer joy of being
in her company, my feelings failed me, yet I was alive in a kind of
pleasant confusion.

I'd never skated on ice then, but I tell myself it's what listening
to Joni Mitchell felt like, gliding on a blade cutting across the face
of a white pond, while snowflakes fell around us, my one hand free
and pushing ahead, the other grasping Alina's, whipping her around
in a saucy spin, carefree except for the portent of chill weather and
crisps of frozen lace gathering over our lashes, blinding us to the
path furling in front of us.

I think we both knew what we wanted once we decided to spend
a day out on the bluffs along the Palos Verdes Peninsula overlooking the Pacific. I'd plunged my body against hers, pressing her back
into a chain-link fence on the border of Alondra Park, where we'd
gone strolling early one evening, pretending to feed the ducks, scattering bits of bread, then hiking up a gentle knoll under the shade
of a stand of eucalyptus trees along where the park bordered the
parking structure of El Camino Community College. I not only
stole a kiss, but let my hands skim and stroke over her lithe torso,
arching under me. In her living room one afternoon, we'd stood
against each other, hips locked, and felt for what mattered under
our clothes, while the worlds of mutability and distraction spiraled
around us. That I took along my camera, a 35 mm Nikkormat with
Nikon lenses, and said I'd snap some shots of Alina with my portrait lens, a mild telephoto that let in lots of light, was only a ruse
and a story we could tell others if they asked. I threw a small tripod
and white umbrella reflector into the back seat too, saying it was to
diffuse the intense California spring sunlight while Alina posed for
me in the field of tall grasses under the lighthouse at Point Vicente.

We'd discovered the spot one afternoon driving down Crenshaw Boulevard to the sea. It was a wide swale of tall rye and sedges

that caught the wind in green scallops and billowing waves as though it were a sea too. There was a tawny trail worn through stems of meadow barley, their russet heads ripening in the lengthening days, and swaying grails of fountain grass we walked through on our way to the wilderland of what was to come. There was the occasional *churr* of cactus wrens flitting up from the steep cliffs against the sea, and Alina pointed out black-vented shearwaters that cut through the air above us. Gulls screeched, hovering and then banking swiftly down past the bluff's edge and out of sight, their cries trailing like peals from skeletal bells along the ridgeline.

When I think back to that time, it's Joni Mitchell's tunes and lyrics that are the soundtrack, accompanying the ache Alina and I both felt, not only for each other, but for a life that held more mystery and surreality than the working-class, suburbanite lives we witnessed every day and that the world compelled us to live out too. When I heard Alina intoning the poems she loved, whether Rilke or Dylan Thomas, when she spun her records and either smiled or wept at the secular altar where she worshipped that music, I felt deeply that there were other worlds expanding in my heart, taking away the rage and boredoms that possessed and vexed me, replacing them, in the words of one of Mitchell's songs, with globes of ambergris and *amber stones and green*. The world could be pretty then, and despair wasn't so much its core, but its trailing lace of fond and sorrowful emotions intermingled in the aftermath of having lived through a miraculous splendor.

Under the bluest ocean of sky, Alina lay down in the scarecrow stems and russet heads of grasses she'd matted around us and her clothing seemed to spiral off in one immaculate gesture, falling to her sides like petals from a bush sunflower. I lay with her and my life has since been lost to a fragile glory.

"Cathy, I'm Lost . . ."

We got found out as a couple almost immediately, but our separate racial worlds were not as gentle with us as Alina's father was. A rival of mine confronted her on the edge of the school parking lot and violently grabbed her arm as she tried to twist away. It snapped a

bone and Alina, stone-faced, wore a cast for the first spring month of school before her parents moved them north along the California coast near Morro Bay. Her father suffered from emphysema and had long planned the move, suddenly accelerated by their dismay that anyone could have hurt their daughter. I felt at fault. But I'd suffered too—goaded into an after-school fight with another Japanese American kid who'd been badgering me between classes. Staged in the backyard of another boy who was eager to see us fight, it looked for a while like I could win it, but I was smaller and shorter than my opponent, and my punches started missing and his landing, hard against my lips and cheeks. He beat me mercilessly and I was not only injured—lip split, eyes closed, a tooth chipped—but humiliated. At school, Alina and I carried our separate signs of abuse marking us as outcasts. We barely spoke after that—her eyes darted at me in the hallways and begged me not to try to—and in little more than a month, she was gone.

I sleepwalked through that last half-semester of my junior year; I remember almost nothing except how I had to grow a carapace of indifference and alienation in order to endure the scorn of my Japanese classmates. I took to making long drives around the South Bay and up along the West Side and to the Hollywood Hills, borrowing my father's car as though I still had dates with Alina. It was a four-speed BMW with perforated black leather seats and a flip-up brodie on the wheel I learned to use navigating the curving roads through the Santa Monica Mountains and hills above Hollywood. I played the car radio a lot, punching the precise keys of the Blaupunkt tuner, catching the FM stations that were starting up then, broadcasting in stereo, no less. One of them played whole rock albums rather than just singles, the style the exact opposite of the incessant chatter of AM DJs, cutting in only between album sides for long disquisitions on the new music, announcing upcoming concerts, advertising hip restaurants, nightclubs, and some chain stores like Pep Boys and Owl Drugs. I drove the freeways, crisscrossing L.A. in all directions, desperately trying to get away from the grief I felt from as many quarters as there were points on a compass. I listened to more and more music, and most of it music I'd just started listening to within the past year—Chicago blues and British invasion blues, acid

rock from Cream and folk rock from the Byrds, and *Bookends,* an intensely wistful album from Simon & Garfunkel. I played Simon's tune "America" on my father's stereo all the time and it circled in my head as I drove aimlessly on weekend nights, no Alina to see, nothing to take her place. Simon's lyrics spoke to the loneliness I felt, to the new, desolate feeling of despair.

On the LP, the tune began with a fade from the song before, then two voices in different registers doubled sweetly on the melody as a guitar accompanied them, at first deftly picked as its strings were bent. Then it was strummed emphatically and a drum kit kicked in, sticks traveling across a set of toms. When Paul Simon's airy tenor voice sang, *Let us be lovers / We'll marry our fortunes together,* his lyrics had me moving with the itinerary of the narrative, boarding a bus with the two idealistic lovers searching for a place *to be* across the wintry Michigan landscape of its setting, where a moon rose over an isolate field.

The song reminded me of Alina and that Rilkean world of poetry she loved, the feeling just over the edge of the bluff we visited so many times, beyond the gray rim of the sea we could barely make out past the brutal haze of L.A. It was like a sanctuary I could fold my waking self into as I drove, listening, the car stereo drenching the cab with its gorgeous sound as though I sat beneath a waterfall of music spilling over me, slicking over my skin in a tender forgetfulness for the time I could hear it. It drew a pain away even as I allowed it to abide, restfully, turning rage and sorrow into the sweetness of regret, fragrant with its own life, a kind of body just being born within the soul. Out of grief. Out of the loss of Alina's lithe torso next to mine, her brown hair peppered with burrs of ryegrass.

Mayall, Clapton, and the Bluesbreakers

My last year in high school was a withering experience of rejection and shunning from most of my peers. I was alone much of the time. I started listening to KMET, an FM rock 'n' roll station that had captured a lot of young listeners, and especially to the impromptu disquisitions and anecdotes that the DJ gave between tracks. His

name was B. Mitchell Reed and, through him, I was listening to Bob Dylan, Stephen Stills and Mike Bloomfield on their *Super Session* LP, B. B. King, and a completely eclectic roster of rock and blues artists. The blues, on records I brought back from my forays out to Hawthorne and Hollywood, were mostly British versions played by John Mayall & the Bluesbreakers and the famed dropouts from the Yardbirds like Jimmy Page and Jeff Beck. I think a record store clerk was the first to point out a Mayall LP to me over in the bins somewhere. It might've been at a place called American Records, a small discount storefront on Hawthorne Boulevard in Torrance, another L.A. suburb a few miles away from Gardena. They sold some of their stock for ninety-nine cents apiece during sales and had an eclectic selection of the new rock music emerging in the late sixties. I heard Elton John's first album there. And Jethro Tull's, before the band had an American label.

At the store, I saw my first psychedelic posters, multicolored, stencil-lettered affairs advertising blues and rock acts at venues far away from Gardena. I drove out to Pasadena to the Icehouse to hear English blues bands Jethro Tull, Savoy Brown, and Peter Green's Fleetwood Mac. I drove up to the Ash Grove in Hollywood to hear a Los Angeles hippie blues band called Canned Heat. I started wearing my hair longer, dressed in Levi's and work shirts instead of gabardine slacks and turtleneck tees, and thought in terms of the ecstasy in music. I turned up the volume loud on the Blaupunkt radio as B. Mitchell Reed on KMET would play Jeff Beck's new album with Rod Stewart singing "Jailhouse Rock," while I drove alone in my father's car through the twisty roads of Las Virgines Canyon from the San Fernando Valley to Malibu and the sea.

One afternoon, listening while I was driving, or maybe standing in a record store riffling through the bins, I heard a track that just caught my ear right—Chicago bluesman Otis Rush's "All Your Love" being played by Mayall & the Bluesbreakers, which drew me in with its heavy beat coupled with the ostinato figure on electric guitar. It bore a close relationship to the Motown R&B tunes I grew up dancing to, hitting a groove and keeping it going throughout the song, the melody clipped and clearly articulated within the measures. But there was also an amazing guitar solo about halfway

through that started slowly, in the same cadence the tune had maintained, yet speeded up just so slightly as it started to scream, distort, crunch a chord, and break off in a final, ascending scatter of rapidly picked thirty-second notes. This was my introduction to the blues of Eric Clapton before he joined Cream.

I bought that record—sometimes known as the Beano album, famous for the photograph on its sleeve showing the band hunkered on a curb with Clapton absently reading a children's comic book of the time. The playlist included classic Chicago blues numbers by Rush, Freddie King, Sonny Thompson, and Little Walter alongside originals by Mayall and Clapton. There were also songs by Ray Charles and Robert Johnson—"What'd I Say" and "Ramblin' on My Mind." I played that record just about every morning before school, waking up and going straight to the living room and my father's Empire turntable, sliding the album out of its sleeve, and nestling it down on the short, brushed aluminum spindle. My mother would have already left for work downtown by then and the house was just me, my awakening brother, and my swing-shift-working, hard-of-hearing father asleep in bed. So I played the music *loud,* drums thunking, guitars screaming and squawking, Mayall singing mournfully in that high, airy Anglo voice of his, trying to sound Black and American. At one remove, aided by the Brit bluesmen, I was trying to *be* Black and American too, though I didn't know it. It was an unconscious thing, pursuing my affections and fleeing alienation, wishing my soul to be let out of its confinements so it could scream in the dives and nightclubs along with my white and Black heroes.

I'd grown up with African American schoolmates since elementary school in what was called "Midtown" in Los Angeles. I was eleven when we moved to Gardena, a town with the highest concentration of Japanese Americans outside of Honolulu; my Black schoolmates were bused to school from Compton, a neighboring suburb that, since the start of World War II, had welcomed the Second Great Migration of African Americans from the South. The aerospace and shipping industries had ramped up and needed new workers, and Southern Blacks from Texas, New Orleans, and Mississippi, displaced from agricultural work, took trains and moved to

L.A. in the tens of thousands, bringing their culture and style with them.

Back in seventh grade, when I was in a boys' choir, Gerald, whose folks were from Louisiana, would show some of us gospel and doo-wop styles before class, how to hold our mouths open behind the notes so our voices would get an extra push from the physical concavity we made. He explained chest voice and head voice and taught us to listen for the overtones we'd produce in the bloom of air around us when we harmonized, a sound conjured by the magical choiring of all of us together. We learned to control the intensity and timbre of each note, how to "swing it" with the rhythms of the song and in unison so that our voices could make a sound as though of one breath and body—two Japanese American boys, a Caucasian guy, and Gerald. We weren't conscious of being that different from one another, but were just kids learning what another knew that we didn't. And it felt then as though the world might be that way forever. I never sang with such joy again.

Fittingly, the first time I'd heard a Clapton tune was because of Gerald, yet again. It was in 1968 during my junior year of high school while I was in English class that a big, distorting progression of notes from an electric guitar launched itself, doubled by booming notes from an electric bass, through the school and into all the open doors of our classrooms. *Boom-boodah-dah-doomp-doomp-doomp-boomp-bwahdoo-boomp. Boom-boodah-dah-doomp-doomp-doomp-boomp-bwahdoo-boomp.* It was Clapton's famous riff from "Sunshine of Your Love" by Cream, a band I'd never heard of until that day. A quartet of my schoolmates, with Gerald as their lead singer, had started up for an outdoor assembly across campus, likely limited to graduating seniors, but everyone in the school was rocked by the blues power of the song and the electric amplification that carried it in the air throughout the campus.

The feeling was infectious, and I quizzed as many of my classmates as I could about the tune, about the band, about Gerald, whom I'd lost touch with since we sang together in junior high. Felipe García, my Mexican American friend, was the only one who could identify the song and the band. He told me about Cream, mentioned "acid rock" and "British blues," and I suddenly had

an entire new world of music to get familiar with. I drove out to record stores beyond Gardena and its restricted catalogue of sounds and found a new quest, a new "underground" to explore. I drove out to the Sunset Strip—a terra incognita to Japanese American teens—to haunt the record stores there.

The music on the Mayall/Clapton album was recorded just after the incident, chronicled in various rock bibles of the time, when the graffito "Clapton is God" was found scrawled on the wall outside a London blues club where he'd just played a gig. But I'd not heard of him until that record fell into my hands, and he was thus a kind of private revelation, a player unlike anyone I'd come across in American music of the time. His style gripped and teased, focused emotions into a resolute ferocity touching on anguish at its bottom, soaring into a transcendent, hypnotic bliss at its heights. It rolled pain and peace into its bent notes and passing chords, invoked syncopated beatitudes just off the beat. It became my morning matins in this way, toast with *liliko'i* butter and coffee my breakfast, and Clapton's version of the legendary Robert Johnson's "Ramblin' on My Mind" my getaway song before shutting my father's stereo down and racing off to catch the first bell of the school day. Before I ever heard Johnson's original, I heard Clapton's cover, Clapton's timing, and his British interpretation of the Delta bluesman's vernacular. It was inspiring, but it was filtered, an impersonation, albeit creative and, I think now, respectful.

School was all right if a bore most of the time. My Japanese American classmates, academically inclined, were all of them on a mission to become doctors, lawyers, engineers, or, at the very least, pharmacists. Economic security was what they'd all been primed to want, their Nisei parents bound and determined to recoup the economic and social standing their own parents had lost because of the wartime internment. Gardena was the suburb, just south of Los Angeles, where housing and land (once dedicated to flower farms) were made available to them when it wasn't elsewhere because of early "red-lining" that restricted where Japanese Americans could buy. And though the internment was a fact and a stain on the psyche of just about everyone in the entire town, there was never a mention of this experience, everyone in the community tacitly agreeing

"to get on with it," trying to wipe away the hurt of remembrance. Consequently, I was among a tribe of dedicated strivers in school—well behaved, stoically silent, and determined to earn excellent grades and good opinion from their teachers. I felt this all repressed too much and acted out in all sorts of ways, voicing opinions left and right, talking with our white classmates about the Beatles and Buffalo Springfield, with Black students about sprinters Tommie Smith and John Carlos and their raised-fist protest on the medal stand at the Olympics. To cover my humiliation and loneliness, I took on the role of the malcontent—outspoken about the Vietnam War, about racial discrimination and civil rights—and just generally became a loudmouth, something anathema to my racial peers.

Once school was over and I got through the front door, I put that Mayall & the Bluesbreakers record on my father's stereo. My mother was away at work, my father was just getting ready to leave for his job on the swing shift at Learjet, and my brother was still walking back from the bus stop where he was dropped after his day at junior high. I snicked the power slide-switches on the preamp and amp my father had built, lifted off the cheap, plastic cover on top of the gleaming Empire 398 turntable, fit the LP's center hole onto its spindle, and settled it onto the black rubber platter mat. I lifted the tonearm out of its spring cradle, positioned it over the lead-in groove of the record, and gently, with the calibrations of my fingertip, let the tonearm and cartridge lower onto the vinyl and the stylus catch the groove.

There was a palpable but soft blast of pre-sound to things even before I heard any music. It was like the puff of a bass note almost out of audible range, like a breath of wind moving through beach-side underbrush that had sprung up under screw pines and coconut palms back where I'd lived as a child in Hawai'i. I sensed it even before the catch of the stylus in the groove of the record, before the shush of it sweeping inside the stereo channels, before any first, brilliant note from a guitar, crunch of a kick drum, or *whomp* from an electric bass. It was like an intake of air before the system began to sing, and I grew used to it, like the shake in the withers of a horse before its owner swung his leg over its saddle and sat himself on its wide back. It was part of the natural rhythm of things, almost a pre-

conscious piece of assurance that the system was well and working, and I learned to count on it before the arrival of any music through the speakers. I've heard it with nearly every good tube system I've powered on and used with any regularity. It tells me the chain of electronics has found its natural starting point and that I'm likely in for a good time.

It's said among audiophiles that aural memory lasts for mere minutes if not seconds, but I recall that the sound of these British bluesmen was strong, emphatic, and thrilling. There was a squawk and bite to Clapton's guitar notes, a crunch to their tasteful distortions, and a kind of soft, oceanic roar to Mayall's organ that flooded the air in the living room. I learned to bob my head to John McVie's bass line and Hughie Flint's kick drum, locked in the pocket they'd laid down for each tune. There was always this amazingly magical sense of space and air to everything, sonic images of the musicians taking up their places left to right across a palpable soundstage, both wide and with 3-D depth. I can recall feeling the sound of Mayall's organ snaking through the living room as though from a stream surrounding me. Mornings, my brother would have awakened somewhere in the middle of the second side, and he'd insist I replay the classic "Ramblin' on My Mind" with Clapton plaintively singing the lead vocal. We'd both sit on the floor, at the level my father had placed his bookshelf speakers in their walnut cabinets and gold-threaded grills. The sound was more immediate this way, and it's my guess now that we sat so our ears would catch the first wave of sound launched directly from the speakers, before the plethora of room reflections softened the snap of Clapton's guitar strings and the overall sparkle of the recorded performance. Plus, it had the added value of mimicking our father's cross-legged pose as he listened to his big band music while he pored over the *Racing Form* and other handicap sheets on weekend days. We lived in parallels this way, unconsciously, my brother and I twinning each other in our new devotion to this fresh music from England, at the same time mirroring our father in his love for a music of his own wartime acquaintanceship with the world wider than our former island home.

Jazz Lessons with Bill Taylor

Before setting off for college my freshman year, I bought an all-in-one, "Circle of Sound" analog system made by Zenith that I probably found discounted at White Front or some other appliance store of the time. It featured a two-speed turntable suspended on springs stiff enough for a Jeep and a built-in solid-state integrated amplifier with a control panel to the side of the table that lit up its silver-dollar-sized plastic dials when you turned the thing on. Finally, there were two cylindrical speakers in a fake walnut finish that made them look like fancy little tubs for trash or side tables for a Munchkin's living room. The speaker wires spooled out from underneath, where they were coiled around a catch-wheel like that of a garden hose. There was a kind of walnut-colored stickum wallpaper wrapped around the outer circumference of the tubs, which had single, supposedly full-range six-inch drivers that fired *upward* into vortex-shaped, off-white plastic reflector cones designed to spread the sound out in an even 360-degree circle. You'd better believe there was some *fancy* engineering to this rig—all mine for $79.95. Because of this ingenious design, which deflected and radiated sound "in the round" and "at the source," you were supposed to be able to place these little walnut tubs just about anywhere, and after I put them well into my dorm room and kept tripping on the speaker cables snaking across my cheap, low-pile rug, I put them where you'd expect any normal, red-blooded, front-firing bookshelf speakers to be—spaced evenly along one of the longer walls and about ear-high to where I'd sit in a wingback reading chair across the room. Though without much bass to speak of (I think the woofers may have gone down to 38 Hz), the system served me staunchly through four years of college and the summers between and after.

Its turntable was the kind that could stack LPs three at a time on a notched spindle contraption that dropped one record while suspending what remained in a heavenly queue above the spinning platter below. Annunciation was what I thought of—life was the spinning LP making its sound; the afterlife was what was silent and

held in the queue above it. And indeed I could have been the imitation of a contemplative Christ as I listened to my angels singing. I checked out a small armful of jazz LPs from the college library—stuff by headliners I'd vaguely heard of—Miles Davis, Charles Mingus, and John Coltrane. After listening for a week, I went to the little record store in the miniature downtown of the village adjacent to campus and carefully chose Davis's *Kind of Blue,* Mingus's *Wonderland,* and Coltrane's *Ballads* from the bins. These became my lullabies, and if I had trouble relaxing and falling asleep after studying all evening, the sounds on these records soothed me like chamomile tea, calmed my worries about making it in the tough, take-no-prisoners, academic school I'd chosen. I'd do German homework, drilling myself on declensions and vocabulary, do the readings for American Lit II (Emerson, Whitman, Melville, Dickinson), then slide these LPs out of their sleeves, careful not to touch the black vinyl, stack them on the turntable's changing spindle, and nick the lever for AUTOPLAY on the controls to cue up the first one. I think Mingus was always first, then Coltrane, and finally the Miles Davis, whose album had a slick way of lofting the phantom of a satin-covered, acoustic pillow under my head. The contraption would then miraculously and automatically shut itself off after the last record, just when I'd drifted off into my jazz-induced slumbers.

The tonearm was a flimsy, S-shaped, aluminum thing, and the cartridge one of those hermaphroditic wonders that you could flip, ass over teakettle, to switch from a stylus designed for 33⅓ rpms to another for 45s. There was a dust brush that hung down in front of the body and stylus like a streetsweeper. Little bunnies and balls would pile up there and I'd skootch a forefinger under the pickup lift on the headshell, plucking the tonearm from the LP, swing it deftly away from the vinyl, lift it some, and then puff like Aeolus (or Dizzy Gillespie) on the broom of it to chase the fuzzy detritus from its bristles.

What sound did I get? It was just good enough to get me by, though nothing like my father's system of choice separates. The speakers were certainly as gimcrack as American ingenuity could make them—more bells and woofers than anything serious. But I could play whatever I wanted without my mother yelling at me to

"turn it down fer Pete sakes," as she had whenever I played blues or rock records back home. I started collecting classical records too—Heifetz and Piatigorsky playing the Brahms sextets and Mendelssohn octet and the magnificent Brahms Double Concerto. I'd heard these bow-tied gents play when I was ten, taken by my grandmother one Saturday afternoon when they performed at the Pilgrimage Theatre across from the Hollywood Bowl. On a Friday evening my freshman year, Zubin Mehta came with the Los Angeles Philharmonic and performed Stravinsky's *Le Sacre du Printemps,* and I went out and got the LP immediately, falling in love with the sound of the oboe, the sound of the bassoon, the sound of a piccolo overblown like the cry of a shrike. I found records of the Budapest String Quartet playing Beethoven's string quartets and they, too, became lullabies for me, soothing me to sleep when Davis's "So What" and "Flamenco Sketches" didn't work. I coveted the richness of string sound, and it whispered lavishly at me like the fluttering chorus of a lover's lashes across my body. I'd been so anxious through the day, my mind and spirits lit up by the heady company of my school peers, who, to a soul, possessed good minds and educations far better than mine. I was an L.A. public school kid among those who'd prepped at Exeter, Horace Mann, Deerfield, and the University of Chicago Lab School. And I drove myself hard to *get there,* so as not to feel dumb in class or insipid at our mealtime chatter in the cafeteria and dining halls. I pushed myself to study most all the time, and though I took no drugs, all the bright foam of my new collegiate life just gave me the all-night jitters. Music was my anti-anxiety pill in those days—*Kind of Blue* my diazepam, Beethoven's Razumovsky quartets my Xanax. And they worked. A lot of nights, after studying, I'd lie down on the bed in my dorm room (I had a single), stack a sequence of three albums on the spindle of my Zenith rig, and let myself slowly drift off, knowing both enspiritment and deliverance were mine if I just let the music take me.

In my waking, leisure hours, though, my curiosities intensified in a kind of potent reduction distilled from just those three jazz albums that had been simmering through my sleep. I thought of it as "America's art music," the small combo (a solo instrument or instruments with a rhythm section) the equivalent of a string

ensemble in the classical tradition, and I pushed myself to learn more. Eventually, I targeted a classmate who lived downstairs almost directly under me; I'd heard recordings of saxophones and trumpets blaring out from his dorm window, carrying out to the courtyard and up the walls to my room on the second floor. This was Bill, a tall, light-skinned African American man who looked like a slim Charles Mingus, his face square-jawed, his frame like a basketball player's. I'd determined he was to be my tutor whether he wanted to or not. I ran into him in the hallways or between classes on campus and asked if I could drop by to listen sometime. A kid from Compton, he was cautious and aloof at first, saying "maybe." But I eventually wore him down, repeating my suggestion whenever I saw him in the lunch line or in the dimly lit hallways of our dorm. We had a good word for each other whenever our paths crossed, each of us one of the few "ethnics" amidst an enrollment that was overwhelmingly white in those days. In fact, as one of the few African Americans, Bill had made it a point *not* to talk to white people, a survival practice that had been discussed and vetted in the pre-term meetings of the Black Student Union. The main thing was that he be solid with his African American brothers and sisters, who mostly believed it was the way to go back then.

Exasperated after one of my incessant pitches, probably in the queue for a meal at our dining hall, he asked, "Why should I let you into *my* room to listen to *my* music?" He meant *Black music.*

Calling on everything I had and trying to sound straight outta Compton, I said, "Because I'm Black myself and you just don't know it."

"Huh?" Bill said. I'd stumped him for a second.

"Black is soul and I've got a lot of it," I said.

He looked at me cockeyed, assessing, his face unperturbed but for a slight flare of his nostrils, then broke out a big smile that showed back teeth. I'd counted on him recognizing that I, like him, came from both a broken and a partly mended culture, one that took from Blackness a lot of its grout. It worked. Bill invited me to drop down to his room sometime for a listen.

When I went over one afternoon after classes, Bill was already playing Pharoah Sanders, an LP from 1966 that I didn't know yet—

Tauhid, which he pronounced *"Tau-heed,"* dramatically elongating and emphasizing the second syllable. It was the magical suite entitled "Upper and Lower Egypt." Its music first weaves a sonic tapestry with tomming drums, thumb piano, a droning bass and guitar, electric piano, and the splashings of varied cymbals. Then, rhythmical improvisation commences on the drum kit, augmented with rattles and a cowbell until the stand-up bass starts plucking out a catchy and insistent pattern. The piano and guitar join back in at some point, playing full, ascending chords from the earth to the skies. Throughout, there is rhythmic play on woodblocks and thumb piano, striking marvelous grace notes that make the progressions sparkle. Finally, about halfway through the sixteen-minute piece, Sanders comes in with a brilliant screeching and honking solo that explores a wide swath of his tenor's tonal range. The tune closes with him chanting in Arabic. Bill and I listened in silence until the side finished. He plucked up the tonearm from the spinning platter of his turntable before the stylus could hit the label, and there was a short moment of silence between us before I broke it with a question.

"What's he saying?" I asked.

Bill gave me a sidelong look, smiling again, but this time tight-lipped.

"Well, he's saying, *There is no god but Allah . . .*" He let that sink in a moment, then elaborated. "*Tauhid* means to conceive of the world as One," he continued, extending his arms and opening his palms upward like a preacher. "*Tauhid of Allah* means both the oneness and the uniqueness of Allah." Then he smiled magnanimously and bowed his head, ever so slightly, his hands coming to rest clasped near his waist, a pose of theatrical dignity and reverence.

Bill's words charged the air in the room. They were of a religious tradition that was foreign and more than a little suspect to most Americans. Black Muslims were unsettling to mainstream culture, and though we'd heard from Malcolm X, Muhammad Ali, and Kareem Abdul-Jabbar by then, only a few outside of its proponents might encounter Islam's teachings, even its vocabulary, in their normal lives. But I had. Back when I'd taught a photography class in Watts for a couple of weekends. Back when I'd taken

a poetry workshop there the summer I was fourteen. One of the counselors was Muslim. One of the teachers would quote Malcolm X's speeches. And the Black parole officers who ran the cultural center where we'd meet would repeatedly expound on the issues of amity and mutual respect among all of us regardless of whether, said he, we were from "a Panther family, a Muslim family, a Christian family, or a family of freethinkers."

I thought carefully about what to say to Bill next. It would be important.

"We have that teaching in Buddhism too," I answered. "*All things possess the Buddha nature.*"

We let these words from the two different religious worlds hang in the air a moment, then Bill got up and flipped the *Tauhid* LP over, saying, "You gonna like this one. It's called *Japan!*"

"What the fuck I know about *Japan?*" I said. "I ain't no Japanese!"

We laughed again, two boys from the 'hood who knew they both understood the ambiguities of origin, displacement, mixed cultures, and the art they produced. That whole year, we connected on Coltrane, Miles, Mingus, and Monk, him loaning me his precious records, one by one, presenting me with mini-lectures before each loan. These were disquisitions on development—the differences between one album and another, what it meant in terms of African American culture, how the music might be "outside" but still based on the blues, the foundation of jazz. Bill also told me not to bring anything back scratched, or he'd have to cut me off. He never did.

American Beauty

For Christmas 1970, when I was twenty, my maternal grandmother, Tsuruko Kubota, sent my brother and me five hundred dollars each "to do with what you like," my mother told us. She'd helped to raise us from the time I was about nine years old until I was fourteen, making meals, accompanying us to doctor and dentist visits, knitting on the sofa in the corner of the living room when she wasn't fulfilling duties and paying attention to us. I loved her with all my heart. But when I was fourteen, she returned to Hawai'i, moving

from Gardena back to Hauʻula on the windward side of Oʻahu, a country town perched on the strip of level land between the green *pali* cliffs of the Koʻolau Mountains and the churning Pacific. In 1955, retired from the plantation store he'd managed since before World War II, my grandfather built a structure on Kamehameha Highway, making half of it their home and half a diner my grandmother ran, catering to police, servicemen, and locals. Few tourists ventured that far—over an hour's drive in those days—from Waikīkī and Honolulu. My grandfather was the sous-chef and waiter. My grandmother was the cook. They made the usual local fare—*shoyū* pork ramen (called *saimin* in Hawaiʻi), pork chops with rice and mushroom gravy, hamburgers, and *makizushi* (rolled sushi) among the most popular. I'd lived there a year with them before moving with my mother and brother to Los Angeles to rejoin my father after his year of trade school in electronics.

There was no question what I'd do with my grandmother's money—I'd fly back to Hawaiʻi, where I'd not been since I was ten, half my life ago, and spend winter vacation visiting her and my grandfather, perhaps see other relatives, and just *be* back home amidst the onshore winds furling through green fields of sugarcane. I'd "go beach" and eat fish and *poi* for lunch. My brother also knew exactly what he wanted—a Gibson Les Paul Custom Black Beauty—and he used his money toward getting it. So, while he stayed in Gardena that winter, polishing his gleaming new axe, learning its action, and bending the shit out of blues notes, I flew back to the Hawaiʻi our family had always referred to as *home.*

When I came off the plane and stepped into the open-air terminal in Honolulu, a huge group of my extended family was there to greet me—more than a score of cousins, aunts, and uncles dressed in their best aloha attire, my grandmother standing in front of them all, beaming away. They were singing a Hawaiian song, two of my older cousins strumming a guitar and *ʻukulele,* and it all seemed a chorus welcoming me to heaven. The music filled my heart and I saw my grandfather bounding out from the crowd. He shook my free hand with a firm and encircling grip, then took my heavy suitcase from me—a cheap, powder-blue Sears & Roebuck one made of heavy cardboard that I'd borrowed from my mother and stuffed

with books. My grandfather laughed as he tried to pick it up, setting it back down on the concrete walkway and then grabbing it again, hoisting the considerable deadweight. Inside were Joyce's *Finnegans Wake* and Faulkner's *Go Down, Moses,* along with some poetry anthologies and books by Kawabata, Mishima, Beckett, Merwin, D. H. Lawrence, and Virginia Woolf. I'd barely brought any clothes, obeying the island maxim that said, *No need! Juss pans and T-shirt 'nuff.*

I spent the next few days eating too well and drinking Primo—the locally produced beer—"talking story" with whatever relative happened to drop by on a given day, wandering around the house and its grounds full of plumeria and papaya trees, going down to the rock scrabble beach nearby, and generally feeling pretty bored. Nothing was happening except the parade of rainclouds trailing their skirts of glory from out at sea, over the canefields, and piling up against the green cliffs of the Ko'olau. I wandered over to the shopping center that had sprung up on the *mauka* (mountain) of Kamehameha Highway opposite my grandparents' place that was *makai* (seaside). There was a place that sold records in bins in one of the mall's storefronts, and I went there one day, flipping through the meager selection, bypassing the old-time Hawaiian music and ubiquitous Elvis LPs, and looking for whatever might be more interesting to me. There were records by Santana (had it), the Carpenters (hated it), and Petula Clark (super-hated). But something else jumped out. Across a background that looked like walnut paneling, its outer sleeve with a design imitating the circular label of an LP, there was a stained-glass image of a dark red rose encircled by psychedelic stencil lettering that looked like razor wire enameled with turquoise and lapis lazuli and spelled out the album's title—*American Beauty.* It was the Grateful Dead's fifth studio album, and being in a mood to take a chance (I was by no means a Deadhead), I picked it up.

I crossed through the rain, the record in my hands still in its cellophane, and went back to my grandparents' home. I slipped out of my flip-flops and went over to where a console stereo stood against the back wall of the large living room (once the open floor of the diner). I lifted the wooden lid over the record changer, swung the

overarm away from the tall spindle, tore away the shrink-wrap from the album, took the record out of its paper sleeve, and settled its center hole over the spindle of the turntable. Then I set it spinning, letting the auto-play function work the tonearm and needle drop.

From the moment the stylus lowered into its groove, the songs of *American Beauty* leapt in my heart like brilliantly colored birds of the rainforest, weaving through limbs and flowering branches of the canopy, and fleeing melancholy for the stately joy released by the notes of the Dead's music. They mesmerized me. Here was a sound I'd not heard before—gentle vocal harmonies (reminiscent of Crosby, Stills & Nash), acoustic guitars and mandolins emphatically strummed, an electric bass played as though Melvin Franklin from the Temptations was singing the part, and deftly chattering drums and maracas played tastefully and on time.

A day or two later, I went to stay a week with my cousin Leanne (a year older than me) and her parents in a seaside cottage on Kawela Bay, a few miles northwest along the shore from my grandparents' place. Most every day, we smoked *pakalolo* as we sat on rattan chairs out on the screened porch overlooking the sea, talked about college and what we'd do afterward (she said a schoolteacher, I said I didn't know, but secretly told myself I'd be a poet), sunbathed on the beach, swam in the lagoon, ate seaweed and *poké,* and listened to *American Beauty* along with records by Joni Mitchell, Country Joe and the Fish, Joan Baez, and Bob Dylan. It became the soundtrack for that time in my life while I read the Moderns, contemporary Japanese novels, and my primers in American poetry. When I think of *The Voice That Is Great Within Us,* that monumental anthology of American poetry from early Modernism to the sixties, I hear Jerry Garcia singing "Look out, look out, the Candyman . . ." When I recall the pre-teen Kawabata Yasunari trekking along a country trail following in the steps of an adolescent dancer dressed in colorful silk kimono and wooden *geta,* I hear Garcia asking if we could hear his trembling voice come through the music. Whenever I see a rain shower gently pockmarking the softly roiling surface of a Hawaiian lagoon, I hear a trilling mandolin and Garcia plaintively singing "Ripple in still water . . ." as the string band of the Dead plucks out a deft accompaniment behind him and a choir

hymns in a warm shroud of near heavenly, subtropical rain that encircles me.

In Memory of Elizabeth Reed

My younger brother, Eldon, had taught himself to play the blues on a cheap electric guitar. A kid in junior high, he'd feigned illness so he could stay home days at a time from school to practice. We shared a room, twin beds on either side of a small dresser where we kept a stereo. Before school in the mornings, we'd play blues records. It was fairly odd then for any teenagers to be into the blues, let alone a couple of Japanese American kids like us. While our classmates argued over the Beatles versus Motown, the Beach Boys versus Marvin Gaye, we listened to ex-Yardbirds Jimmy Page and Eric Clapton playing a shuffle blues duet together called "Freight Loader," to Albert King's "Crosscut Saw," and B. B. King's *Live at the Regal* every morning.

Eldon had already outgrown my old acoustic nylon-string guitar, that cheap Spanish-style thing I'd picked up at Wallichs Music City when I thought I could teach myself to play. I snapped strings, over-tightening them. I put the guitar aside and went out to play baseball one afternoon and came home to my younger brother, all of thirteen then, fooling with it. "I got it tuned for you," he said, and strummed a perfect chord. Later that week, he could play everything on the Al Caiola record—not perfectly, mind, but recognizably. I think our mother gave him a cherry-red Harmony electric guitar the Christmas after. He was playing that when he heard something on a Mike Bloomfield LP he thought he could teach himself to play. He took our old Wollensak cassette recorder and taped one of Bloomfield's solos from the record, then played it back, starting and stopping the tape, again and again, rewinding it in small, incremental measures that he could imitate, one by one, reconstructing the solo. Each day, after I came home from school, he'd demo his progress for me. After the first day staying home, he had one of the licks down. And before the week was out, he could play the solo, note for note, in its entirety. I was flabbergasted. He sounded just like Bloomfield. What's more, he'd taught himself the

fretting and bending the strings all on his own. He recorded his own rhythm track for T-Bone Walker's "Stormy Monday" on the Wollensak and then played a screaming lead over it, asking me to sing the vocal. Three years later, at sixteen, he was the lead guitarist in a local blues band, playing gigs at colleges, festivals, and parties all around Los Angeles.

It was over the summer vacation between my sophomore and junior years in college when Eldon first played me *At Fillmore East,* a double LP by the Allman Brothers Band. I'd moved out of my dorm room and brought my stereo with me, the Zenith "Circle-of-Sound" unit. It was compact and fit into our tiny, shared bedroom adjacent to the front-door entrance to the house and sharing a wall with the two-car garage. I placed the record changer on top of the dresser Eldon had decorated with magazine cutouts from *Life* of Woodstock and from *Rolling Stone* of blues artists like B. B. King, Albert King, Clapton, and Bloomfield. The turntable had a clear-plastic flip-up cover latched with metal hinges on the back. We'd lift the lid, settle a three-disc stack of LPs onto the spindle, flip

the carrier arm over the top of them, switch the rig to automatic operation, then carefully lower the lid back down. An ingenious device embedded in the spindle (called a "Collaro spindle" after its patent) would retract, sliding the bottom LP a precise measure away from the notch-lock still suspending the other LPs, and the first record would drop onto the platter. Robotically, the tonearm would swing over to the lead-in groove of the first LP that dropped, stop miraculously in midair, and then helicopter down softly onto the vinyl, making a shushed and scouring sound that said the stylus had found the groove. After this mechanized ritual of manual and sensor-controlled operations, the music started.

What struck first was that snarling rendition of Blind Willie McTell's "Statesboro Blues" with Duane Allman's screaming slide guitar lead energizing the air in our dank little tract-home bedroom. The yellow wallpaper (original to the home when our folks bought it) had a pattern of little stemmed roses in diagonal lines all over it and the song's soulful notes suddenly drained them of their absurdity and drabness, electrifying them all in a brief, blues-induced, hallucinatory dance. Allman's slide sounded as powerful as a wailing harmonica but snakey and supple like a Hawaiian lap-steel guitar played raunchily, roaringly, tuned to jubilation-in-pain and not colonial tourism. It made me cry. I couldn't help it. So much to it. So much of what my brother and I had felt about the blues and the hurt and rage it called for us to release.

And what did two Japanese American boys know about hurt and rage? Maybe nothing, but we were the inheritors of two and a half generations on Hawaiian sugar plantations, killing labor and regimented lives, and the poverty that accompanied it all. Our parents had escaped to Los Angeles, the New World for them, and we were about getting a new foothold on things, including new American identities. Three generations removed from southern Japan where our ancestors had come from, Hawai'i no longer providing us with livelihood or lifestyles, we post-plantation teenagers struggled in our new surroundings, cherry-picked and bricolaged ourselves into whatever called to us. And what did was the Southern blues of white freaks like the Allman Brothers and Johnny Winter and the Black

musicians like Albert King, Willie Dixon, and T-Bone Walker who were their models.

As we continued to listen to *At Fillmore East,* the stereo played Willie Cobbs's "You Don't Love Me," Thom Doucette's harmonica honking on the LP with chromatic richness, in bluesy screeches and tweets. Duane Allman—on his Gibson Les Paul played with a bottle slide—snarled, screamed, and moaned like a Pavarotti of the blues, sometimes letting the notes linger, at others biting them off, damping them with the heel of his palm, then wailing swiftly into another rapid run of blues roulades. At one point I thought I heard guitarist Dickey Betts's entry solo just lag the beat for two or three measures before catching up. But when both guitars played in unison, yawking and twanging together, it made me want to get up and move my Macon.

The tune I remember most is "In Memory of Elizabeth Reed," a composition by Betts, their second lead guitarist. It begins with an achingly elegiac melody played by Betts on his guitar (a Gibson Les Paul too), as he elegantly swells and diminishes his notes by deftly spinning the volume control on his pickup. It was a sweetly melancholic sound I'd never heard before, and at first I thought he was using a wah-wah pedal and somehow controlling the voicings of his notes that way. But Eldon said no, that Betts was using his pinky wrapped around the volume knob mounted just below the strings on his guitar.

"Like Stephen Stills did on 'Wooden Ships,'" Eldon said, miming the action with his right hand, pinky extended and curled. "It's stronger and more direct than using a tremolo bar or wah-wah pedal. And at the same time, it's more subtle and sophisticated too—almost like you're bending the note with your fingers on the fretboard, only this is with your little finger on the volume knob. It's cool."

The technique created a wondrously haunting sound that's stayed in my mind ever since, a paper flower of memory, a capsule of colorful Chinese tissues, shrunk to tight pellets, that blossom gloriously into slowly unfolding petals when tossed into a bowl of water. I remember the song for Betts's lead, but if I play it back

in my mind, the harmonizing lead guitars then emerge, Allman's accompaniment sweetly twinning Betts's melody. And then the whole band races into an extended blues jam on the theme, Gregg Allman's insistent organ and Berry Oakley's electric bass supporting and driving the two guitarists as they each take expressive solos. It struck me then as a kind of chamber music—a Razumovsky of the blues, as Beethoven might have written had he grown up poor and in Georgia. The tune had progressions like that, seemed through-composed (albeit with measures dedicated for improvisation), was performed with tight cues and precision, and emanated and flowed and evolved from an instigating andante of mordancy into scream-ing glorias of allegro and presto guitar-playing by tune's end. Near the finish of the song, during the double drum solos, Jai Johanny Johanson and Butch Trucks played in tight unison on the beat, with light variations, the rimwork of one simultaneous with the press rolls of the other, producing a delicious complexity of well-timed, complementary strokes. The simultaneous work of the two pro-duced a tapestry of rhythm not unlike the harmonic complexity of a polyphonic choir, one drummer's jazzy polyrhythms against the other's steady, bluesy *thunk* and *chunk*. Here was American blues-rock taken to another level of taste and power. "Somethin' else," as Eldon said.

The backstory of it, though, is yet another kind of haunting. I read somewhere, maybe in *Rolling Stone,* that Dickey Betts wrote the tune in homage to Elizabeth Reed, a Southern woman born in the nineteenth century who married a Confederate officer near the end of the Civil War, lived a long life with many children, and passed away in 1935. Her body was buried in a cemetery in Macon, Georgia, where he and the Allman brothers had grown up, where they and other members of the band had spent a lot of time drink-ing, talking, and thinking of the music they would make *one day.* Betts had woven his inspired melody out of homage to their *spiritus loci,* this adopted ancestor, and composed it in remembrance of that fiery earnestness he had in his younger years with all his bandmates.

Back at college the following year, the double LP of *At Fillmore East* played in regular rotation, alternating with my Miles-Coltrane-Mingus sequence and stack of Beethoven quartets as I studied Japa-

nese ideograms, medieval English poetry, and twentieth-century art movements. I'd do my German homework, speed through readings for American Modernism, then slide these LPs out of their sleeves, stack them on the turntable's changing spindle, and cue the first one, most often the Allman Brothers kicking me off into jubilation and a progress of reverie into the night.

Whitman's Barbaric Yawp/
Coltrane's Equinoctial Song

The summer of '73, I was on my way to Japan after graduating from college in California. I won a fellowship sweepstakes of sorts, one of the seventy graduates from thirty-five colleges across the country who'd been tapped for a swank non-academic fellowship. What were my plans? "To fuck around and write for a year in Japan," I said, smugly hip. I'd never been, nor had anyone in my family been back in the hundred or so years we'd been Americans.

Bert Myers, the Los Angeles poet who was my first teacher, had constantly and consistently mentioned several great poets to me. Although my own tastes had run to Pablo Neruda, Gary Snyder, Philip Levine, and whosoever might have been on the cover of what we called "The Rolling Poem"—the *American Poetry Review*—Bert had enjoined me to make studies of Yeats, Baudelaire, Dickinson, and Whitman above all. All were mysteries to me—I'd managed to read none of them during an entire undergraduate curriculum of English Literature and Far Eastern Studies. I'd traded an education in nineteenth-century American poetry for Japanese and Chinese literature in translation, for T'ang dynasty Chinese poetry and painting. I'd given up a semester of Yeats—something I'd planned and wished for—and instead shored up some German and social sciences in order to graduate on time. I'd done myself in with choices, lower division electives, and courses not taken.

So I was reading Whitman the summer after Commencement, driving around my little L.A. suburban hometown from drive-in to Chinese takeout to the branch of Sumitomo Bank that would receive my fellowship check in America and dispense cash to me in Shinjuku, Japan. I had a shitty Signet edition, a "pocket" Whitman,

with a hard-glue spine and a cartoon cover that made our poet of the open road look something like a Musketeer from a Hollywood costume drama. It was hard to get into the stuff, frankly. I mean, the spiritual claims, the grandiose rhetoric, its speechifying. I'd glance through the poems, trying to imagine an audible lyric voice for them, and it was impossible. I preferred Georg Trakl, Attila József, Han Shan in translation.

I had no "training" in Whitman, but only Bert's praise for him to guide me. Bert had said Whitman was "the poet of American spiritual optimism, *before* America had turned to shit." I remembered snatches of classroom commentary about Whitman that other English majors—distant cousins to me by my senior year—had mentioned over dorm dining-hall dinners. Democracy. Barbaric yawp. Indian religions. Hegelian *Spirit*. Driving Redondo Beach Boulevard in my car, negotiating highway dividers and islands and left-turn signals and merging traffic, I asked myself, "What the fuck does all *that* mean?"

"Spontaneous me," "My limbs, my veins dilate," "I am the poet of the Body and I am the poet of the Soul," and "I depart as air" were all lines that cracked me up. Well aware that derivation and influence doesn't work in reverse, I still caught a whiff of Bob Dylan in the line about "boot-soles." When the poet yelled about having a feeling in his big toe, I thought of Stevie Wonder. Simplicity. Directness. Bold speech. And no cunning.

Whitman was sailing by me and I was sailing by him. He meant a corny book stuffed into my Levi's back pocket. He meant a conversation I couldn't get into, another of my teacher's "greats" who eluded me. "Not my *thang*," I told myself. And I planned to forget him.

On the way to the beach one night, driving around L.A. to console myself with the dark emptiness that sand and ocean could give, without my college girlfriend who lived in Oregon, I switched on the car radio. It had FM radio capability. I punched the tuning button to a station I liked. Jazz. KKGO, it was called most recently (before it folded); its call letters were KBCA then. The tune that came on was John Coltrane's "Equinox":

Drums: *Booom. Shboom. Whatchaw. Whatchatchatat.*
Chacaww. Whatchatchatchat. Chacaww.

Piano: *Boomph, boomph, boomph, a-boomph, bawum,*
baww. Silence.
Boomph, boomph, boomph, a-boomph, bawum, BAWW.
Silence.
Cha-boom, boom, boom. Silence.
Cha-boom, boom, boom. Silence.

Tenor sax: *Bwee-daaah, daaah.*
Bweeup-daaah, daaaah. Bweeup.
Daaah, daaah.
Bweeup, daaah, daaah.
Dwee, dah-dah, daaah, daaah, bweeup, daaah, daaah.

Bass: *Boomphum-Boom,*
Bwaawhoom.
Boomphum-Boom,
Bwaawhoom.

This was near the beginning of Coltrane's "modal" period. He took a simple tune, a melodic line, and gave it a bit of syncopation. Then he repeated it. Again and again. The "head," it was called. A "vamp," my musician friends called it. "My Favorite Things" was elaborate compared with this. "Equinox" was a chattery drum, traps and cymbals; a presto piano vamp; then a reverberating bass plucking out a walking line, vaguely reminiscent of bluesy, big-band Mingus except it was made minimal. Well, basic. Nuttin' dere *but* da bass. *Paah-doom, doom-dooom.* Vibration. *Paah-doom, doom-doom.* Vibration.

Then Coltrane's tenor line. A rip. *Bwee-daaah, daaah.* On the Selmer. Big, bell-like notes over the buoyant bass. From the grave, man, as Coltrane had kicked the night of the Newark riots in '67, and this was '73. Big noise. Simple noise. And it was beautiful. Like a work song. Like a chorus of angels made to do the work of man. Fucking *American* music.

It had the beauty of the spoken word. Like a verse from Psalms or Lamentations. I heard in it the dithyrambs of Ecclesiastes. *Listen to the Preacher,* I told myself, *for he hath the Word of God.* The Coltrane tune proceeded like anaphoric verse. It set up a rhythm. It used a phrase that repeated. Again and again. And then it made it change. It made variations that spun and twisted back on themselves, but always returned in some way to the first one. If it made crescendo, there would again be a crescendo a line or two down. Or it would have decrescendo—the opposite. If an accelerando beginning, a like accelerando or complementary ritardando ending somewhere else. A kind of speech. *Phrase. Repeat. Repeat. Variation. Super variation. Return phrase.* A sequence something like free-verse poetry.

Walt Whitman's poetry. For Coltrane had given to music the prosodic and emotional secret of Whitman. And through Coltrane I *learned* to hear Whitman, finally. I carried Whitman throughout Japan. Throughout the sojourn year in Japan, walking Matsuo Bashō's narrow road through the deep north of Tōhoku, visiting temples, partying with six-foot Australian disco women, sneaking out of the monastery, and practicing my poet's identity in the triply-funktified collision of cultures that was my soul, I carried *Leaves of Grass* around with me like the bass-bottom of a melody, like 'Trane's tenor line set to the rhythms of my step. I had a kosmos in my big toe, Jack.

Three years later, in Seattle, working on a poem in the parlor room of my apartment in the University district, I was playing a tape of Coltrane's "Equinox" on my Sony cassette player. I was trying to write a poem about the open road, about Highway 99 in California. About taking some kind of car trip like Jack Kerouac celebrated in *On the Road.* I wanted a rhythm. I wanted a compositional structure. I wanted a musical rhetoric of form.

Boomp-boomph shBWOOM BWOOM BOOM!
Bwee-daaah, daaah . . .

I felt Coltrane in me then. I felt his phrases. An upswelling of music like sexual potency. I heard Whitman. Whitman's "Starting from Paumanok." The barbaric yawp fit like Coltrane's saxophone

over Paul Chambers's bass line. You try it sometime. It's the same. You may surge with the ocean of life. You may surge with American music.

If you ask me again, I'd say Whitman means Coltrane to me, that Coltrane means Whitman. Nineteenth-century optimistic ofay runs into twentieth-century reformed-drug-addict cool Negro saxophone genius. They depart as air. Look for them under the bass line. Of all our voices they are with us, camaradoes. Step and toe. *Bweeup.*

Part Three

Tubeworld, 1

Falling in Love with the Vacuum Tube

During my first few years in the world of high-end audio, I was no longer young, but middle-aged, a tenured, full professor with a new passion. I had an insatiable urge to hear *everything* that this expensive equipment could provide—the finest attenuations of piano notes having taken their full shape, decaying in air; the sinuous palpitations in the melismas of an operatic soprano; the silken warmth of orchestral strings articulating theme and development in a Mozart symphony. I was so obsessed with finding equipment that could provide the best sound that, on a recruitment visit to another university where I might teach, when asked if I'd like to see anything special in the area, I wanted to shave some time away from investigating real estate listings so I could visit the showroom of JoLida, an audio company based in Maryland and known for its affordable tube gear. I was taken there by my faculty host and welcomed by its owner and chief engineer Michael Allen. We spent a long session listening to CDs of Alfred Brendel playing Haydn piano sonatas that my host had brought along, trying them with different amps, speakers, and CD players. I didn't get the job, but I'd heard the JoLida sound and expanded my scant knowledge of audio gear.

On trips to see friends and family in Los Angeles, I'd look up audio dealers nearby and quickly make up a roster of them, routed for drive-time efficiency, running from Redondo Beach up to Palos Verdes to hear Von Schweikert and Nestorovic speakers, from Santa Monica to Pasadena to audition Conrad-Johnson tube amps, from Irvine to Upland to listen to a BAT and then a Cary CD player. Even on a research trip and retreat in Hawai'i, I'd leave my beach-side cottage on an isolated, reef-protected bay on the North Shore of O'ahu and drive an hour to the bustling, subtropical metropolis of Honolulu just to hear more speakers, amps, and CD players.

With each visit, I learned just a little bit more of what audio

could do. I took notes, was doing exactly what my SoCal friend had recommended (just to listen awhile and learn what different sounds components could give) and felt I was building both knowledge and awareness, hearing a bit more deeply and acutely with each session, sharpening my memory for sound, for all its features and nuances, and, like cool dew beading upon the grasses of a garden at dawn, its evanescent, faceted, and sometimes crystalline beauty. I could not stop myself, it had become such an obsession.

I come, in fact, from a line of obsessives. I saw it first in my father, of course, before I knew what it was—his quest to build a perfect do-it-yourself stereo to listen to that music of his, mixing big band with Hawaiian pop, changing tubes and constantly fiddling with the circuit. Later, his younger brother, who was a boxer in high school, then a gambler and a writer himself (he authored the novel *Hey, Pineapple!*, a Korean War saga famous in the islands), told me over coffee and slices of cream pie in a Gardena diner that my father had cultivated bonsai in their backyard when they were teenagers.

"Your father just misted and twisted those tiny trees," my uncle said, "until he had what he wanted."

"And what was that?" I asked.

"A tree just like a skeleton and old man's fingers, moss on the roots. Every time we came home from boxing or football practice, he'd spray his trees, wind wire, and make knots with wax string. He was dedicated."

My younger brother, too, had these traits. After he got going on my clunky Spanish guitar, he became dedicated to the blues. By the time he was sixteen, he had his own Gibson electric guitar and a band to play it with and was so fast and fluid, he could play "I'm Going Home" (infamous for its devilish flurries of notes) by Alvin Lee and Ten Years After. By eighteen, he'd switched to the Allman Brothers and could imitate Duane's solos too. It was a marvel to witness, and at sixty years of age now, steeped in the blues lifelong, he can make a grown man cry on a tune like "Stormy Monday." He played it at our father's funeral and wore dark glasses so no one could see he was weeping.

As I aged and found out more about our family, I learned that

our start in America came when two brothers, one being my grand-father Torao Hongo the younger, arrived in Hawai'i during the early part of the twentieth century from Kōbe, which was likely simply the port from which their ship embarked. His older brother Torakiyo became a businessman and bought up a lot of shorefront property in Hilo, on the Big Island. Torao, meanwhile, ran through a lot of things on O'ahu, starting as a taxi driver, then a car sales-man, then the owner of a teahouse for the nighttime entertainment of his Japanese clientele. When the teahouse went under, he fled to Volcano, the village near Kīlau'ea on the Big Island, and, staked by his brother, opened a general store up there in the rainforest com-munity of truck farmers and orchid growers. The brothers were said to be "educated": they were literate, had bookkeeping skills, and knew things like tea ceremony, flower arranging, and even painting—all very unusual for immigrant Japanese. A relative told me my grandfather could read "fancy books" in a Japanese no one else could read and practiced calligraphy, holding classes, selling scrolls to the villagers to dress up their living rooms, and ordering his ink, stone, and brushes from Japan.

Why was this? I asked myself, until by deduction and a few more clues, I wondered if the two brothers were descendants of the dis-enfranchised samurai who'd led a rebellion against the shogunate in the late 1800s. History said these swordless and erstwhile warriors, dispossessed of income, status, and fealty to any lord, took it upon themselves to march from the southernmost provinces on the island of Kyūshū to the island of Honshū and Edo (what became Tōkyō), asking for the restoration of the emperor and, by extension, their own annuities. Of course the plan was a foolhardy one—to gain supporters from the countryside and spark a rebellion to undermine the government as they wended their way north. They were cut down by the superior forces (and guns) of the shogunate way before they reached their destination. Those left still alive were executed and their families disgraced. It's from this history that I think the two patriarchs of our family in America fled, making their way from Kumamoto in southern Japan to the port of Kōbe, smuggling in their traits of pride and obsessiveness, and gifting us, their descen-dants, with a tendency to take ourselves out of the flow of workaday

life and spend time in a solitary pursuit. My aunt in Hilo, Torao's daughter, arranged flowers. Her daughter is a weekend painter of seascapes. Other cousins breed and grow orchids (and, maniacally yet pleasurably, play round after round of a game called *golf*). My bluesman brother trains champion bird dogs for a living. And I write poetry and retreat into the sound caves of high-end audio.

Despite the beauteous music I heard emanating from my system, I kept churning through gear for two years. I bought a French CD player that upsampled the sixteen-bit digital CD signal into a twenty-bit signal, hoping it would add more refinement to orchestral string sound and choral music, more liquidity to my beloved opera singers. It sort of did but it sort of didn't too. The sound was just *different* from the magical Cal Audio Labs player I'd gotten from my first audio friend—more refined, perhaps, so when the violin sections in Jascha Heifetz's recording of the Mendelssohn Violin Concerto played, they sounded more like individual violins playing together and not a gigantic flute or set of plumbing pipes playing the music for the violins. But it also punched under its weight (about fifteen pounds), as when Jack Bruce plucked his bass on Cream's "Born Under a Bad Sign" and I could barely hear it, let alone feel it boom and blow against my pant legs. I wanted bigger sound, better sound, a sound that would fill my soul as well as the living room around me.

I made pilgrimages to audio boutiques whenever I got to bigger cities like Portland and Los Angeles, and I listened to tons of gear. I once rented a car just to drive a circuit around the Bay Area and visit a Romanian man in Hayward who modified Denon universal players, adding tubes to their solid-state circuits. Then, I drove to Oakland to listen to a troika of expensive CD players in the condominium of a home dealer. Next, I crossed the Bay Bridge to a retail palace in San Francisco and heard nearly a dozen other CD players there. I drove up from Eugene to Salem one Saturday to hear an SACD player in the living room of yet another home dealer. He made his own audio racks fashioned from mahogany and shaped like a trilevel medieval assault tower. He let me borrow one for some weeks. It worked really well but I didn't buy it, and when I took it back, his wife got mad and told me off in their driveway.

In Hawai'i for vacation, I drove from the windward side of O'ahu to Honolulu, having arranged to meet a man who ran his dealership out of a decrepit garage of blackening wooden planks and siding in Makiki. When I got there, I found a rotting wooden building at the address he'd given me. I parked on the street and followed an arrow painted in white and rough block letters that spelled out ENTRANCE. I walked to the side of the building, passing an old Maytag washer, a tower of stacked tires, a couple of old hi-fi consoles, and the ancient shells of a few cathode-ray television sets. A light rain steadily came down. When I got inside, a bearded man, Japanese American like me, dressed in overalls and a long-sleeved T-shirt, greeted me. He was about my age—mid-fifties then—and we managed to exchange a few awkward pleasantries while studying each other cautiously. He asked what I came to hear and I told him there was nothing specific, but that I just wanted to explore his sound. The man then launched into a short lecture on what he termed "audio criticality" and praised the sumptuous sound of the Philadelphia Orchestra recorded by Columbia as an example of that, but only if played back on the right equipment.

It was dark and musty in his shop, and I felt I'd stepped into a cave entrance to the underworld, asking to drink of its shadows. There was audio gear, incongruously pristine and new, on open metal shelving all around us. I saw bookshelf speakers, Conrad-Johnson electronics, and vintage receivers from the sixties, stripped of their cages and showing their tubes. The man, calm and self-possessed as a master of martial arts, played me a recording of Mercedes Sosa and the Estudio Coral de Buenos Aires singing the South American folk mass *Misa Criolla*. In that little shop of oddness and horrors, the gorgeous music transcended its setting, and I brushed aside my misguided scorn as Sosa's plaintive alto, accompanied by the creole choir, an indigenous drum, and varied folk instruments, roused my soul.

During the late aughts, back in L.A., in my teenage hometown of Gardena one fall, I looked up Bruce Edgar, a speaker maker I'd read about who lived nearby in Torrance. I found him in the phone book. I called and he invited me over. He lived in a modest tract home in a lovely neighborhood lined with mature sycamore and

maple trees, the residential street paved with an immaculately black asphalt. He answered the door and waved me in, smiling behind a white walrus moustache. Heavy-set like me, he got around awkwardly, explained the circulation in his foot was hampered from diabetes, yet still toured me amiably through his ranch-style home, showing me an extraordinarily tall radio antenna mounted on a pole next to his house, speakers in his garage that looked like empty bookcases, unfinished horns of shaped and sanded wood as big as buckets laid down on a worktable, and, in his living room, a row of framed magazine covers that featured his speakers. We sat down on his couch, and he played a lot of rock 'n' roll that sounded wonderful; then, sensing I seemed bored with the system or music, he suggested we drive across town to his workshop in Lomita where he kept his reference system. We caravanned over.

His workshop turned out to be a huge industrial space that could easily have housed a body and fender operation, a custom chop shop, or a fleet of vintage eighteen-wheelers. There was that kind of echoing noise of an immense space contained by an overhead roof. Blasts of sound flew through the air—a compressor starting up, an acetylene torch being lit, a pile of wooden boards collapsing on the concrete floor. I wondered how an audio system could compete with all that, let alone make a decent sound in that vast, almost hangar-like space.

Edgar pointed to his speakers—huge behemoths that looked like floor-to-ceiling bookshelves with hand-sized jet engines attached to them. Those were his JBL compression tweeters, shark-shapes of gleaming aluminum encircled with ribbed halos of like silver metal. Edgar showed me his tubed amplifier that combined power output for the speakers with a preamp stage for the incoming signal, which was sourced from a modest Magnavox CD player he could have bought at a big-box store. After he powered everything on, he put on a CD of orchestral music—Jascha Heifetz, the violinist of the chamber group my grandmother had once taken me to hear. It was his recording of the Beethoven Violin Concerto. The orchestra's sumptuous strings seemed to fill the space of the entire building, drowning out all the tappets and small thunders of stray industrial noise, each section sending their separate, silky notes across the air

from the huge speakers. The place sounded as though an orchestra was in there with us. And Heifetz's violin sang sweetly, pianissimo trills and forte notes emanating effortlessly, nimbly, across the incongruous setting.

Over the next few years, I went through three more tube amplifiers and even tried one that was solid state. I was restless about the sounds I wanted from my audio system, moving away from my initial passion for that lush, enveloping tone I'd treasured as a remembrance of my father's system and toward something more "transparent," or true to the live sound of music. With the Prima-Luna amp, choral voices were superb, but it quailed from rendering the most powerful operatic singers and when an orchestra built to a crescendo, it lost its resolve and the sound collapsed in that all-too-familiar hash that I assumed meant it was underpowered for the speakers and the music I tried to play through them. The grandest notes of sopranos seemed to distort when they should have sounded their loveliest. Missing was a delicacy and ease I knew from live performances, the supreme clarity that was captivating at La Scala when I'd been there.

A friend happened to mention, while I was visiting him once at his home in Carmel, that he had "a tube amp" stashed in his closet.

He was telling me a story about visiting an audio store when he spotted something that just caught his eye—a Cary SLI-80 integrated amplifier made in North Carolina. It had an output up to 80 watts and looked to him like an old fifties hot rod with cans of capacitors sitting right on top of the chassis next to its tubes. "It looked like the Batmobile," he said. On the spot, he bought it for its looks. He collected things—luxury watches, a BMW M3 sports car, a twelve-cylinder Volkswagen. He even had a cabin cruiser he kept in Monterey that he took me out on one summer day before its engine (idle for months perhaps) quit and we had to be rescued and towed back to harbor by the Coast Guard. He was like that—collecting things but not keeping them up. He loved the fascination even more than function or performance—the quick rush that evanesced and then gave way to yearning for the next toy or bauble. I couldn't believe he had an amp like that. I made him drag it out from under his suits and spare slacks, a bucket of shoes beside it spilling over, and we slid the dread Batmobile of a thing out, a single tan Ferragamo stuck between the caps and tubes. After removing the shoe and taking a moment to admire the amp and re-seat each of its tubes, I insisted he let me set it up. I went rummaging for his cables, commandeered a Musical Fidelity CD player, and hooked the amp to his speakers—a pair of Bowers & Wilkins towers. I flipped switches on, admired the glow from the tubes, shoved an opera CD into the player (I'd brought along a wallet of them to play while driving), and pushed the start button. A glorious sound bloomed. It was Renée Fleming's rendition of "Quando me'n vo," Musetta's waltz from the Café Momus scene in Act II of La Bohème.

It's a saucy number, the character Musetta playing the coquette, bragging to her bohemian friends about how she manipulates older, sex-starved men and has them eating out of her hand. Her acting evident in her voice, flirty and lighthearted, Fleming sings the aria with an amazing vocal flourish at its climax, and rendered by my friend's amp, her voice was more gorgeous than I'd ever heard it before—smooth and creamy through her midrange, thrilling at its top end, lifelike and expressive as though real. Her voice quavered in the air between the speakers. I thought of all those juicy tubes—the complement of exotic electronic factors making for a powerful

presentation of sound. My PrimaLuna was weaker—every top note clipped the amp, collapsing operatic highs into a staticky hash. The Cary amp was *better.*

Leaving for Oregon the next day, sitting at the Monterey Airport waiting to board, I went on the Web, found about three SLI-80s for sale on AudiogoN, and made an offer on one just before I went through the security line. By the time I landed, the seller had accepted, and the amp was mine. It arrived about a week later, accompanied by a fat multimeter in a yellow plastic case, bias prongs to measure its current settings (you had to monitor these to make sure it wouldn't blow up), and an extra set of tubes. I was in hog heaven—poking with the prongs, measuring with the multimeter, adjusting the bias screws on the amp with a snubby screwdriver I'd found included in the shipping box. The jolt I felt doing these things was as though the electricity flowed through me and not the amp.

The SLI-80 used a different power tube than the PrimaLuna—a KT88 rather than an EL34. The KT88 was barrel-shaped, bigger, and, in the Cary's ultralinear circuit, put out twice as many watts (80W) as the EL34 did in the PrimaLuna (40W). While the EL34 was tubular, a fairly slender glass cylinder, the KT88 was broader, chestier, and fat around its middle. The EL34 was originally created in the Netherlands by Philips in 1949 as a pentode (five internal electrodes) but subsequently produced by other companies as a beam *tetrode* with four electronic elements inside its glass bottle. The KT88, invented by the Marconi-Osram Valve Company in England, was a "kinkless" tetrode, also a tube with four electronic elements, yet with something mechanically and electronically special to its insides that averted a sonic flaw inherent to the original EL34. Their published ratings, released at the time they were each invented, said their respective output powers at peak were about the same, both rated at 100 watts maximum.

When I asked around as to why their sounds were so different, I got a lot of different answers, and few of them were simple. The most reasonable view I got didn't point to any power rating of the tubes or even the amplifiers they were in, but to the relationship between the amps and my Italian speakers, which possessed an impedance curve, a shifting electrical resistance to the current

sent from the amps that itself depended on the relative pattern and spread of frequencies of the signals being sent. Orchestras might be full range, covering most of the audible range in the collective of their instruments' capabilities. A choir, though, might produce a signal from the higher midrange up to the trebles, with a possible intensity of harmonics as well. Opera singers not only might range from the mids to the highest trebles, but they also produced prodigiously continuous signals that never lapsed and could cause certain amplifiers to produce corruptions of sound due to clipping (running out of power) or even ragged discontinuities in the electronic wave within the signal itself. Opera singers *drained* power like thirsty horses drank at a water trough. Electric guitars, with their rapid bursts of signal, weren't anywhere near as tough on amps as Angela Gheorghiu (a soprano) or Rolando Villazón (a tenor). Essentially, the speaker presented a varying electrical load to a given amplifier, the output of which over the curve of frequencies might or might not be compatible with the characteristics of the speaker. For choral singers, even forty-voice choirs like the Huelgas Ensemble singing a motet by Thomas Tallis, the PrimaLuna performed wonderfully with my Italian speakers, perhaps "clipping it softly." But, for an operatic soprano like Renée Fleming singing arias from Puccini and Bellini, the Cary amp worked much better.

"What it comes down to is, does your amp *like* your speaker?" a dealer once told me. "An engineer could probably quote you numbers, but the two either match or they don't. And that's the heart of your system—the match between the amp and your speakers."

I thought there was no way to definitely tell what would go with what without trial and error or listening to advice from an old hand, which did and did not serve me well.

"Oh," the old hand said, "it also depends on what music you like, I guess, especially you 'cause you like *opera*—something *nobody* likes and nobody ever tries to play. You're on your own."

Oh, great, I thought. But I already knew that and bumbled along as best I could, buying and listening for a while, then buying something new and selling the old, churning through gear like a manatee moving through mangroves, chewing omnivorously through water hyacinths, turtle grass, hydrilla, and lettuce. I was insatiable. I went

rapidly through an amp from Musical Fidelity and CD players from Audio Refinement and Conrad-Johnson. I almost reveled in this new arcana of audio brands, my spectrum of familiarity growing to encompass Vandersteen and Alón speakers, Herron and deHavilland electronics, then Lector, Ayre, and ModWright CD players.

Another year went by, and I moved on to yet another tube amplifier, an Air Tight ATM-2. This one was from Japan, its casing painted battleship gray with heavy transformers in the back and a bias meter on its control panel to measure output current at start-up and after warm-up, too. It sported a dance line of six miniature signal tubes arranged along the front of its deck, above a row of smallish, silver-colored control knobs. It was said to be more linear, airier, and provide a more resolving sound than the Cary. I'd been reading around the online audio forums and e-zines, and reviewers and enthusiasts who posted on the forum threads were critical of gear like mine as "overripe," plummy in the midrange, dark in character (emphasizing bass and midrange), and lacking in upper-end sparkle, treble extension, and, worst of all, *neutrality and transparency.*

Neutrality meant an even balance throughout the frequencies, while *transparency* referred to a piece of gear's ability to be true to the recording itself. I thought I liked my retro-sounding Cary amp fine, particularly on jazz, rock, and chamber music, and I even liked it on CDs by my favorite operatic singers—it rendered the micro-details of vocal performance like vibrato, glissando, and melisma so well—but, chasing perfection, I'd started worrying there might still be a sonic shortcoming. Orchestral violins could sound like wind instruments sometimes, obscuring the inner details and shifting dynamics of phrasing, and, even worse, string instruments could sound as though they were made of glass and not wood. My ear was developing, and my audio sensibilities started trending toward fussiness and discrimination.

What I think I was chasing was a level of *fidelity* to the real thing. The romance of the lusher tube sound I remembered from my father's stereo days slowly fell away as I demanded more of what I was hearing, a closeness to the sound of music in life, more perfection in my audio gear so that it could provide, not the *strangeness* of

an artificial presence of sound, the beguilement of my initial love for audio, but a closer imitation of music's sound in the world. That eeriness of Edison's scratchy doll's voice had no spell. The command of Alexander Graham Bell summoning his assistant Watson from the other room had no authority. And Arthur Lyman's shimmering vibraphone on "Ilikai" could no longer capture me in its treacly thrall. The luscious ghost in the machine had lost its allure, and what I wanted was something as close to the real thing as possible.

On the Vacuum Tube

I never learned as much about the vacuum tube as I did when I blew one up. It was by accident and the fault of over-anticipation, excited as I was to power on my first stereo amplifier (one without the preamp stage), a step up from the Cary and PrimaLuna inte-grated amps I'd started with. The Japanese-built Air Tight ATM-2 was supposed to deliver an even more resolving and refined sound and present my precious operatic and choral voices in a more lifelike manner. I'd done my research, choosing the Air Tight from a small raft of options in my price range (setting the bar of cost ever higher), and identified a used unit from a seller in San Jose, California—a deputy district attorney, as it turned out—who was easy to deal with and eager for the sale. I paid him about three thousand dollars, and waited for my new treasure to arrive.

I had developed a passion for orchestral music combined with full choirs—magnificent pieces like Mozart's *Requiem,* Beethoven's *Missa Solemnis,* Bach's Mass in B Minor, and Brahms's *Ein Deutsches Requiem.* The deep and sonorous song from their choirs matched with stirring orchestral music, and the fact that the compositions each had numerous movements that progressed through changes of pace, tone, and emotional affect stirred me and lifted me with thoughts I felt were worthy for a poet to dwell in. I'd long been con-soled by a double-cassette tape of the Brahms piece that, with a kind of superior Sony player, had accompanied me in my itinerancy as a graduate student, from Berkeley to Michigan to Seattle and Cali-fornia. Now, long settled as a university professor in Oregon, I'd

come to explore more in the genre, trying to teach myself not only its splendors and musical complexities, but its fine focus of human feeling, its tradition of honoring human grief and transforming it into exultation, devotion, and humility in the realization that there were grander schemes to the universe. I still mourned for my father and for his suffering and isolation that never was redeemed.

But not all my pursuits were noble or poetic. I'd also become sort of a fiend for "tube-rolling," the practice of tweaking your electronics by substituting different vacuum tubes with the same electrical values as the values of the stock tubes that came with a given piece of electronics. These weren't the big bottle output tubes that sent signals to the transformers to be amplified, but the miniature, nine-pin triode tubes that picked up initial impulse signals sent from source gear—a CD player, a phono cartridge, or even your DAC or streamer. Usually, with the latest equipment, manufacturers installed cheaper, readily available Russian or Chinese signal tubes from some of the only contemporary factories that still made vacuum tubes. Though once ubiquitous throughout the Western world, vacuum tube manufacturing had all but disappeared since the early 1970s. There had been numerous huge factories in Holland, France, Denmark, Germany, and England, as well as in Japan and the United States, but they had all shut down and had their machines dismantled and tools dispersed when transistors had come to dominate manufacturing and the audio world in the seventies. Famed manufacturing facilities that had produced vacuum tubes for decades, like Siemens and Telefunken in Germany, RCA and Tung-Sol in the United States, Genalex and Mullard in England, Philips and Amperex in Holland, Matsushita in Japan, and Radiotechnique in France had all closed their doors by the time I got fascinated with audio tubes early in the new millennium.

I learned all this browsing online, finding the websites of various tube sellers—the anonymous Tube Depot, Tube Store, and Tube World, with their plethora of current production tubes from China and Russia. But the websites that attracted me most were created by eccentrics and maniacs about tubes—Brent Jessee Recording in Chicago, Upscale Audio in California, and Vintage

Tube Services in Michigan. Each of these had the phone number of a *real guy* you could call and talk to, and each of them would talk your ear off about whatever you wanted to know, not only about the sonic characteristics of whatever tubes you were interested in (how they might sound in your amp), but about how they *found* them in a big lot in an estate sale down in Saint Louis or down a windy road in some warehouse in Serbia. These guys were tube hunters and sought out stocks of old audio tubes stuck in moldering military installations, on dusty shelves of boarded-up electronics stores that had gone out of business, auctions of the property from small companies in receivership, and even in your grandmother's crazy brother's garage.

Used tubes could be scrounged from old, often unused electronic equipment stockpiled by commercial enterprises. Old Baldwin and Hammond organs used vacuum tubes. The early IBM computers and later Hewlett-Packard computers used tubes. Guitar amps, test equipment, helicopters, and old electronic control panels of all kinds had vacuum tubes. But for unused, "new old stock" tubes, the largest caches came from big finds in military and industrial warehouses (particularly in Eastern Europe). And private collectors had them. These all reached the consumer market via tube hounds like Andy Bouwman. His online business, Vintage Tube Services, advertises a boatload of brands and types of vacuum tubes. Like a sportsman's ammo and field trial patches, Bouwman's website (I'd almost say it was "handmade") sports the commercial logos of a dozen companies on its splash page—Amperex, RCA, Sylvania, Telefunken, Mullard, Siemens, and Tung-Sol "vibration-tested electron tubes."

Tubes You Can Trust!

The central image of Vintage Tube Services is a grand pyramid made of four power supplies and a central meter stack arrayed on a workbench, their white sweep gauges spread like shaving cream on the upper lips and chins of each dour measuring instrument.

"DEAR MUSIC LOVER," Bouwman says on his website,

Welcome to the Vintage Tube Services. VTS was founded to provide properly selected tubes for those seeking the very finest in music reproduction. At VTS, art and science are combined with old-world personal service as an alternative to the so called "premium suppliers" that use a high-speed computerized testing system to select from a large number of highly questionable candidates. The madness started 22 years ago with absolutely every Tom, Dick & Harry on the block with a pile of tubes and a tube checker getting a web site or going on the bay as big-time expert tube dealers. 98% of those guys don't know anything about tubes, the history of tubes, Electronics, the history of electronics or even audio or the history of the same! From what I have seen, they are more interested in finding the highest price a tube can sell for, rather than how or if it will give YOU pleasure in your music system year after year! VTS was first to provide properly selected & prepared Vintage tubes for Hi-Fi, musical instruments & studios, and that tradition continues with ongoing research & development. VTS starts with the longest lasting, highest quality tubes that were ever made. These tubes have stood the test of the last 100 years! And despite the ongoing sham in advertising, no amount of salesmanship or "consulting with European audiophiles and engineers" is going to bring back the greatness of the real thing.

If you called him up, you'd find that Andy talks exactly like that—voluble, crusading, and making claims that his products not only are the best, but test the best and are proven to be the best over time. He loves to help. He loves to say, "Yup, yup, yup" as he

listens to you describe what you think you're looking for, trying to put a succinct question to him. Then he loves to describe the sound of the exact tube he tells you you're looking for. And he loves to set you straight on any "piece of junk" you might buy from some other seller or website. I've talked to him dozens of times and have kind of fallen in love with his voice and mettle over the phone. His personality seems forged, blown, and shaped by the same manufacturing forces that created audio tubes themselves. Blessed with idiosyncrasies, causing occasional consternation, and magically capable of shimmering disquisitions on sound, Bouwman and his tubes are a national treasure. Anyone who doesn't think so is a destroyer of the universe.

Bouwman's website gives a short history of each of the brands of the particular tubes he carries, along with pithy descriptions of their sonic characteristics. I browsed his site for hours, creating a kind of orchestra of miniature signal tubes on a list as though they were musical instruments of a particular tone and frequency range. I'd call him in Michigan on my cellphone, discussing which 6DJ8 tubes I should add to my little collection. He advised pairs of RCAs made in Germany by Siemens & Halske, along with Mullards and Amperexes, each providing a bit of a different general tonality—the RCAs clear and resolving, the Mullards warm and slightly lush, the Amperexes warm too, but a bit more open, transparent, and clear, sometimes even "pretty energetic," he'd say. To make things even more obvious to a neophyte, Bouwman places the image of Boston's long, rectangular Symphony Hall at the bottom of his webpage. Superimposed over its range of seats, from stage to orchestra to mid- and rear hall, he places the logos of each tube company over where he's determined its characteristic tonal balance would best physically fit. The big "S" from Siemens in Germany is first, taking up the rows of seats adjacent to the stage. Telefunken's diamond logo follows immediately behind that. These positions mean an "up-front" presentation. Tung-Sol, Sylvania, Amperex, and RCA follow to the mid-hall seats. Then, way at the rear, it's Mullard, whose sound has "the deepest, darkest tonal balance." You can easily see why I bought my first tubes from him, starting to assemble my own small collection of the glass bottles. In

shoeboxes and cigar boxes, in plastic lure boxes, and in Tupperware containers, I built up a little storehouse of vacuum tubes—12AU7s, 12AX7s, 5814s, 6922s, 7308s, and the 6DJ8s my father had adored. They all added up to my own repertory company of electronic actors—understudies, small variants, near equivalents, and radical substitutes all vying for the same placements in the two to seven sockets for miniature tubes on any given amp. Two years into the hobby and I'd already become an addict for small differences in sound.

Bouwman's site displays images from old industry ads touting the virtues possessed and manufacturing breakthroughs reflected by many of these tubes. The Amperex 7308, made for computers, featured a "ruggedized ampliframe." The Mullard "10M" 12AX7 was guaranteed to last ten thousand hours "of effective performance . . . within two years from date of purchase." There was "less hum lowest noise" with Mullard's 7025. But RCA's 7025 was "quiet as the whisper of a butterfly's wing," and its ad featured an imago hovering over the nippled blossom of a radio tube. And Tung-Sol's "completely new beam power" 6550 was "first in its power range . . . designed specifically for audio service . . ." These slogans were splashed across stark graphics—silk-screened images of tubes held by ghostly hands, half-tone photo images of jazz artist Louis Armstrong clowning with two white musicians wearing striped coats, out-of-scale images of tubes beside factory gear, airplanes, and workers bent over machinery. The ads could also sport vintage colors—dark reds over light manila backgrounds, stark cut-out white lettering over patches of black, or all letters and images in shades of monochrome red. Lettering could be jagged or blocky, and the images of people, though rare, showed them managing test equipment and wearing white lab coats. The world they came from was closer to manufacturing, more machine-crazy, clunkier, as dazed with the era of hi-fi electronics as we today are cavalier and breezy about our democratic republics of digitized communication, portable hard drive libraries, and smartphone-enabled lifestyles of convenience and confusion.

Over the two weeks I waited for the Air Tight amp to be delivered, I did some research on vintage American and English output

tubes. Though Bouwman's site had everything to say about signal tubes—the miniature, nine-pin ones that handled the audio signal first in the chain of electrical handoffs, functioning as input phase inverter, voltage amplifier, and driver devices—it had little to say about power tubes, the big bottles of the family, which sent signal to the heavy iron output transformers that then provided the electrical current that carried the final signal to the speakers. And Bouwman didn't much sell power tubes, either. I found, instead, that there was a hobbyist and specialty magazine called *Vacuum Tube Valley* that published a ton of articles on just what I wanted to know—what different brands of old power tubes existed, what they sounded like, how they stood up to prolonged use, and even where some of them might be had other than eBay. There were articles on the history of the KT88, the very tube in my Cary SLI-80 and in my newly arrived Air Tight ATM-2 as well. Other articles covered the EL34—the tube in my PrimaLuna Prologue One that sounded so good with choral voices. I read about even older tubes like the Marconi-Osram Valve Company KT-66, the Western Electric 300B originally used in movie theater amps of the 1930s, and the RCA 6L6, popular for use in guitar amps since the early sixties. There were "shoot-outs" galore—articles comparing the sounds of the various power tubes by listening to them in the same amplifier, with the same source equipment, on exactly the same music. And nearly all the tubes I read about could be had through Vacuum Tube Valley, which happened to be the name of the retail business run by another old hand, a tube hound named Charlie Kittleson, owner and publisher of the magazine of the same name. He had a website too, and there were power tubes galore on it, not only with prices, but with test ratings and thumbnail sonic characteristics (derived from the shoot-outs, I suspect).

All the vintage power tubes were said to have sweeter, silkier, and a more refined treble sound than current production tubes made in Russia and China. Why that is has been argued variously—differences in metallurgy, differences in structure, differences in quality control, lost tooling, and lost artisanal knowledge among the workforce, etc. Typically categorical, Andy Bouwman at Vintage Tube Services has this to say on his website:

For the entire first half of this century, the very finest minds in the world were applied to making these tubes, starting with Thomas Edison. To give you an example of what goes into the making of these tubes let me relate an event that happened a couple of years ago. I had a visitor from Germany who is a super sharp chemist and hangs out with the aristocrat types in Europe. He is a member of the Royal Academy of Science but doesn't know, or care much, about tubes or audio. So, he is sitting in my living room having coffee and he looks down at some tubes lying around and launches off about how one of his elderly friends at the Royal Academy used to be one of the chief chemists at M & O. They used to talk about corporate culture, trade secrets and how closely guarded many of the chemical & metallurgical processes were, as they had been worked out over the preceding half-century. He mentioned that the old guard, Edison & Marconi, (and the next generation after) set the example of guarding important formulas & processes because of the patent wars they had fought all their lives. Even at that, he said with all of the equipment in front of you and everything running well, that the ticklish nature of making the finest tubes (inconsistencies in raw materials etc.) is more like cooking than anything else and they knew that someone without their experience could never make tubes quite the same. Well all I can say is that these guys had it because it is now 2015 and damn near of all of the hundreds of real KT-88s, 77s & 66s I have distributed over the last 26 years are still making sweet music for their purchasers! This includes 20 KT-88s that run in a pair of Jadis JA-200s that were tubed up in 1995. Not only have they not missed a single beat but they still all measure and look almost NEW! I have had my own personal set of 4 for 16 years and they are still perfect.

The two most highly touted of these powerful output tubes were made back in the fifties—the Genalex KT88, made in England, and its rival, the Tung-Sol 6550, made in Newark, New

Jersey. The two types, though structurally different, had only slightly varying electronic specifications and were virtually interchangeable "drop-in replacements" for each other, according to the lore in the magazine and on Kittleson's website. But the Genalex tubes went for astronomical prices on eBay—about $300–$450 per matched pair at the time—and I balked at shelling out so much. The vintage Tung-Sols, on the other hand, were relatively affordable at only about $75 apiece. So, for $150 a pair, they were bargains in my quest for the best possible sound.

I read that the Tung-Sol 6550 was an improvement over the RCA 6L6, an earlier, less powerful, but very successful American output tube. While the 6L6 was initially employed as a tube for power amps back in the fifties, its current usage was mainly in guitar amps made by Marshall, Ampeg, and other makers. But the 6550 tube was, as an ad of the time proclaimed, the "first in its power range . . . designed specifically for audio service." It was a tube *made* for hi-fi and put out more watts (in a matched pair run Class A/B, 100 at peak) than the 6L6 and could withstand more voltage and carry more current than its predecessor—both good things for the fine, sustained stereo sound I needed to reproduce operatic voices. Going by the numerous descriptions and reviews, I thought it would be the best possible output tube for my music. I quickly went on eBay and bid on a few pairs, snagging a couple for very good prices, I thought. And, after they arrived, wrapped in sheets of toilet paper and layers of Bubble Wrap—a common practice of great charm—then nestled in popcorn in their shipping boxes, I admired them lovingly.

These were the famed "no-hole, black plate" tubes with single halo-getters inside (getters are a kind of heater for oxidizing the excess compounds left over from the manufacturing process).

Their shape was that of a cartoon weight lifter's—handsome and V-chested, with aluminum bases stamped "Tung-Sol" in faded black lettering, the once gleaming metal spotted with rust or mottled with shadings of corrosion I thought were immensely attractive, a kind of antique patina of authenticity. I tried them out in my Cary SLI-80 amp, and they did indeed produce a sweeter and more liquid sound with clearer highs and crisper string attacks, more languidly trailing decays. I took renewed pleasure in the sound of an aria sung by Renée Fleming, in choral pieces by Palestrina and Dufay, in the jazz ensembles of Miles Davis and Bill Evans. Though piano was never more crystalline, voices never as sweet and soaring as with those vintage Tung-Sol tubes, I pulled them from the Cary after only a short trial. I wanted only to confirm their quality and then set them aside, awaiting the arrival of the Air Tight.

Once it landed, double-boxed and surrounded in gray foam padding that was particularly cushy, I put aside the contemporary, Russian-made output tubes that the seller had boxed and shipped separately from the amp (a normal precaution), and installed in their stead my coveted quartet of vintage, American-made Tung-Sol 6550s. I anticipated a pure, resolving, and highly clean sound, especially in the treble range, and absolute sonic glory when it came to singers.

The Air Tight looked grand, its casing painted battleship gray with a troika of heavy, Japanese-wound output and power supply transformers in back. After seating the vintage Tung-Sol tubes in their tube sockets (each audio tube has an array of metal pins that make electrical connection to the equipment via designated sockets) on the top deck of the Air Tight amp and then wiring all the components together with my speakers, I powered the amp on and pushed PLAY on my CD player. It was so, you know, *the moment I'd been waiting for.* I think gamblers and audiophiles especially prize this brief wormhole of anticipation when the future is fraught with the promise of unknown fortunes and an enriched world of shallowly imagined splendors encroaches on the flattened, threadbare existences where we often find ourselves confined.

There were twenty seconds of beautiful orchestral sounds with period instruments (string players eschewing vibrato) and then ten

more precious seconds of a glorious full chorus. I'd chosen "Kyrie" from Mozart's unfinished Great Mass in C Minor in a new recording by a French orchestra, choir, and soloists, and I anticipated the coming of that particular swoon of intermingled intellectual and emotional rapture this music always gave me. But then I heard a half-stifled *pop!*—as if from a wet ten-cent firecracker—and, through the little circular perforations to dissipate heat machined on the amp's top, I saw the briefest burst of yellow light flashing from within the amp's inner carriage. This was followed by a gradual implosion of sound, as though a sonic balloon had just been pricked and was deflating slowly, the grand choral voices of Le Concert d'Astrée congealing in a Doppler swirl, the orchestra contracting like an anemone around its adventitious prey. A miniature thunderhead, a cartoon chanterelle of gray-white smoke rose up from the chassis then, and the lightest touch of a carbon stench spread through the air. All outside the house was silent—it was a summer afternoon in August, the morning light was already diffuse, radiating, seemingly without purpose, from a nearby window—and then I heard the softest and unchanging upper midrange hum, like an old TV test tone from the Emergency Broadcast System. Though my blood had run cold, I still managed to shut the amp down and stood in shock, trying to comprehend what had just happened.

This was my first audio disaster (there would be others, but none more outwardly calamitous), and I stood on the carpet of my living room like a Scottish stone megalith and imagined a cold, North Sea wind streaming around me. For minutes, I just stared at the system and collected myself. *What had gone wrong?* I'd hooked everything up properly, I thought, as I cast my mind over every knurled and pin-jack connection of wire, the seating of each tube in its designated socket. Could it be that the vintage tubes could not hold up under the amplifier's powers? I thought back to biasing the proper current settings for each tube in the amp, remembering that I had to be sure to set the current within very specific parameters of milliamperage. I had been especially careful, following the gauge indicators for adjusting the current level of each output tube as the amp warmed up. The gauge was on the front panel and had a little "sausage balloon" in it to indicate the recommended settings for the

sweep needle, adjustable with its own dedicated control dial. I'd set both pairs of tubes at the lower end of the curved balloon, staying at the conservative level of recommended current. Or so I'd thought.

I made two phone calls after that, explaining what happened and asking what had gone wrong and what could I do? I called the seller, who disavowed any responsibility, naturally, as I'd crucially changed out the Russian tubes that came with the amp for the vintage Tung-Sols I'd installed. I called the shop that had sold him the amp, and the tech there guessed what had happened. He explained that the Air Tight amp "pounded" the tubes with high voltage at start-up and that it was obviously too much for the delicacy of my vintage tubes to withstand. He recommended I find a local shop to repair it, which, he said, would not be that hard as, more than likely, the flash I'd seen was the flare from a bias resistor as it flamed out under stress: "Your white smoke says prolly only one burned up." *These guys know everything,* I thought. A resistor is a small device, often about the shape and size of a capsule of extra-strength Tylenol, made to slow, at a rate determined by its electrical impedance value, the free passing of electricity through a circuit. The amp was designed, he said, to take out a resistor in overload situations (like the perfect storm of electronic stupidity I'd created) rather than a fuse. "Don't ask me why they didn't use a fuse," he said.

I found a local stereo repair shop, and its owner was able to locate the exact problem—a resistor and a single blown tube—and fix the amp quickly and cheaply, as it turned out. I was saved.

Inside the Tung-Sol 6550 Tube

When the resistor inside the Air Tight amp flamed out, I didn't care that much about what exactly happened inside the tube that had also blown—the old Tung-Sol 6550 tube from the fifties. Instead, I remember wanting to turn time back to the moment when I first heard the Mozart Mass come on, period strings thrumming ominously, mysteriously, chorus chanting *Kyrie,* elongating the vowels in sweet and portentous descants, timpani thumping away. The glory of the music was so short-lived, I hadn't even gotten to hear the crystalline soprano voice of Natalie Dessay, sinuously palpating

through the ornaments of her solo. I'd shut the system down after the smoky blowup and gotten only silence.

But, once the amp was fixed, I asked around and found things out, piecemeal at first, then getting a more complete picture in a couple of telephone sessions with audio engineers who worked on tube equipment. From what I gathered, the blowup of the tube and failure of my amp was a sequence of cascading electronic and metallurgical events. These altered the mechanism within the tube itself. This resulted in both the tube's failure and the almost simultaneous burnout of a resistor through which electricity in the amp flowed. The burn-through of the resistor and the tube's failure produced a dead short in the amp, cutting off the audio signal and creating that high-pitched hum I heard. I have to admit, explaining the electronic mishap can sound much like the "Who's on First?" routine of Abbott and Costello. A lot depends on following sequence, and, once your mind gets mixed up, you lose track of who's on first, and even the explanation seems a comical mash-up of all that's gone on.

The 6550 tube (invented by Tung-Sol in 1955) is a fairly late development in the twentieth-century technology that created the vacuum tube, starting from early in the century. In contrast to its ancestor, the incandescent bulb oft said to have been invented by Thomas Edison, which has only one major element (the filament), the 6550 tube has four essential elements, or electrodes. It's what's called a beam power tetrode. The elements are the cathode, the anode, the grid, and the screen grid. The cathode supplies the electrical current. The anode accepts that current. The grid controls the rate of current flow from the cathode. And the screen grid makes the current flow more efficient, creating more output power. The screen and control grids are made of two concentric coils of wire (resembling springs), each supported by two rods, one on each side, and they lie between the cathode and anode inside the tube. They help the tube function more efficiently in terms of electrical flow and power output. The "beam" part consists of two flat pieces of metal, bent in two wings like a folding screen, that confine and channel the flow of electrons into directional sheets of current. The turns of their inner grid coils are also aligned to help cause the

electrons to flow in sheets. As it transfers no charge itself, the beam electrode isn't counted as an electrical element along with the other four, so the Tung-Sol tube can be called a tetrode rather than a *pentode* (a tube with five electrical elements). In all the advertising when the tube was released, Tung-Sol announced that the 6550 was invented exclusively for audio application and made to provide the necessary current to drive a hi-fi loudspeaker. But in order to perform this task, the 6550 tube needs to be inside another electronic device—the amplifier.

My Air Tight ATM-2 is a two-channel (stereo) amp that uses three transformers, one as part of the power supply for the amp (and its tubes) and the other two for output current feeding the two channels that go to the speakers. In the particular type of audio circuit used in this amp, called *ultralinear,* a circuit partially invented in the 1930s by Alan Blumlein, two of the elements inside each 6550 tube (both the screen grid and anode) are connected to an output transformer that itself (through the whole chain of the amp's circuitry) is connected to the power supply.

At start-up, my amp's power supply sent a tremendous amount of voltage (around 600V DC) through both the screen and control grids, quite friable, actually, inside of each tube. That high amount of voltage likely passed between the screen grid and the beam plates in one of those tubes and caused electricity to arc—to leap from the screen grid as a small lightning bolt of uncontrolled electricity freed of its designed pathway. This instantly vaporized much of the metal in the screen grid's coils, producing a plasma cloud of uncontrollable electrical current. It was this plasma cloud bursting alive that I heard as a small wet pop. The cloud created a sudden burst of light inside the tube, like a small nova inside the envelope of glass. This succession of events disrupted the orderly flow of the electrical current, shortening the proper channels of electronic travel so that the tube failed. Just ahead of tube failure, though, while circulating from the tube to the power transformer, the burst of high voltage at start-up almost simultaneously caused a resistor (that little Tylenol-capsule-sized part) to overheat, melt, burn, and then vaporize its own internal electrical elements from the inside of its

ceramic casing, flaming out in that cloud of smoke from inside the amp's understory, shorting the circuit, and cutting the tube out of the electrical circuit completely. Fortunately for me, because of this, no further arcing could occur, saving me the even more calamitous disaster of wrecking the power supply transformer (a serious and expensive piece of iron) itself. When I heard the dwindling Doppler tornado of choral sound emanating from the speakers, it was the flow of current fed to them from the amp's storage capacitors slowly dying away. The resistor failing (by burning out) effectively meant it cut off the current, saving the precious transformers from overheating and getting damaged. Having the tiny resistor flame out and a tube short itself was much cheaper, and it was easier to replace a tube and resistor than a heavy transformer with all its windings. I was lucky, relatively speaking, and the event was a wake-up call. It was also my slapstick qua panicky introduction to understanding the vacuum tube and appreciating its role in the history of audio.

Kevin Deal of Upscale Audio

I once spent a day hanging out with Kevin Deal (who has a completely Dickensian name), the owner of Upscale Audio, where I'd bought the PrimaLuna amp, a Cary CD player, and the Italian speakers I first loved. It's rare to say this about a retailer, but I like him and I trust him. He is a shortish man with receding blond hair and a neatly trimmed chin beard that bristles when he smiles. And he smiles a lot when he talks, amused with his own expressions, and grimaces too when he says something disparaging about a promising "whack" of rare tubes that turn out to be "pure shit." We met in Upland, near the Claremont Colleges where I'd gone to school decades before. His home is up a cul-de-sac that abuts the nice rise of a hillside tawny with wild oat and rye grass. At the top of the hill, there's a water tower where Deal used to go drinking with his high school buddies. His home is a large ranch-style place with a big garage, two motorcycles and a Porsche parked in its driveway, and a double door in the front by some century plants and a rock garden. He is obviously prosperous. We met in his office, a small room off

the entrance filled with audio gear—rare and old tubes stuffed in stationery cubbies, a Cary SLI-80 amp painted an automotive red resting on a file cabinet, a corner desk covered in papers and single tubes in boxes, and a big sign on the wall that says WILL WORK FOR TUBES.

"Had that made in Paris," he says, catching me looking.

Deal wears a headset while he works, answering the phone when customers call, scooting around on his swivel chair, pointing to stacks of orders when his assistants drift in, answering questions about the transconductive readings on power tubes, matching triodes in a big order for signal tubes, and touting gear, gear, gear every chance he gets.

Deal talks fast, like a guy peddling hubcaps at a sixties swap meet—or tubes by the lot at an auction. He speaks with the broad, somewhat nasal emphatics of his Southern California roots, tossing off surfer and skatepunk terms with panache and verbal English. I'd have expected him to have worn a mullet in a previous incarnation, dressed in Dickies and chinos, and maybe Chuck Taylor tennis shoes. He weaves sentences like a bipolar spider might a web—a little woof here, skip a big space, a little warp there, skip, and come back to the woof of the web two sentences over. He speaks with gaps and references like that, weaving a piecemeal verbal tapestry that makes sense only after you wait awhile for everything to fill in and knit together. His voice has the bite of a carnival barker and pattering speed of a used auto auctioneer, but he has a way of throwing a glance at you that is totally sincere when he makes a point. *He ain't lyin'.*

"I go out to the desert in Parker, Arizona—a trip out by the Colorado River. When you're out that way, to find tubes during that period of time, you had to go where nobody else went. That would be out in the middle of nowhere. That would be in the desert. That would be up in Central California in little gold rush towns up there in the Sierra foothills. Just anywhere you never want to go. And I walked into a TV repair shop, 1995. That's where I used to find stuff. Along with very expensive Altec 604A speakers in a cabinet. Absolutely perfect. Would get them used once every year at the Parker Valley Fair. Out there to go jet-skiing on a three-

day weekend like the Fourth or Labor Day. I found a lot of good stuff. I'd just go, *Hey! You got any old junk you wanna get rid of?* Sometimes people would just give you things. Sometimes, if they asked me what I'd pay for it, I'm paying them a fair price. They'd say things—*Here, you want this?—fifty bucks.*

"One of the biggest finds I ever made was in Newport Beach of all places. You wouldn't think I'd find a thing there, beach city, crawling with weekenders, surfers, spring breakers everywhere. Lotta Asian audiophiles—Korean guys live in the OC. They are really good at digging things up. Tenacious. With my wife, supposed to be in a hotel in Newport for our anniversary, I *smelled* it. There's just something here. I just smell it! *Okay*, she says. *Go ahead.* Along 55 in Costa Mesa. Old TV shop. I go in and say, *Hey, ya got any old tubes you wanna get rid of?* The guy says, *Oh, yeah, we've been tryna find a way to get rid of 'em* and blah-blah. They had a big old box of tubes. He said, *Oh, and I have these up there.* Imagine you're in a shop and I'm on this side of the counter and there's a thing up here. A shelf full of cubbyholes. And from where he's standing and there's tubes right up here. He says, *Well, I've got these too.* So I peer around the corner and *Western Electric 300B,* there you are! But not just any Western Electric 300Bs. Engraved base ones—which are the earliest. They're worth thousands and thousands of dollars. And metal base EL34s. Had not been touched. Just thick with dust on top. 'Cause they put them up there sometime in the thirties I would guess. Then they put the EL34s in up there in the late fifties and never took them down. Never dusted them. Never did anything with them. I said, *Okay, I'll come back* (and I knew my wife would be ticked at me). I said, *Okay, I gotta get in the car or I'm gonna be in deep shit.* So I go back to the van breathless, I go to my wife, *Okay, yeah, he's got some stuff but we'll come back.* And we start to pull away and I said, *I CAN'T COME BACK!* I put it back into drive and I parked and I went in and I got the most important stuff out of there and I said I'll be back day after tomorrow to get the rest of it.

"And so I'm supposed to have a nice romantic weekend at this hotel and I'm at the beach and I'm with somebody who means so much to me which is absolutely true but I kept having to go down to the hotel parking lot back out to the van and look at them because

I was afraid that somebody was gonna wake me up. Something's gonna happen when this didn't happen when this doesn't exist or something it's so weird it's gonna be gone because it can't happen that way but it did. So I went down to the van I think *four* times during the next two days. I think I got six 300Bs and I got a perfect brand-new quad of Amperex metal base EL34s. I got some Western Electric rectifiers. And I still have those. It was quite the deal. And they were just happy for me to get 'em out 'cause they were fed up with it and they were gonna throw 'em away!

"There was a guy who talked with a high, squeaky voice. Somewhere from the Midwest. 1992. Old ham radio operator. He worked at a place where they had government things there. They had tossed Tung-Sol 6550s aside at his workplace—government airport. Things were getting rusty. They've buried over a thousand! Just buried them! He says, *If I find a way to excavate, are you interested?* To the best of my knowledge, they've never been unearthed. 'Cause I've never heard about them coming on the market.

"What really got me into tubes? High school. Best friend Steve Shiffler had this receiver, the Sansui 1000A, which was tubed. My other friend Andy had the Sansui 990DB, which was the new solid-state, transistor super receiver which was replacing tubes. It was so powerful—125 watts! Everyone wanted to have one. But when push came to shove my friend Steve had this down in his basement where we go to smoke dope—endlessly. And we knew that his sounded better. It wasn't even close. And we knew it had something to do with tubes, but we didn't quite know why until later. It wasn't even close. Older one sounded better. We knew it.

"Five years old. Neighbor built kits. Hi-fi kits. He had an oscilloscope, would be playing stuff in his garage. I go over and watch him. Told everybody I wanna be an electronics technician. First big awakening after that—1968. I was ten years old. Friends listening to surf music. It's just what you did. Either listen to Motown or surf. Just heard the Doors' first album and Jimi's *Are You Experienced* and Big Brother's *Cheap Thrills*. Sister had them. The '68 epiphany was with music. Then, when I went to Montclair High School, there was a lot of racial tension. It was difficult. But I was in the mentally gifted program. Got to go at my own speed. It was the dumbest

thing they could do. My own speed was I used to go to Straw Hat Pizza Parlor and smoke dope, do laser experiments, play games. When I got out of high school I started working at a store. Sold electronics and sold tubes. 1977. Old guys would come in and buy tubes for their McIntosh or whatever. I said, *Why are you doin' that?* Old guy just looked at me and said, *You don't know anything, do you?*

"Never finished college but always had my own hi-fi system. Had the nicest. Got a job. Bought my own stuff. Then I got the job at a parts place—Mission Electronics. Then I went from there to Cal Stereo. Then went to Roger Sound Labs. Biggest Audio Research dealer in the nation. I had an original SP-9. Wanted to get some 6DJ8s for it. So I go to this parts store down on Holt Boulevard [in Upland] called MarVac Electronics. Bought an International Service Master 6DJ8. And I plugged it in and I go, *Oh my god!* I couldn't believe how good it sounded. Looked at the tube. Tube said *Made in Great Britain.* Sounded so good. Started putting pieces together. I found out that that first tube was a Japanese tube. They were not gonna be popular in the seventies. So what they would do they would conceal the identity of who really made the tube by marking it *Made in Great Britain.* Counterfeit! Started buying Mullards and Tesla [tubes]. Playing with it. Curious about sounds of different types. I found that the important thing—not small, but giant.

"I was married at the time. Mainly buying stuff. I'd sit on sofa at night. I'd hold tubes up to light and note structure. This one's got a hoop that's mounted on a stick. This one's got a hoop that's mounted on a flat bar. There's a buncha little codes on the bottom. This one's got little pointed things. You know, just do that over and over again and you do it for a few years and you start connecting the dots and figuring out who does what. The giant thing that's in the structure of the tube that makes it different and different *good.*

"My first years in this business I worked every day from morning until night. My living situation allowed me to do that. I had a wife who was disabled. She had MS. It's not sad. It was an incredible experience. She passed away in 2005. My wife got so sick. And she had MS when I met her but I didn't care because when you love somebody that much that's it. She got sicker and sicker. When it got to the point when she needed care, business exploded, where I was

able to have three girls and all they did was take care of Lisa. It's funny how things work out. Just hanging out with her, looking at tubes, and making phone calls was okay.

"My early Upscale Audio ads—the one that put me on the map—had her in it. I did an ad called *I've got a woody!* I placed in *Stereophile*. And people flipped. Thought it sexist. Controversy. People talked about it so much a competitor accused me of sending in the first complaint letter.

"I don't wanna be stodgy. I don't wanna go out that way. My dad was like that. He was a Beatnik. When he was much older, before he died, he was wearing a *kufi*. He was a jazz musician. My mother was a singer and my father was a bass player. That's how they met in bars up north. I was born in Roseville, California. 25 March 1958. By Sacto [Sacramento]. Moved down here because of my father. Father worked in the prison system. I grew up in Chino prison. It was pretty wacky. Played catch, games with guys who robbed banks. My father got into the prison system as a social worker because he liked helping people. He definitely had his own thing.

"Me too. I got a lotta customers I deal with that are professionals—doctors, attorneys. They don't want me to talk about what they do for a living. Because I'm the pressure relief guy, so when a doctor calls me I don't call him doctor I call him *Dude!* And they like that. Guy calls up about that little Aja! CD player that we sold for $750. He says, *Kevin, it's Dan.* I thought it was this other Dan. I go, *Hey, Dan, what's happening?*—*Well, I'm thinking about getting that CD player and it's $750, you know. I don't know if I want to spend the money on it.* I go, *Dan, you cheap motherfucker! It's not that much money! Are you crazy? Especially for what it does, what else is gonna compete with it?* He goes, *Yeah, you're right.* Laughs at me. *Okay, write one up.*— *How do you spell your last name?* He's a major judge that you see these big federal cases in front of him. He's a guy in the newspaper—they named a freeway after his father—but they don't want to be treated like that. They want to just talk about music and have a good time. And being a judge is not a good time. Maybe sometimes it is, but a lot of times it's not. What's-his-name from Steely Dan was a customer for years and I didn't even know it was him—Walter Becker was buying stuff from me.

"I'll tell you what happened. Back in the seventies, I was looking for tubes again. Around that time, there used to be an old radio shop. In Pomona. Called Kitron. It had been there since the thirties. It had closed down. Covered up. It had rain damage on the building. *I bet there's some good stuff in there.* I kept leaving notes. *Call me, I'm interested in tubes. Call me.* Stopped by for six months. No one ever called me back. So then, I'm driving by, this old guy comes

out of the door. *Hey! I've been tryna get ahold of you.* He was a little cagey. But we went in and I bought some stuff and I got his phone number. Got in, bought stuff. He had a lot of tubes. Bought more. *You know what I've got? I've got twenty thousand tubes,* he says. *See those Olson boxes? My son was VP of Olson Electronics. After they shut down, we ended up with all their tubes. And I wanna sell for a buck apiece. But you can't go through 'em. You can't do anything. You just gotta pay me a buck apiece. Gotta buy all. No peek. Twenty thousand dollars.* I thought about it, and I knew what some of them were worth so I knew I could do fairly well with it. I got a cash advance on four credit cards. Took 'em all to the max. I made the purchase.

"Found a guy named Bill, elderly, who'd go to electronics flea markets. *You think you can sell these? Absolutely! We'll do this and do that and we'll pop a bottle of champagne 'cause it's gonna be party time! This is the deal, Bill, first I've gotta get my money back. So the first tubes that you sell, it all comes back to me. After I recoup my 20k, then I'll split the rest 50/50.* He said, *Cool.* Money starts poppin' in. Five . . . 5k . . . then it kinda stops. I get a phone call from a guy, *You're Bill's friend, aren't you? I bought 5k worth of tubes last week and he was s'posed to bring me more.—What's the problem?—He was s'posed to bring me tubes—When did you give him the 5k? Last week?* I hadn't seen that money. So I found out that Bill had ripped me off. And he lived in Central California. And my head was spinning because I still owed 10k plus he had my tubes. So I didn't even tell him I was coming. I got a big van and I drove it all the way up there to Meadow Vista where he lived and I showed up at his door and he was shocked to see me. We said things and he tried to bullshit me about stuff and I knew it wasn't true and he wasn't gonna give me the tubes back. It was a nightmare. So I left for a little while and I talked to the police about what happened and they said it was a civil matter—they couldn't help me get my stuff back, it's my property and they can't help with that. Then I went back to Bill's house and all my stuff was piled in the front yard. So I loaded it into this big old cargo van and I came driving back down here and now, I had no choice! *I was IN the TUBE bizzness!*"

Part Four

Tubeworld, 2

The Vacuum Tube and Electronic Amplification

A lot can be said about the vacuum tube and its history, but it boils down to a few basic steps, a handful of inventors (both staid and eccentric), and a kind of "march of technology" sequence through the latter nineteenth century and early part of the twentieth century. Though one might go back as far as primitive man noticing lightning strikes, Benjamin Franklin's kite and key experiment in a lightning storm, or a string of European scientists and inventors throughout the eighteenth century (they tinkered with wires, heat, current, and pickling jars), it's handy to start from the gradual perfecting of the incandescent bulb during the latter part of the nineteenth century and the significant additions and changes to its basic structure in the early twentieth.

Contrary to the American myth that holds that Thomas Alva Edison (1847–1931) was the inventor of the light bulb, it was actually a string of scientists from all over the world that contributed to its development beginning late in the eighteenth century and into the early twentieth. English chemist Sir Humphry Davy (1778–1829) came up with something first. He was searching for a safer means to provide light to coal miners who'd been using torches and other open-flame lamps. Obviously dangerous, these caused numerous explosions when they came into contact with what was called "firedamp"—a combination of methane and other gases prevalent in the mines. Concerned about the loss of life in these conditions, Davy created a device that placed a wire gauze (a kind of mesh) over an enclosed flame, and made an incandescent lamp that operated somewhat like the contemporary Coleman lantern. Later, in 1801, Davy took a platinum strip and passed some electrical current through it, causing the strip, a kind of filament, to glow with light in the open air. The light didn't last, but this is how the notion of electrical incandescence got its start.

Then, in 1841, Frederick de Moleyns, another Englishman despite the flamboyant Francophone name, put a platinum filament in an evacuated glass bulb, creating the first light bulb. He used powdered charcoal heated between two platinum wires to make his lamp. But it, too, didn't last long, as a curious blackening would build up at the top of the inside of the glass, eventually obscuring its light. Finally, in 1878, English physicist Sir Joseph Wilson Swan (1828–1914) and American inventor Thomas Alva Edison, working independently of each other, both came up with light bulbs that could last awhile, taking carbonized cellulose as the filament. Swan used a paper material and Edison used plain sewing thread. But Edison found a more efficient vacuum pump to evacuate the air from his bulb, and consequently it lasted a much longer time—up to six hundred hours, according to historians.

It had long been known that the better the vacuum, the longer the filament material would last, but coming up with an efficient means to create the vacuum had been a stumbling block. Edison found the Sprengel pump, an 1865 invention by German chemist Hermann Sprengel. It used drops of mercury falling through a small-bore tube to drain air from the system it was evacuating and was able to create the highest level of vacuum in its time. But even after he'd switched to a longer-lasting carbonized bamboo filament in 1881, Edison noted that his bulb suffered the same syndrome of unwanted blackening inside its glass that de Moleyns had encountered. In this historical period before parts of the atom were completely described, Edison regarded this blackening as a vexing and very curious phenomenon. He hypothesized it was the deposit of carbon from the charred bamboo filament migrating to the glass wall in a process that was later labeled (by yet another scientist) the *Edison effect*. It was the earliest citing of the movement of electrons freed from a heated element (a cathode) and attracted to a separate surface inside a sealed tube of glass. Edison was describing the effect of an electrical current not only moving across a conductive filament (that produced light), but also causing the transfer of atoms of carbonized material across space, moving particles of a substance from the surface of the heated bamboo filament to the inner surface

of the glass light bulb. This movement of electrons became the basis of later vacuum tube technology.

The Fleming Valve

In 1904, John Ambrose Fleming (1849–1945), an English scientist working for the Marconi Company, demonstrated that this same Edison effect could be used to detect radio waves. He took the basic structure of the incandescent lamp and placed a metal plate (another electrode) within it beside the filament. This simple addition created a *diode,* a tube with two inner elements instead of just one. In addition to the detection of radio waves, this diode had the ability to *rectify,* that is to transform AC voltage into a pulsing DC current, creating the potential for power applications (actual amplification) when used with tubes that were invented later.

The diode operates by having its element, the cathode or filament, heated to a super-hot degree so that electrons boil up on its metal surface (at first made of tungsten), then are attracted by the more positive voltage of electricity applied to the anode, or plate. Electrons travel the relative vacuum of the tube over an electrical field created between the cathode and anode, producing a current that flows only toward the anode. As the anode isn't heated, current cannot flow back across the gap of space to the cathode—the heated filament itself. The diode is a one-way electronic street. And, since the current flows only one way, only half the waveform of the AC current is reproduced, resulting in a pulsating DC current in a process called "rectification." The diode was patented by the Marconi Company as the "Fleming valve." The English called it a "thermionic valve" for its valve-like ability to control the movement of electricity. Heat the cathode and electrons flow. Shut off the heat (like a valve) and the flow stops cold.

If you look at early photographs of this Fleming valve, its form factor presents as a kind of space spider—the alien being from the discontinued TNT television series *Falling Skies.* It has the bulbous head of a Martian atop a robotic-looking, three-legged spaceship. The parts of the Fleming valve were a globe of glass peaked with

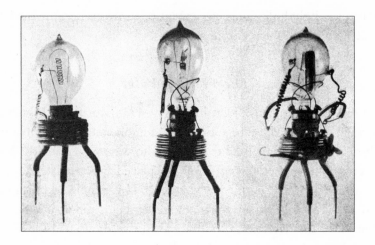

a nipple (the residual connection to a blowpipe used in glassmaking), its two interior electrical elements (one a loop of carbon or a fine wire of tungsten, the other a flat or cylindrical metal plate), a stem or collar of sorts, a base or cage of horizontal ribbing, and three thickly jacketed wires emerging from below this cage (positive, negative, and ground wires).

I saw some Fleming valves up close myself once, in a display of several such diodes in a glass cabinet of an old schoolhouse building in Gold Beach, Oregon, one day in early fall. I was there for the annual meeting of the Tube Collectors Association as a guest of its president, Ludwell Sibley. I'd been corresponding with Sibley about tube substitutions—what variants of which tubes might stand in for one another and where I could get them—and the various handbooks he had for sale and also available free online, when he invited me to the meeting not far from where I live in Eugene.

It's a sweet drive down to Gold Beach, not more than a couple of hours and change, mostly on I-5 through the south end of the Willamette Valley and up through woodsy hills and dales of the Coast Range and Cascades folding into each other. You get to a point where old Highway 99 crosses I-5 again and it's the turnoff to Gold Beach, a former stop on the rail line built for logging back in the nineteenth century. There's just a patch of shops and bungalow dwellings there, but the roadside trees in the fall are gorgeous— maples turning golden and coppery, shaking in the midday breezes

like blaze-skinned salmon jumping up a waterfall. And there's a quality of seasonal light that's unforgettable—like it was luminous matter bestowed by parting hands over the landscape, saying *Let us feel fucking good together.*

The meeting was held in a converted schoolhouse, a big single building, complete with a broad concrete staircase leading up to its double-doored entrance beneath fake Greek columns. I paid my registration fee ($15), for which I received a thin catalogue and a ticket for a sandwich lunch (it turned out to be a choice of tuna or egg salad sandwiches on white bread with a bag of chips and a pickle), and spent most of my time cruising the display rooms on two floors. The other attendees were mostly single elderly men or elderly couples dressed in summery clothes—shorts, chinos, and polos— and carrying canvas bags they filled with brochures, souvenir hats and visors and coffee cups, books and pamphlets, and, occasionally, a vintage tube they'd discovered that they just had to have. Most of these tubes did not function and were only for display—for my fellow registrants were mostly dyed-in-wool collectors on the prowl.

In one of the rooms on the second floor, I found a display of Fleming valves, anchored vertically with what looked like twisted pipe cleaners up one side of a wood-frame, glass-paneled display case. There was one example of the first Fleming tube—that skitter-looking creature previously described. But there were several others, each marked with approximate dates of invention, along with the name of its creator or, in one instance, company of origin (Marconi Company). To me the things mostly resembled Erlenmeyer flasks stuffed with wires and plates, or bulbous test tubes that sprouted inner latticeworks of delicate wires. Some looked like the bottleworks inside the lab of Dr. Henry Frankenstein from the eponymous movie by Universal Pictures (1931). I imagined bolts of blue electricity snaking and sparking from inside their glass, twin electrodes of a Jacob's Ladder channeling thin arcs of freed juice climbing upward from the case to the ceiling. I wasn't very realistic. I was full of fancy and romance, as it was still my early years in the hobby, fueled by curiosity and fascination with the evolution of this technology. I had gaps to fill—just like the dream of a ladder Jacob saw in Genesis as a staircase heavenward, emptied of man but spot-

ted with the angels of God, stepping and arising and descending and burned like strikes of lightning in his memory.

De Forest's Audion, or the Invention of the Triode

The invention of the *triode* (a tube with three electrical elements), the first thermionic valve capable of amplification, was accomplished by the curious, ambitious, and dogged American inventor Lee de Forest in 1907. I say "dogged" as he worked tirelessly on a number of inventions, including one for simultaneous sound to accompany film, but he never felt he achieved the success he envisioned for himself. From his youth, de Forest had the fantasy that a life of inventing would bring him fame, riches, and standing in a world he saw as out of his grasp somehow. He'd grown up the son of the first president of Talladega College in Alabama, a school primarily for African Americans. As both a Northerner and a white man, de Forest was ostracized by his Southern schoolmates and, though perhaps happy among them as a child, increasingly distanced from the Black students at his father's school. Although he was eventually educated at Yale (with three degrees, including the Ph.D.), he was nonetheless a man always out of place, viewed as an upstart by his Ivy League classmates and a muddler by fellow engineers. His creations appeared to them as mash-ups of the inventions of others, and he initiated, suffered, prevailed, and lost many patent battles throughout his lifetime, not the least of which concerned his invention of the first triode. He called this tube "the Audion," combining the Latin root *aud,* "to hear," with the Greek *ion,* meaning "something that goes." (The term *ion* was likely first used in electronics during the nineteenth century by Michael Faraday, an English physicist and chemist, who came up with it as a name for the electrically charged particles he hypothesized were formed in gas when an electric discharge was applied to it.)

I read in Tom Lewis's book *Empire of the Air: The Men Who Made Radio* (1993) that de Forest took the momentous step in fall 1905 of ordering a glass bulb "about the size and shape of a small pear" from Henry W. McCandless in New York City, a manufacturer of auto-

mobile lamps for Westinghouse and General Electric. The bulb had a brass candelabra screw base and a carbon filament—a construction very similar to Fleming's diode. De Forest had been studying Fleming's work and became interested in a paper Fleming had authored on the "oscillation valve." The de Forest bulb had a nickel plate placed very near the filament with a short wire leading out of it through the tube. On December 9, de Forest took out a patent on a "static valve for wireless telegraph systems." Five weeks later, he took out another patent on a similar tube that ran wires from a small battery to both the filament and the plate. But he wasn't through.

In November 1906, de Forest ordered yet another tube from McCandless, this time with specifications that called for *three* elements within it: the familiar plate and filament of Fleming's valve, but now, interposed between the plate and filament, the order called for another nickel wire as close to the filament as possible. All three elements had wires leading out of the side of the tube. When the third wire was positively charged, de Forest found it would attract a stream of positive electrons flowing from the filament and *accelerate* them toward the plate. The more positive the charge, the more the charge on the plate. Here was finally the moment—the technological beginning of electronic amplification.

An assistant to McCandless, one John Grogan, suggested de Forest add a zigzag bend to the third wire to create a greater surface area to attract electrons from the filament, amplifying things further. The Audion worked. De Forest could hear the result through earphones. *More* sound came when they added what he called "the grid" of this third element. In effect, the action of the third element *amplified* the wave of electrons flowing from the filament to the plate, increasing the efficiency of flow, resulting in stronger electronic waves. This was a triode, the first significant improvement to Flem-

ing's simpler diode tube, and an advancement in early audio technology that would have far-reaching implications.

The grid controlled the flow of electrons across the vacuum between the heated cathode and anode. When a negative voltage was applied to the grid, it slowed or closed the flow of current, depending on its level of charge, by repelling electrons away from the anode. When a positive voltage was applied, it increased current by accelerating the amount of electrons that flowed to the anode. The triode tube was thus a voltage-controlled device. Since it was also capable of carrying a small radio-wave signal from the grid to the plate, converting it into a much amplified signal and with exactly the same waveform—*boom*—this commenced the electronic age of sound.

In a 1907 paper presented to a meeting of the American Institute of Electrical Engineers, de Forest claimed his Audion was "a new receiver for wireless telegraphy." He'd taken Fleming's diode valve and introduced a grid-shaped electrode into it, making it more efficient for conducting current. The earliest versions of this tube were round globes (like clear tennis balls or the unfrosted, 20-watt light bulbs around a makeup mirror) with a coil of wire wrapped around a neck that protruded from one end and two thick wires sticking out of the other end.

At the Tube Collectors Association meeting in Gold Beach, I saw that the earliest triodes looked like a glass model of the Soviet Soyuz spaceship, an elongated spheroid without the solar wings. I could see the three different electrodes inside it—the heater/filament, the plate, and de Forest's new addition, a squiggle of wire that served as the grid. Later, more "lamp bulb" type versions were round and had screw-cap ends (where they'd fit into a socket). These were fashioned from the spherical envelopes of auto headlamp bulbs like those McCandless made, which were then commonly available. A version marked "R 1922 Triode Marconi-Osram UK" was particularly handsome—a round bulb with a top nipple and a silver-collar base that wrapped around its stem. I'd see a replica of this one at the KR Audio factory in Prague some years later. The triode tube could amplify an electrical signal, and later versions of it were used for telephone and radio systems in North America and Europe.

De Forest's showmanship manifested in the way he decided to get his device better known to the public. He got the director of the Metropolitan Opera to stage a special event that he transmitted to various stations scattered around New York City. On a night that Enrico Caruso was singing the lead roles in *Cavalleria Rusticana* and *Il Pagliacci* (two Leoncavallo *verismo* operas with especially hammy tenor parts), de Forest hung a microphone over the stage, rigged up a radio transmitter backstage, strung up an antenna on the roof, and threw a switch. Using earphones, receivers picked up the signal only faintly, however, and drift and distortion kept the experiment from being a popular success. Yet, here was the beginning of radio broad-casting in America—an Italian tenor singing arias full of bathos to faraway listeners straining to discern every note, hoping for sublim-ity but receiving a signal that was full of static. How much like me they were in my beginning days in audio, devoted to the art and yet disappointed in the technology that tried to reproduce it.

A Visit to the KR Audio Factory

"De Forest invented audio amplification tube," the Czech engineer Marek Gencev declared when I visited the KR Audio tube factory in Prague. I was in the city to teach a poetry workshop at Charles University in the summer of 2013, and I'd taken a couple of after-noons to see how audio tubes got made. Despite his youth (I guessed he was in his mid- to late thirties), Gencev is the chief engineer of KR, a small, independent company that manufactures fine audio tubes in an old Soviet-era factory. He is baby-faced, his blond hair cut uniformly short, and has a whisper of a ginger-colored beard on his chin. Over two separate afternoons, he showed me around the KR factory, and both times he wore only a pair of thin slacks and a plain white T-shirt. It's hot in Prague in the summer, and in the rooms of that small factory, it got even hotter.

Founded in 1994 as Vaic Valve by the initial partnership of Alesa Vaic and Riccardo Kron, KR was reconstituted under the new name by Kron in 2000. It is now headed by the "K-G-B" triumvirate of his widow Eunice Kron, Gencev, and Jindra Boube-likova, a slim and deeply tanned middle-aged woman who handles

its administration. I met them and the production team in a small manufacturing facility once operated by Tesla. The current KR factory had been the research and development lab for the Soviet government during the Czech Communist period.

After class one Tuesday afternoon, from a building called "the Faculty" at Charles University on the banks of Vltava River, I took a taxi to the industrial district at the far reaches of Prague where KR is located. We rode along the riverside for a mile or so before the cab headed inland through poorer commercial areas, then down a broad thoroughfare past an ice rink and soccer stadium to a narrow industrial lane of two-story offices and shops. These lay in the shadow of an imposing multistory building with T-E-S-L-A spelled out in yellow lettering against a blue background. I dismissed the cab, walked up to the kiosk entrance, and was directed under a passageway for cars and then up a flight of concrete stairs to a suite of rooms.

Down a battered hallway, I found an open office door and heard voices. Inside the room were Eunice Kron and Jindra Boubelikova talking in a rapid exchange of Italian. Eunice was expecting me, and she turned and gave a charming greeting in an accent that was unmistakably that of a native New Yorker. A corpulent, shortish woman with brown hair and pearly skin, Eunice is from Brooklyn but has been living in Europe most of her adult life. After I met the multilingual Boubelikova, a thin-lipped ringer for Donatella Versace, we went quickly down the hall to the vacuum stage of tube manufacture, where Eunice introduced me to Gencev. They took me through a quick tour of the sprawling factory, consisting of eleven rooms altogether. There were two administrative offices, a listening room, five rooms dedicated to tube manufacturing, and two more for amplifier assembly. The glassworks and vacuum rooms were among the most fascinating, both hot from torches and other tools of fabrication.

"No new tube since *forever*," Gencev exclaimed in his insistent Slavic accent, with a clipped syntax that charmingly abbreviated his English. His speech was forceful. We were walking down a hallway, and he stopped to point into the testing room at a row of triode tubes on a long table, their heaters glowing an amber red. There

was a parallel row of simple light bulbs beside them. Gencev said the light bulbs, hooked in parallel, were the way they controlled overall electronic impedance to the tubes. It's always *science*.

"De Forest invented grid. Grid sent moderate current to anode plate. He made magnificent Audion tube. We are only following. What is new?—*Only mateeree-alz*." In place of the classic metals, KR uses mica oxide supports, nickel cathodes and anodes, molybdenum for the grids, and thoriated tungsten or nickel filaments. At KR, there are no modern machines—they work with twentieth-century artisanal skills and a lot of tube-making experience.

"Only knowledge and old machines produce these tubes," Gencev said, handing me an 845 triode. It was big as a pony's penis. I felt its serious heft. "No one is teaching vacuum tube," Gencev explained.

KR learns from studying old designs, first testing and measuring old stock tubes for their electrical characteristics and dynamic parameters. It's also important to find out their design geometry, noting the physical dimensions of each tube under scrutiny, its spacings, size of grid, etc. When data sheets and a working tube aren't available, Gencev cuts into an old tube, opening it up, and analyzes it—he practices reverse engineering.

"It's basic science," Gencev pointed out. "We make calculations to determine gain qualities of tube."

Talking later with Eunice Kron, I found out that KR Audio started as an idea one day in 1991 when she and her husband, Riccardo, a collector of old radios, were browsing through a flea market in Scandiano, Italy. The Berlin Wall had just fallen and there was excitement in the air—"like a new age was dawning," Eunice said. Behind a wooden table covered with old tubes was Alesa Vaic, a "scarecrow of a man" as described by Eunice, and an inventor and tinkerer who'd come up with his own working version of the original Marconi radio tube. It sat there among an array of new old-stock radio tubes. Riccardo Kron immediately recognized Vaic's accomplishment.

"He got a sensation," Eunice told me.

The tube Vaic had fashioned was a working replica of a 1916 Marconi-Osram R, a high vacuum triode manufactured first by

TM in France and then in the United Kingdom as a radio frequency amplifier for use during World War I. A very simple design, the R was close to de Forest's Audion bulb but had an axial grid, no flashing, and no getter.

"If you can make a Marconi tube for radio," Kron said, "can you make an audio tube?"

There were no audio tubes made in the West then, Eunice told me, as its factories had long been shuttered and abandoned, and in the East the Russian manufacturers were retooling. Vaic accepted Kron's challenge of creating a tube from scratch, and quickly the two men formed a partnership. They created the company known as Vaic Valve and hired a glassman, the team beginning experiments with the goal of making a better 300B triode tube, originally invented by Western Electric. Two years passed before the Vaic Valve 52BX VU got developed and released. It was Alesa Vaic's startling version of a 300B that finally gained the attention of the audio world in 1994, when Kron and Vaic were asked to upgrade a vaunted Japanese amplifier called the Kegon, made by Audio Note (now Kondo), which had previously been using Chinese tubes.

In 1996, Vaic and Kron relocated their manufacturing facility to the Czech Republic, taking over the old Tesla research facility. As a prototype lab for Russian and Soviet military tubes, it already had gas lines convenient for audio tube production, Eunice said, and although they still used the old Tesla machinery, they did not imitate Tesla's technology. But the partnership, perhaps as fragile as glass, had started on a downward spiral by then. By 1997, the partnership fractured, Vaic forming AVVT and Kron branching off into KR Enterprises. "So, in 2000 we reformulated the company as KR Audio."

In 2002, Riccardo Kron died, and the company seemed destined for closure, except for Eunice's determination. Marek Gencev

had just signed a contract to be the company's engineer a few months before, and Eunice felt a responsibility, not only to him, but to administrator Boubelikova, the glassman, and various other artisans and mechanics on the KR Audio staff.

"KR was an obligation to people both living and dead," Eunice told me.

Gencev had worked for the company since he was seventeen, soldering boards for amps, then signing on as engineer after graduating in 2001 with a master's in electrical engineering from the Czech Technical University. He is Boubelikova's nephew, and the glamorous woman raised him. KR Audio is indeed a family.

"My husband built the foundation," Eunice said, lovingly. "But Marek built the house."

The KR Audio team now produces numerous triode output tubes, and they have just started making their own signal tubes as well—those miniature, nine-pin, twin-triode devices I'd been collecting to tweak the sound of my amps. Marek Gencev had resurrected the knowledge of how to fabricate these tubes once made in Germany by the tens of thousands, devising a new manufacturing technique by analyzing the electrical properties and spatial geometry of a few samples that still existed. It was artisanal reproduction of a lost industrial gem, and I was fascinated not only that it could be done, but that the art of tube manufacture had revived itself at KR.

At the factory, I saw there are roughly nine steps to the tube manufacturing process: (1) glasswork wherein tube envelopes are cut, heated, and shaped into their characteristic bottle profiles and the glass stems made with their lead-in wires; (2) electrode parts manufacture and assembly, wherein various metal parts are fabricated and then assembled into the cathode, anode, and grid of each tube (making up the electrode); (3) tube assembly, in which the glass stems are connected to the inner electrode system and then put together with their appropriate glass envelopes; (4) tube exhaustion, in which the air inside each tube is evacuated so that a vacuum forms within the tube; (5) inductive heating, when the anodes of each tube are heated via a coil that surrounds the bottles, burning off impurities; (6) cathode activation, when the barium compound

that coats the filament is super-heated so that the cathode can produce active electrons; (7) closing the tube via melting the thin glass
pipe connector to each tube; (8) gluing each tube to a base and
burning it in; and (9) testing and storage for final shipping. These
steps take place in five different rooms of the factory: a glassworks, a
vacuum room, a chemical room, a mechanical room, and a testing/
packaging room. The entire process to make a single tube out of
roughly 128 separate pieces takes about 120 hours.

In the glass room, while observing the shaping and cutting of
the glass envelope, I was introduced to Ladislav Krouzel, the glassman. I took him to be around seventy and about my height (5'8").
He wore a dark blue polo shirt, denim trousers, boots, and no work
apron. After introductions, he picked up a four-foot-long tube of
glass in one hand and grabbed a blowtorch in the other, shooting
flame at the open end of the glass, heating it until it glowed a light
red. He heated the glass until it softened enough to form a rounded,
somewhat molten cap. Then he cut the tube about a foot below the
cap and mounted the part he cut onto a lathe, where he put a torch
to it again, further softening the glass so he could reshape it as it
spun within a metal mold, transforming it into the characteristic,
tapered shape of a KR 300B XLS—a kind of super-300B.

Next, he took the glass envelope out of the lathe and walked to
the other side of the room, where he mounted it on a shaft of met-

alwork connected to a glass stem—the inner electrode assembly of the triode tube—that was sticking up from a vise. He clamped the envelope from above and heated it up again, this time from the bottom, blowing pressure into it through a small surgical tube so that glass extruded from the bottom, sealing the tube's neck to the glass stem. Excess glass drooped of its own weight away from the nearly finished stem, and then he drew the entire tube off with a deft lift of his hand. This was a critical stage, a weak point of manufacturing, and his timing had to be perfect, drawing the glass tube and connected inner assembly away from the vise and the extruded excess glass while it was still slightly viscous. Gencev told me there must be no mechanical tension as this was achieved or the hardened glass of the tube neck could break, causing potential leaks, and resulting in a tube that had to be junked.

Langmuir's Hard Vacuum

Lee de Forest, for all the significance of his invention, didn't quite understand how his Audion worked. He was as much huckster and inventor as he was a pure engineer, and thought that the small amount of gas left in the tube after it had been pumped out was the crucial manner of electrical conveyance. It was not. It took Irving Langmuir (1881–1957), a chemist and physicist working at General Electric (he was later to win the Nobel Prize), to develop de Forest's Audion into a more workable amplification device. Around 1913, Langmuir developed the "hard vacuum" inside the tube, which evacuated nearly all of the remaining gas left inside after the shape of the glass was formed, adding a crucial step in the manufacturing process. While de Forest had speculated that all the gases still trapped inside his Audion were imperative to its electrical functioning, Langmuir theorized that an audio tube should have, as much as possible, all oxygen, water vapor, and other gases eliminated. And he found a way to do it. He employed a condensation pump of mercury vapor, used elsewhere in industry, to evacuate the tube to a high degree of vacuum very rapidly and created the procedure still in operation today.

When I was in Prague, Marek Gencev showed me this process.

I witnessed a phalanx of big 845 triode tubes having the living air sucked out of them via their glass stem straws, each of them hooked to an ancient mercury vapor pump left over from Soviet times. The tubes were connected to an exhausting machine and pumped free of air to a relatively high level of vacuum (less than 10-2 Pascal). The vacuum is necessary, Gencev told me, in order to direct the free electrons emitted by the cathode toward the anode of the tube. Regular atmosphere creates obstructions to the movement of a tube's emitted electrons, but a vacuum eliminates these obstacles (essentially, the electrons of atmospheric gases). KR first uses a rotary pump that creates a rough vacuum, then follows it with the diffuse pump I was seeing that works by blowing mercury vapor from silicon oil heated to 200°C. Consequently, the vacuum room was a very hot place. The line of gleaming tubes looked like a nursery of glass jars hooked to a large ventilator that made just a whisper of noise as it worked away, raising the temperature of the room. The imposing machine was mostly a big, gray tank with a valve and a long tube running out of it. At one bulbous end, a white-faced gauge stuck up above it, the needle steady.

"We get Langmuir's hard vacuum this way," Gencev said. "Down to 1E-5 Pascals [a billion times below atmospheric pressure at sea level]. Much superior vacuum for long life tube."

Langmuir named his tube the "Pliotron," after the Greek *plio* for "more" and *tron* for "device" or "instrument." It was capable of amplifying more clearly and at higher frequencies than the Audion tube, revolutionizing radio communication. It's from Langmuir's Pliotron that the term *vacuum tube* eventually evolved and came into common usage over the course of the twentieth century.

Yet one of the most important factors in the development of the vacuum tube as an amplification device had nothing to do with the number of electrodes or quality of the vacuum. Instead, it had to do with the chemistry and metallurgy of the cathode, or filament, inside the tube and involved the switch from using tungsten wire as the heater material (of the cathode) to another kind of metal.

One of the problems with using tungsten was that it needed to be made red-hot (800°–1000°C) in order to stimulate its molecules to throw off electrons inside the tube—the electrons that streamed

from filament to plate in the current flow that created amplification itself. Thoriated tungsten, a metal coated with a layer of thorium, was a much better source of emission and was later in wider use, but it needed to be heated to an even higher temperature—about 2400°C and white-hot—in order to operate. Being a friable metal, the tungsten wire would eventually break and the entire tube would have to be replaced after only a few hundred hours of use. The solution came in 1913 from H. D. Arnold, an engineer at AT&T whom de Forest approached for help with his Audion. Arnold, more of a chemist and a better scientist than de Forest, conceived of an Audion with a filament covered with an oxide of barium that would inspire thermionic emission at a much lower temperature (about 800°–1000°C and only orange-hot), making this tube more stable and longer-lasting. It is the metal oxides of barium, aluminum, strontium, and calcium that give audio tubes their touch of romance even now, as they glow in the dark like tiny embers of sound just on the verge of being born.

In Prague, Gencev said that KR uses molybdenum, a silvery-gray metal oxide, to make their grids. Filaments, anodes, and cathodes are nickel, and cathodes are coated with oxides of barium, pretty much the same stuff Arnold discovered in 1913. Then, they put their tubes through another process in what they call a "cleaning station," where tubes are attached to an inductive heater that operates via radio frequencies emanating through a metal coil. The coil slips down over the glass bottle of the tube and is activated, heating the anode inside the tube until it glows red. Impurities and excess gases are burned off and the characteristic flashing—that silver coating around the bottom of a tube—appears. Gencev explained that the flashing is the residue of vaporized material from within the circle of the getter of each tube. If there is no flashing or the flashing is transparent, he said, the tube vacuum is bad. And, again, you have junk. The point is not to make junk, Gencev said. The point is to "make the good tube for classic audio sound."

Various tubes followed de Forest's and Langmuir's first inventions—new, more powerful tubes, like the Western Electric 43A, designed to play the audio track for early talking pictures during the thirties; the Philco 71A, for its Model 20 "baby grand"

cathedral radios of 1930; the Genalex KT66 in the Marshall Blues-breaker amp, used in the sixties by Eric Clapton; and the GE and RCA 6L6GC tube for the Fender Twin Reverb, used by rockers Chuck Berry, Jimi Hendrix, and Jerry Garcia of the Grateful Dead. And, in an interesting application used during the Cold War, it was a Russian 6C33 triode tube that the Soviet air force chose to depend on for telemetry and guidance systems in their MiG-25 fighters. Finally, in the early days of computers, the Univac 1101 computer used twelve Tung-Sol 6550 beam power tetrode tubes (the same ones I'd collected) in its tape drive.

When the transistor came along in the late 1950s and early '60s, widespread industrial use of the vacuum tube died away, and by the early '70s, solid-state amplifiers took over most of the consumer market. But a curious rebirth occurred in hobbyist and specialist audio almost simultaneously with the tube's demise in consumer and industrial applications. Emerging from the small clan of do-it-yourself hobbyists who had been placing old tubes in stereo ampli-fiers all along, others in audio (electronics and sound technicians, engineers, and designers) began to realize that these old "glass amps" sounded noticeably better than the "sand amps" created by main-stream companies that used the transistor. A new, niche market for leftover industrial tubes blossomed during the 1980s and '90s, and an amazing, artful array of new consumer audio equipment was designed to make use of them. This was the point at which I entered the hobby, finding my PrimaLuna, Cary, and Air Tight amplifiers excellent machines for making the world's finest recorded voices sound as sublime as the angelic orders of heaven.

Listening in Triode Mode

In May 2008, in a corner hotel room at the Vacuum State of the Art Conference in Vancouver, just across the Columbia River from Portland, I heard an astonishing demonstration put together by Jef-frey Jackson of Experience Music. Jeffrey had driven a moving van full of his exotic gear all the way to the West Coast from his home in Memphis, Tennessee, just to make the demo. He brought Goto Unit tweeters and compression-driven horns, cavernous Baltic birch

bass horns, and a raft of custom-built electronics he'd made in his shop. One was a version of a vintage Western Electric amp originally designed for movie theaters that Jackson housed in a trilevel antique walnut cabinet. It looked like a pie safe lit from within by extremely low-watt Edison and Fleming bulbs. The amp was basically three tiers of glowing electronics, based on the Eimac 75TL triode transmitting tube for its output stage, and it had Western Electric 287A mercury-vapor rectifiers, direct-heated 112A driver tubes, and vintage Weston meters. Jackson told me it produced a flat-out maximum of 10 watts. And it was a one-off—a Jackson-signed original piece and the only one ever built. He took a CD I'd brought, my copy of Mahler's Fourth Symphony performed by the Los Angeles Philharmonic with Esa-Pekka Salonen conducting, and spun it on a contemporary player. The speaker's woofers were black, half-oval wooden caves resting on the floor, big orange pillows stuffed in their mouths for damping. Behind them, the horns of their midrange units, big as sousaphone bells, were suspended from two large frames. The Goto tweeters were stuck on planks below them, looking like matching Mini-Me's of the horns. When the music came on, the system sounded prodigious, movie-scale big, and had a soundstage that hung the moon. Images were indeed big-screen with an amphitheatrical depth: huge grunt in the bass, a pure sweetness to the strings. Jackson's system attracted quite a crowd—I noticed a few Asians among the twenty or so American audiophiles that gathered—and they all flattened themselves against the windows and walls of the big hotel suite, preserving the travel of fine sound waves from diffusion by the obstruction of their own bodies, as the sublime music blasted away.

Then, in the late fall of 2011, I visited Jonathan Halpern of Tone Audio in his home in Huntington, New York, out on Long Island. He told me he preferred 78s to everything, so he pulled out his collection of old shellac records, and we listened for a couple of hours. He had two speaker systems set up, one with a German-made Klangfilm Siemens two-way, the other with a pair of German-made Auditorium 23 speakers of a new design that incorporated Chinese-made replicas of the Western Electric 555 driver. Both speaker systems shared the same electronics—a Shindo preamp and amplifier

that used the Western Electric 300B tube, a modified British Garrard turntable, and a Japanese tonearm that was twelve inches long. His specialty monophonic cartridge was a German EMT OFD65.

Halpern shared his philosophy, explaining he thought mono was a much better way to listen to music, that stereo is already distracting—"You start looking side to side rather than listen." Much more trickery comes in. "All that sparkling and tinkling," he said. "With 78s, you cannot look away."

And he was pretty much right. He played "Mood Indigo" by Duke Ellington and His Orchestra and the sound wasn't up in your face, but more natural—the clarinet sweet, the tenor and alto both with strong schwerve and sashay, and Ellington's piano briskly sparkling. When the trumpet came on, worked with a plunger mute, there wasn't that brassy shout you sometimes get that pierces your ears. The plosives had more warmth and a soft shimmer, the growls more grunt. But when Halpern played an old Julie London recording, "Cry Me a River," one that I must've heard dozens of times before in stereo, I at last understood what all the fuss was over the old triode tubes, shellac records, and sexy singers of yore. I didn't think about the system so much as feel that London was singing in a small, intimate lounge just for me. Barney Kessel strummed his big-bodied electric guitar, his choice chords vibrating, Ray Leatherwood's big double bass stalked the bottom of the tune, and then I imagined a gray helix of smoke from the cigarette in the torch singer's hand twisting up beside the long curls of her red hair like it was the thinnest of ladders to heaven. I couldn't look away.

Armstrong, Sarnoff, and the Radio Era

The great age of broadcast radio was made possible by the work of Edwin Howard Armstrong (1890–1954), an American electrical engineer educated at Columbia (and later a professor there), who took the vacuum tube and early amplification technology to another level. Before Armstrong, amplification was mainly confined to experiments, the development of prototypes, and telephone relay systems—more practical communications than the widespread leisure activity that audio became.

In 1912, while on summer break from his studies at Columbia, Armstrong hit upon the idea of a regenerative audio circuit he called the *heterodyne,* which took the output from a triode tube and fed it back into its own electrical operation at its input in a kind of round, multiplying the amplification factor thousands of times over (15,000). Imagining this, I thought of the "hall of mirrors" scene at the climax of Bruce Lee's *Enter the Dragon,* when the villain, a kind of kung-fu Satan, lures Lee into a maze of mirrors that multiplies an image ten thousand times—well, an infinite amount—such that Lee finds himself fighting myriads of Satanic masters rather than the original he seeks to defeat. I thought of audio amplification this way. I *still* do. It's easier than doing the math.

As Tom Lewis tells the story in *Empire of the Air,* when Armstrong returned from his vacation, he tried his theory on his own amp and headphones in the attic of his parents' home in Yonkers, New York. His sister described him bursting in her room late at night, dancing like a cat and screaming, "I've done it!" The sound from his headphones, left on his workbench across the hall, was so loud, the signal could still be heard. His new heterodyne circuit, applying "positive feedback" to the audio tube, enabled the triode not only to amplify radio signals, but to transmit them as well, making possible the powerful national radio network that soon developed—RCA, or Radio Corporation of America.

RCA became the monolith of American radio under the tough and visionary stewardship of a Belarusian Jewish immigrant named David Sarnoff (1891–1971). Sarnoff started as a telegraph code operator in the Marconi Company and then quickly rose from office boy to the chief executive and czar of RCA. Early on, he foresaw that radio could rise out of its shell of confinement from pure communications into commercial broadcasting and urged the development of inexpensive, vacuum-tube-powered "radiolas" that could be purchased by the average consumer and be the cornerstone for a national network of listeners.

One of the earliest Radiolas is a 1924 model, named Radiola III, that looks like an instrument of GI issue in a war surplus store from my childhood. There's one on the Internet. It's a squarish brown mahogany box with what appears to be a top plate of Bake-

lite. Two clear glass bottles (WD-11 Radiotron vacuum tubes) rise up from inside of the box and protrude above it in the upper right corner. There are two dials on the left—a bottom one that sets amplification level and a top one that looks to be an ON/OFF sweep switch. On the bottom right is the selector for station frequencies. The amp level and station selector dials are marked with simple numerics from zero to ten. Along the upper right edge are what look like taps for connecting earphones.

There are many stories about Sarnoff, but two stick out for me, one true and the other, at best, a legend. Sarnoff once claimed to have been the sole telegraph operator to stay at the key when the distress signal from the *Titanic* went out in 1912. The truth is that he was there, but was only one of three such operators who kept in touch with the ship after it ran into the iceberg and before it sank. The other story, true rather than embellished, is that Sarnoff set up the broadcast of the heavyweight boxing match between Jack Dempsey and Georges Carpentier in 1921, which resulted in the first serious boom in the demand for home radio equipment. Radios were everywhere after that, and broadcasting succeeded in creating audio listening as an activity of pleasure, families gathering at night around the new hearth of glowing vacuum tubes inside their Radiolas and consoles and Tallboys. Sarnoff had taken de Forest's invention of the Audion and Armstrong's repeater and heterodyne innovations (there came a several-years-long patent battle in the courts over whether Armstrong's were original inventions or "derivations" of the Audion) and applied their amplification technologies to radio broadcasting, making possible the magnificent phenomena of remote listening to live events and limitless reproductions of recorded sound, along with a new era of leisure.

I think of my grandmother in Lāʻie, on the windward side of Oʻahu, crocheting something as she sits beside a brown Philco cathedral radio during the late thirties, and I wonder what music she might have heard, what program might have reached her in that village surrounded by the sea on one side and the green cliffs of the Koʻolau on the other. A radio would have been very unusual, but I'd expect my grandfather, who was a high-earning storekeeper, to have gotten one to entertain the family. Would she have listened to

American jazz or pop tunes? Would she have stood for the inanities
of East Coast comedians? I think there must have been a Japanese
show from Honolulu in those days, a station that would've played
ryukōka, early-twentieth-century Japanese pop songs based on folk
melodies and recorded in a Tōkyō studio, pressed on 78s, shipped
to Honolulu by steamship, and spun on a rig by a DJ in a station on Nuʻuanu Street near the *onsen* teahouse that was there. The
soprano on the record would be singing plaintively about a soldier
leaving her and their home village, stepping on a boat, spending
his lonely nights in China far away from Japan. My grandmother
would have had jasmine flowers in a celadon bowl beside where
she sat listening, their fragrance as redolent as the radio waves that
brought the music, both amorphous and magical in their transmission. I think it was the sound of the human voice that the vacuum
tube was designed to amplify,
not only its measurable audio
characteristics, but things like
emotion, inflection, and body
in the voice—that which is
ineffable and yet thoroughly
recognizable as human.

A few decades after Arm-
strong's discovery, under the
charged ether of this new
atmosphere that inspired significant invention and breakthroughs from yet another
genius or two, we get hundreds of different kinds of
audio tubes made by Western
Electric in New York City;
RCA in Menlo Park, New
Jersey; Mullard Radio Valve
Co. Ltd. in Hammersmith,
England; Amperex Electronic
Corporation in Brooklyn,
New York; Marconi-Osram

Valve Company, also in Hammersmith; La Radiotechnique in Lyon, France; Tung-Sol Lamp Works in Newark, New Jersey; Telefunken Elektroakustik in Berlin, Germany; and Sylvania Electronics in Emporium, Pennsylvania. We get radios made by RCA and talking pictures and movie theater amplifiers and horn speakers made by Bell Labs. RCA and NBC soon cover the world in broadcast waves, and radar from the Royal Air Force saves London and wins the Battle of Britain. We get *The Mercury Theater on the Air, Fibber McGee and Molly, Challenge of the Yukon, Speed Gibson of the International Secret Police,* and actor Frank Readick Jr. intoning "Who knows what evil lurks in the hearts of men? The Shadow knows!" as the signoff for his radio show heard in homes from Brooklyn all the way across the plains to Portland, Oregon. To create a portentous echo effect and that peculiar nasal sound of an acoustic horn, Readick used a water glass pressed next to his mouth. We get Orson Welles in "The War of the Worlds," a radio enactment of an imagined invasion from Mars that scared the bejeezus out of women knitting in a circle in a hardware store closed for Halloween on a Sunday night in Lawrence, Kansas. We get Franklin Delano Roosevelt delivering his famous "Day of Infamy" speech from the floor of Congress direct to America's living rooms, calling for war with Japan on Monday, December 8, 1941. We get walkie-talkies for American GIs at Anzio, Saipan, and the Battle of the Bulge. We get Arturo Toscanini conducting the NBC Symphony Orchestra from NBC Studio 8-H in New York performing Samuel Barber's *Adagio for Strings.* We get Caruso singing "O solo mio" and "Una furtiva lagrima" on lacquer 78 rpm discs. We get "Black Bottom" by Joe Candullo & His Everglades Orchestra on Silvertone Records, "My Mammy" by Al Jolson on Brunswick, "Cooking Breakfast for the One I Love" by Fannie Brice on Victor, and "On a Little Street in Honolulu" by the Hilo Hawaiian Orchestra on Victor too. We get "Sitting on Top of the World" by the Mississippi Sheiks, "Preachin' the Blues" by Son House, "Razor Ball" by Blind Willie McTell, and "Somebody's Been Using That Thing" by Big Bill Broonzy. Something my father must've heard on the way to basic training at Fort Bragg, North Carolina—"A Nightingale Sang in Berkeley Square," sung by Ray Eberle with the Glenn Miller Orchestra. He

might've heard Bob Wills and the Texas Playboys singing "My New San Antonio Rose" as he boarded a troop ship bound for Naples in 1944. Maybe the sweet melody from a baritone clarinet on Duke Ellington's "Sophisticated Lady" ran through his mind as he crept uphill through an olive grove in Castellina-Marittima, German rifle shots slapping into the trees all around him. I like to think he might've heard Frank Sinatra singing "Say It" with the Tommy Dorsey Orchestra blaring out of a barracks radio while shaving with one hand on the bone handle of a straight razor, the other cupping a metal mirror, early one morning near Scandicci along the southern bank of the Arno as it flowed west from Arezzo to Pisa.

Moonglow

I imagine my father at seventeen, crouching outside his pup tent in Vada in July of that hot summer of '44. The regiment had just finished a tough campaign from Suvereto to Collesalvetti in the Tuscan foothills, driving the Germans out of several hill towns and villages just up from the seacoast where they'd come and bivouacked for a break. Their tents were spread out over vineyards adjoining several farms near the Vada River, and there was a powdery dust drifting from the floodplain, covering everything in a slick film that was hard to escape without constant washing. That day, most of the men had gone down to the pebbled shore, looking to get relief from the intense heat and rinse themselves off with a dip in the sea. But Al Hongo expressly stayed behind to listen to a record player he'd paid a supply sergeant to scrounge for him. He was overjoyed it came with a small stash of ten-inch records and that one of them was *Moonglow* by Artie Shaw and His Orchestra, one of his favorite big bands ever since he'd heard them in Honolulu, when they'd stopped off in the islands on their way to Japan for a tour of "the Far East."

After shaving, he settled down and crouched near the front of his tent and out of the light wind. He'd placed the heavy record player on a stool and flipped open the jointed cover-top that held a single speaker on its underside, which now faced him. He unhooked the heavy tonearm from its anchor, carefully slid the shellac record out

of its paper sleeve, and set it down on the rubber mat of the metal platter. He'd gotten a generator going beforehand, filling its tank with a little gasoline, and it made such a noise he'd have to push the volume on the record player pretty high. But he'd have had to do that anyway, as his hearing had been weak since childhood, and though the noise of the generator might've bothered anyone else, he heard it only as a soft pulsing in the background. He picked up the tonearm with light fingers and lowered its stylus onto the outer grooves of the spinning disc, then turned the volume dial on the player up slowly till he could hear Carl Maus's tomming drums and Shaw's sweet clarinet start up on "Frenesi," the tune on the first record's A-side. But he was just using this track as a gauge of sorts so that he could test it against his own hearing.

One reason he stayed behind was because he knew he wanted to play the music loudly enough so that he could hear it, aware that the volume would attract too much attention. He wasn't secretive; he just wanted a little bit of privacy so he'd not be self-conscious about how hard of hearing he was. He didn't want to have to put an ear up as close to the speaker as the white RCA dog did to the bell of the Gramophone on the RCA record's black label. Though the boys in his squad were indulgent, he couldn't be so sure of others in the company, and he simply wanted the solitude to explore the music in some peace and without questions, complaints, or kidding. He wanted to hear the music travel in air and reach him sitting a ways away, but for that, he'd have to turn it up loud. He adjusted the volume higher and higher until it reached a level at which he could hear Maus's rhythmic drumming, the call-and-response of Shaw's clarinet and his band's horn chorus. And then he grabbed a stool and sat down to enjoy the rest of the sides spinning out that captivating swing-time music he'd come to love as a younger teen in the McCully district of Honolulu.

He put on "Stardust" next, a cover of Hoagy Carmichael's composition from 1927, with a trumpet introduction from Billy Butterfield. He could hear its lower register, but the brilliance of its top notes faded as the dust in the air surrounding him settled to the ground. The violins were mere wisps of sound he almost had to

guess were there. But when Shaw soloed, he could hear most every tripping note, high or low, something about the instrument's overtones a grace of hearing that he was granted. Following the melody and the flourish of improvisations with so much glee, he smiled and leaned back on the stool, his hands resting on his knees, smacking them in a rhythm that moved from his unlaced boots up to his bare fingers. He wanted to dance, but that would come later, after he learned how, in USO clubs once he returned home; that's when he met my mother, who was in business school studying stenography. For now, though, he was immersed in music, feeling the 4/4 beat of the double bass keeping the bottom, then the plosive pleasure of Jack Jenney's trombone. When the music faded at the close, it was a sweet and easy outro, a goodbye unlike the staid, silent aloha from his father on the day of his departure for boot camp. In this music, nothing was cold as his father's spiritless handshake; all was aglow.

In the tuneless moments between spinning each record in turn, there came a grind of gears and the chug of a truck clutch slipping a few yards away, but Al didn't care. He was in the groove of the music and it carried him in a way nothing else in his young life had done, more than fun with guys back in high school when he played on the football team, more than shooting craps on the weeks-long passage from the East Coast to Naples, where he'd landed with the others in his regiment. He thought he heard a few guys squawking and light commotion coming from near the camp's mess, but he ignored the stray sounds and put on another record, the one he'd most been looking forward to.

"Moonglow" begins with a short flourish of violins I think my father could maybe sense but barely hear. Then the orchestra and Shaw take up the tune, Shaw's clarinet stating the melody, bursting into an explosive phrase to announce his solo, tasteful and smoothly surprising. The horn chorus restates the melody and syncopates it, playing with close harmonies and tight cues. Bobby Sherwood's guitar keeps on the steady dance beat along with the bass, and the whole thing "swings like hell," as my father would say. Yet, musically, the band swings only lightly and mainly on the four. I think, besides Shaw on clarinet, he loved Jenney's trombone playing the

most, as it was all so palpable to him, alive in the small range of his best hearing, most sensitive to baritone radio voices and comfortable only with calm, resonant speech. "Moonglow" was like that to him, I imagine, a slow and easy foxtrot, light on its toes, with a beat discernible to his weak ears and a succession of solos as clear to him as Euclidean geometry. Its pleasures reached through the salty air of his bivouac in Vada and eased into his soul.

When I was cleaning out the living room in my parents' old house in Gardena after both of them had died, I found, among a small cache of my father's records, a twelve-inch RCA mono of Shaw's *Moonglow* from 1956. On the cover, dancing in the bell of Shaw's clarinet, are a decorously posed couple in forties formal garb with just a hint of giddy romance in the woman's upturned face, cocked to one side as if about to receive a kiss. I doubt the smitten couple in the photo are Kim Novak and William Holden from *Picnic,* dancing at night on the boat dock in Joshua Logan's 1955 film from the William Inge play, but the same aura of unobliged *eros* encircles them like the array of tasseled Chinese lanterns reflected against the dark waters of the lake in the moving picture. When

I play the record on my mono rig, an Ortofon CG25 DI Mk II SPU cartridge with dedicated ST-M25 step-up and via the moving magnet setting of my phono preamp, I sometimes stare at the cover and put my folks, Albert and Louise, there inside Shaw's ebony bell, dancing on the floor of the USO in Honolulu, the world to come signified by the light purple wash of the album's cover, the gaud of yellow letters in its titling a forecast from the glow of their final abode in the modest lights of the urban night sky over Los Angeles.

The Tetrode and the Pentode

During the first half of the twentieth century the technology of the vacuum tube developed quickly. Walter Schottky (1886–1976), a German physicist working at Siemens in 1919, took the triode and inserted yet a fourth electrode into its bottle, creating the *tetrode* via the addition of a screen grid between the control grid and the plate (anode), making possible higher voltage gains within a single tube than could be had with the simpler triode. The electronics explanation involves decoupling the anode and control grid from each other, thus eliminating a specific kind of capacitance problem. Tetrodes were used in radios, televisions, and audio equipment until the 1960s, when transistors took over.

But the tetrode also *created* a new problem. Because of the way it is wired, the screen grid acts as a secondary anode (or positive plate), attracting electrons, causing unwanted oscillations and dropping the level of current the tube puts out. This is called the tetrode kink and is an undesirable effect, compromising stability as well as performance. To counteract this kink, Gilles Holst (a physicist) and Bernard D.H. Tellegen (an engineer) introduced a fifth element into the evolving vacuum tube and created the pentode in 1926. Working at the Philips Physics Laboratory in the Netherlands at the time, Tellegen and Holst created what was called a *suppressor grid,* which was placed between the screen grid and the plate (anode) to electrostatically repel electrons bounding from the anode and prevent them from being captured by the screen grid (and lowering current output). Their suppressor grid deflected and captured secondarily emitted electrons (which had a tendency to bounce off the

anode) *before* reaching the screen grid. The Dutch-invented pentode was quickly put to use in movie theater amplifiers, radios, and (by the 1930s) record player amplifiers designed for the home.

My first tube amp, the PrimaLuna Prologue One, used a quartet of pentodes (the EL34 tubes), and I loved them so much that, within the first month I had the amp, I went on eBay and bought another quad of these, electronically matched by the tube seller to work well together. The tubes were made by RFT/Siemens in East Germany and sounded so good when I put them in the amp, I bought fourteen more, unmatched this time, and from a different seller.

Yet another new output tube was the kinkless tetrode created in England in 1932 by Cabot Bull and Sidney Rodda, two EMI engineers. Due to international agreements between EMI and RCA in the United States, the invention languished for a few years, but Marconi-Osram Valve Company finally produced the first commercial iteration, designated the KT66, in 1937. Then, almost two decades later, in 1955, MOV released the KT88, which became the most powerful output tube of its time. The Russian-made tubes that came with my Air Tight amp (and that I initially set aside) were contemporary reissues of this original KT88.

The Tung-Sol 6550 tube (introduced in 1954) that I blew up in my Air Tight amp was its American competitor, though not quite as powerful. It was called a *beam power tube,* that is, a tetrode with beam-forming electrodes around the main electrode assembly. These were two shieldings of metal that directed the beam of output current so that its electrons would move in a steady, flat, fanlike flow toward the anode rather than bouncing away and potentially losing output power. By the late fifties, the KT88 and 6550 tubes, relatively late inventions in tube audio, were the most powerful output valves for amplification.

I eventually created a fair collection of vintage 6550s, about two dozen, all American made by Tung-Sol in the sixties and seventies. But Tung-Sol supplied the tubes to many other companies in the business, and my tubes are stamped with labels that say Tung-Sol, RCA, Sylvania, GE, Baldwin, and Leslie. The Leslie Company used them in the vacuum tube amp installed inside their famous whirling speakers, which in turn were used in Hammond, Baldwin,

and Wurlitzer electric organs. The Leslie speakers used a rotating drum and horn to imitate the sound of a pipe organ, whose spatial characteristics came from individual pipes that were spaced apart. And, whether you're listening to Lawrence Welk and His Champagne Music organist Jerry Burke playing "Ain't She Sweet," or to the Incredible Jimmy Smith playing "Back at the Chicken Shack," or bluesman Al Kooper on Bob Dylan's "Like a Rolling Stone," or rocker Gregg Rolie on Santana's "Oye Como Va," you are listening to the Hammond B-3 electric organ. And if you are listening to the Hammond B-3 organ, you are listening to the Tung-Sol 6550 vacuum tube.

The Heathkit A-9 Amplifier

The Heathkit A-9A was first introduced by the Heathkit Company of Benton Harbor, Michigan, in 1953—maybe a decade before my father put an A-9C (the last iteration of the model) together in our living room from a mail-order kit. Made on the same chassis, both were integrated amps, which meant they provided not only the output current to drive speakers but also the preamplification necessary to amplify the tiny analog signal coming from the phono cartridge of a turntable. Amps drive speakers, preamps boost signals from an analog source. The A-9A put out 20 watts, a whopping amount for its time, and had a wide frequency response even by today's standards (20 Hz–20 kHz), with a vanishingly small amount of harmonic distortion (the bane of any amp's existence). In opposite corners of its chassis, like two prizefighters on their benches before the bell, there were two big, potted transformers, both painted black. The bent aluminum chassis was a baked gray hammertone enamel that kind of shone in the right light. Sticking up in front were three signal tubes, two 12AU7s and a 12AX7, the latter on the front left encased in a protective, brushed aluminum, silo-like sleeve that had a cutout on the top (for heat to escape). Between them were two tall capacitors sheathed in shinier aluminum. In the back row next to the output transformer were three glass bottles—two 6L6G output tubes and a 5U4G rectifier tube (for changing AC current to DC). There were four black Bakelite knobs on the faceplate for source

selection (phono or tuner), volume, treble control, and bass control. It cost my father maybe thirty-six dollars when he bought it—about half his day's pay as an electronic technician in 1962.

It must have been one of my father's first treasures. He was a man who had few personal possessions—some flashy clothes (silk aloha shirts and Italian loafers), non-Sears briefs and undershirts my mother thought extravagant, and his tools were most of what he had. But his stereo was the big indulgence. I think he must have pored over the catalogues of the day, dreaming up acquisitions, exercising expert discernment based on his electronics training, able to interpret all specifications down to a probable final result of sonic performance in the mix and match of components he planned to compile.

A friend who made two of my own amps gave me copies of *Allied Electronics for Everyone* editions from 1959 and 1963, their pages yellowing and brittle to the touch, but still blessed with shaded depictions of speakers, microphones, radios, tape decks, tuners, phono cartridges, lavish stereo consoles, amps and preamps. Black print, half-tone pictures over a light sienna background field occasionally darkened to a bolder brown, the items floated up at you under trademarks in distinctive fonts, presenting images and succinct copy in newsprint folios of modest splendor. Just about all the hi-fi companies listed their gear in the Allied catalogue—there were Garrard turntables, Scott and Fisher tuners, Hickok tube testers, Marantz and McIntosh electronics, JBL and Jensen speakers, immaculate pieces of equipment made by Bogen, Sony, and Barzilay. I spied in it the Empire Model 398G Troubadour turntable my father bought for "Net $175," maybe taking advantage of the monthly payment deal of $9 per. I found the Dynaco PAS-2 stereo preamp with its "Time-Saving Printed Circuits" he'd assembled, which cost $59.95 "With Case." The vaunted Dynaco "Stereo 70," a 70-watt basic amplifier kit, was there too, "Net $99.95," also "With Case." Everything under the hi-fi sun was included and each ad published specifications along with a brief paragraph of hype. Like Sears and Monkey Wards catalogues in their day, this was the "dreambook" for hi-fi in its own Paradiso of heavenly lights, each called in their turn to speak their stereo song of wisdom to the earthly pilgrim.

≈⅄⌇⌇⟋

In 1955, when I was four years old, things had already run dry for
a life in the islands for my mother and father. We were living in
Kahuku, in the old Castle & Cooke sugar plantation village on the
North Shore of Oʻahu where my matrilineal family had gained a
toehold after three generations. My mother worked as a secretary at
the plantation hospital, a series of green-painted bungalows up on a
small rise and across Kamehameha Highway from the main village.
My father had been working as a lab tech at an agricultural station
of the University of Hawaii but got laid off, eventually finding a job
as a laborer on construction crews to build the Marine Corps Air
Base in Kāneʻohe. But their lives were going nowhere—no future
except to skip from job to job, filling the lower rungs and margins of employment as King Sugar (as it's often called by historians) slowly wound down, closing mills and plantations all over the
islands, the old economic engine of island capitalism becoming less
and less profitable as cheaper sources for raw sugar started to hit the
global markets. So, young and looking forward, my parents hatched
a plan for my father to use the GI Bill to go to trade school in Los
Angeles while my mother stayed back with my brother and me. She
moved us in with her parents in Hauʻula, where we shared living
quarters next to the Crescent Café, my grandmother's roadside restaurant. The plan was for a year of separation and then we'd reunite,
perhaps in the islands, my father hoped, but likely on "the Mainland." While my father lived with relatives near Echo Park and
took electronics classes at Trade-Tech, I sought out a small stand of
bamboo growing on a hill opposite my grandmother's house and
sang made-up songs with the wind as it thrashed lightly among the
tossing green arrows of delicate leaves.

Nights, we'd sit in bent bamboo-framed chairs or on the floor,
gathering around the small, green screen of an old black-and-white
television set to watch a series of shows that brought me into another
world—the slapstick skits and lavish big band music of *The Jackie
Gleason Show,* Sgt. Phil Silvers and his hilarious schemes and gambits
on an army base, *The Ed Sullivan Show* with its jugglers and its plate-
spinning, tuxedoed clowns, its talking Italian mouse and Spaniard

ventriloquist Señor Wences. It brought me the sound of Bronx and Brooklyn accents, the weeping "Man on the Street," crooners and Carusos of the day, the dizzy routines of Burns and Allen I didn't quite understand, and then Elvis, who rocked the whole world.

In July 1957, my mother, my younger brother, and I flew from Honolulu to Los Angeles on a Pan American Airways propeller plane to join my father, who met us on the windswept tarmac of the old airport. He was grinning like a pumpkin at Halloween, happy to reunite, proud of his new job at Packard-Bell in Santa Monica, and driving a lime-green Edsel he'd bought on a payment plan on the used lot at Felix Chevrolet just the week before. He'd found an old house in Midtown on a crowned street lined with palms and moved us in amidst the accoutrements and evidence of his new trade—an oscilloscope and trays of tools on the dining room table, cathode-ray television tubes extracted from their cabinets, radios and tuners and receivers all lying mute on the cheap carpet of the living room, a soldering gun and vacuum tubes out of their colorful boxes idle on a card table he'd set up as a workstation by the unused fireplace. It seemed to me a bounty like Christmas toys scattered around the house, each thing a mystery with its own secret ability and delight that would be revealed once opened and given my touch. But it was only my father who was allowed, as he gathered it all up and consolidated. They fit at one end of the dining table and on the card table that wobbled under the weight of the plethora of tubes and tools he put there.

By 1963, we'd moved to our three-bedroom, ranch-style tract home in Gardena. It was a palace compared to what we'd been used to. Before then, to make extra money, my father had regularly taken in broken TVs, radios, and stereo gear for repair. But the clutter of dismantled electronics had dwindled as color TVs started to come in and my father's skills lagged (his expertise was limited to black-and-white). Our living room that used to be crammed with gear had become a space free to enjoy. My father filled it with stereo.

One night, I think it was during summer, my father got an old white, double-sized bedsheet from the closet (with my mother's approval) and laid it on the carpeted floor of our living room. Then he spread out the different parts for putting together an amplifier—

the Heathkit A-9C or Dynaco Stereo 70, I can't rightly recall now. He put the screws, nuts, and washers in a teacup on one corner of the sheet; the transformers were in opposite far corners, and the rest of the numerous electronic bits were vaguely grouped like clusters of toy soldiers positioned on the white field of the bedsheet. In front of him, as he sat on a *zabuton* on the carpeted floor, was spread the printed guide—a step-by-step set of instructions, a schematic drawing of the circuitry, and an exploded illustration of the entire assembly.

I recall no music played while he worked diligently at this for two or three nights. I wasn't allowed to ask questions or interrupt, but he let me watch, sitting quietly at the right edge of his work maybe an armspan away from the magical bedsheet that covered the carpet. I saw little wisps of blue-gray smoke rise as he soldered away, a little burst of beaded silver erupting from the pliant coil of solder wire he held against the tip of the Weller gun. I saw him douse it on the yellow sponge from time to time, a little susurrus of hot metal on the wet sponge escaping as he did it—a sound I'm sure he could not hear. He used a magnifying glass and a little flashlight. He used screwdrivers and tiny tongs like tweezers. He used a silver penlight. He used pliers called "needlenose" that, locked on one of the resistors, looked like the jaws of a honeycreeper digging a beetle from under the bark of an *'ōhi'a* tree. He checked off each step in the manual with a pencil he'd kept tucked over his ear. When he was done, he set the tip of his Weller gun down on a wet dish towel that he kept beside his knee. He leaned back, then, slapped his thighs while still sitting cross-legged, and said, simply, "We go try 'em!"

Heathkit and Dynakit were two companies that brought many of my father's generation of returning veterans into hi-fi. Heathkit was the older, founded as an airplane company early in the twentieth century by E. B. Heath, a barnstormer from Chicago who created plans to make lightweight planes from a kit. He died in a crash, testing one of his own planes in 1931. The company changed hands a few times, and after World War II, the new owners had

the thought to buy up surplus airplane parts and electronics equipment, selling off the airplane equipment to another company and keeping the electronics. Heath Co. relocated to the shores of Lake Michigan and started producing kits for electronic gear, like an oscilloscope that could be had for only $39.50, then moving into components for the new home hi-fi market. In 1947 and for more than a generation, Heathkit marketed a two-piece amp, then single-chassis amps and preamps, ham radios, transistor radios and vacuum tube receivers—all to be had for prices about 30 percent less than factory-assembled equipment. Returning GIs and their wives loved that—more money for sliced bread every week and fur coats on layaway. All you needed were a few hand tools, the easy-to-follow (a *numbskull* could do it!), step-by-step Heathkit instructional guide, and the kitchen table as a space to lay out and solder together all the parts. Most used empty egg cartons to organize and separate all the loose bits. My father used teacups emblazoned with white cranes and a red sun and hand-sized rice bowls called *cha-wan* that gave off a ceramic tinkling when he poured screws and washers into them.

Heathkit had started a craze—the idea of do-it-yourself, home assembly—and in 1955, Dynaco Company (founded by David Hafler and Ed Laurent) got into the business, launching their own popular line of DIY electronics called, inventively, Dynakit. The achievements of the Dynakit line perhaps culminated in 1959 with the release of the Mark II Dynaco Stereo 70 amplifier that, starting at $99.95, sold over 350,000 units during the course of its run—a number more than any other amplifier *ev-vurr*. After building his Heathkit A-9, my father built a Dynaco Mk II S-70. You can pick up one that still works on eBay for about $750. I got my own S-70 from my friend Kevin Shuler at his stereo repair shop in Eugene. You plug it in and the cigar-shaped bottles of the EL34 pentode output tubes start to glow red from inside, their filaments alight, their aura as suggestive as mariachi static.

The revolution in hi-fi might be traceable not only to the DIY craze, but also to the release, in 1947, of the schematic of an amplifier circuit published in *Wireless World,* a British electronics journal. Popularly misnomered as the "Williamson amplifier" after the article's author D. T. N. Williamson, it was actually a design,

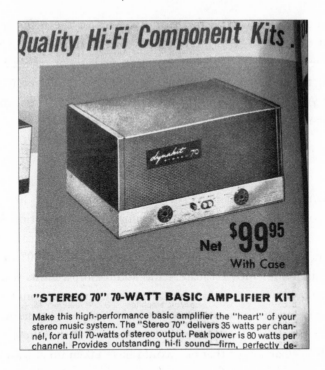

Quality Hi-Fi Component Kits

Net $99⁹⁵
With Case

"STEREO 70" 70-WATT BASIC AMPLIFIER KIT

Make this high-performance basic amplifier the "heart" of your stereo music system. The "Stereo 70" delivers 35 watts per channel, for a full 70-watts of stereo output. Peak power is 80 watts per channel. Provides outstanding hi-fi sound—firm, perfectly de-

or much of it, created by the electronics genius Alan Blumlein a decade before. It employed global feedback, a way of mathematically altering the electronic signal by doubling parts of it back on itself, minimizing the irregular waves of output distortions, making the amp sound more natural and sweet once hooked up to speakers. It also used quality output transformers and the newly invented, high-performance KT-66, the British-designed kinkless tetrode. Impressively, it also put out a whopping 15 watts—a number huge for that time, dominated by movie-theater amplifiers producing watts merely in single digits and requiring horn speakers as big as a Packard sometimes. This increased amp power while reducing speaker size, making home audio a better possibility. Within a few years, American companies made it their business to adopt the circuit too and began producing their own versions of the Williamson amp. Out came the Altec Lansing A-323 from California, the Brook 12A3 from New Jersey, H. H. Scott's 210A in Massachusetts, and the Radio Craftsmen C-500 out of Chicago. There came more magazine articles with variations on the Williamson circuit, using

the American 6L6 valve for output power or the 807 type tube instead of the British KT-66. These designs subbed in U.S.-built machinery too—the Chicago, Peerless, or Stancor transformers— making the amp cheaper and quicker to build for the growing number of American home enthusiasts. Know-how, American parts, and the new ardor for DIY took off from there.

The Mini-Flex Speaker

The University Mini-Flex speaker was one of the first book-shelf models that was part of the home stereo revolution in hi-fi back in the early sixties. Before these compact ones came along, speakers were huge affairs built on older, acoustic-horn technology designed for movie theaters and auditoriums. Costs were high and sizes were more than any average family man could accommodate in a standard postwar three-bedroom tract home like ours. Just two speakers like those built by Altec Lansing, Klipschorn, or Stromberg-Carlson of that era would have taken up at least a third of my parents' living room, dominating the corners like the tuba-shaped ventilation ducts on a battleship. The bookshelf speakers were boxlike and much tidier, and they brought the world of hi-fi to the American middle class, introducing stereo as a new form of home entertainment and pleasure. It celebrated leisure and had the promise of making home-life "aesthetic," cultured. After my father built his own system from mail-order kits he could assemble in his spare time, he made a retail splurge on a fancy belt-driven turntable (also a new invention), and he got compact speakers at Fedco, a membership discount store that targeted civil service workers (my mother worked for the City of L.A.).

These were the Mini-Flex, a three-way speaker system built by University Loudspeakers in White Plains, New York. Only fifteen inches wide, about ten inches tall, and less than six inches deep, one fit on our lamp table in the right corner of the living room and the other on the fireplace mantelpiece near its left corner. That they were on different levels didn't seem to matter to anyone, so long as they both fit tidily into the space—and so they did, for the decade between my going to middle school and graduating from

college. And their look was fairly chic—oiled walnut cabinets, dark brown and beige checkerboard grill cloths interwoven with gold thread, and a secure feeling of heft and density if you lifted them. They were completely unlike the flimsy, flip-top speaker that had come with our old Silvertone phonograph, a tinny gizmo of unadorned simplicity that wrecked whatever real music you tried to squeeze through it. The Mini-Flex, by contrast, played wonderfully, sounding superb and sophisticated, creating a deep sound field that enveloped me whenever I heard it, its bass ample by comparison, supporting the higher frequencies so they could ripple and shine in a way that reminded me of the surface waters on a lagoon back where I grew up in Hawai'i.

Likely designed by Victor Brociner, an engineer originally a partner with Avery Fisher at Philharmonic Radio Corporation, the University Mini-Flex was a later, compact version of the larger Acoustic Research AR-1, the speaker that rocked the audio world into the hi-fi era back in 1954. A New York inventor named Edgar Villchur, trained in art history and largely self-taught in audio engineering, came up with a new speaker design, tackling the problem of poor bass response, uncongenial size, and frustrating distortion that made the big, tuba-like speakers of the day sound like shit in the lower registers. Before Villchur changed everything, bass distortion was a huge problem, the large speakers of the era sounding woolly and full of unnatural boom due to the mechanical suspension of the main woofer cone. Villchur came up with the idea to use a loose suspension, and instead of an open box that allowed the cone a wide excursion, he used a sealed enclosure that counted on the natural air compression inside the box to keep the woofer under control, employing air like a spring that checked the cone's horizontal movement, making its acoustic performance more linear, even across its frequency range. The sound was smooth, tighter, more accurate and Villchur's solution was simple and elegant—like the pulley and winch designed by Brunelleschi to build the Duomo in Florence, like his curved, terra-cotta tiles that used their overlapping, mutual lines of stress to make the eggshell shape of the dome itself.

A woofer cone gets moved by electric current in a voice coil embedded in its membrane. The voice coil interacts with a pow-

erful magnet suspended behind it that itself is held in place by a mechanical basket, often called a *spider*. As the woofer moves, it excites and compresses the air in front of it, creating sound waves. But in the prior designs, the woofer's excursion had often caused it to move outside the control of the magnet, creating distortions that made for unrealistic sound. To counter that, prior tech mounted the woofer in an open box and used stiff, mechanical suspensions that, though they curbed excursion, still resulted in other irregularities of performance. Villchur saw that the open box and stiff suspensions created more problems than they solved, so he did away with them, suspending his woofer inside a sealed box that used the compression of air behind the membrane of the woofer cone to control excursion. *BAM!* He'd created the sealed-box, air suspension, bass-loading speaker. It was cheaper, smaller, and easier to make. Hi-fi could move from mansions on Park Avenue and in Beverly Hills to tract homes in Levittown, New Jersey, and Paramount, California, bought on VA loans by men like my father.

"I used a different kind of elastic restoring force, one derived from an air spring, instead of the mechanical springs of the suspensions," Villchur said in a November 1993 article by Steve Birchall in the *Journal of the Audio Engineering Society*. ". . . After all, the speaker has a cabinet which encloses a cushion of elastic air," he explained. "All I needed to do was to decimate the springy stiffness of the speaker suspensions, and reduce the size of the enclosure until the air spring was strong enough to replace the mechanical springs that we threw away . . ."

To make his prototype, Villchur used a standard twelve-inch woofer made by Western Electric for their own loudspeakers: "I cut away part of the spider of one of them, making it more compliant. I also cut away the entire rim suspension and replaced it with a suspension made of mattress cover material (because it is very compliant, and adequately impervious to air) . . . The acoustic suspension speaker was substantially superior [to the Western Electric speaker] in fullness of bass, and especially in lack of distortion."

The story goes, Villchur hatched his idea in the early fifties while teaching a night class on audio engineering at New York University, the first class of its kind anywhere. He tinkered, he studied

books on electronics and physics, he repaired audio equipment, and he built custom systems for the rich, all the while dreaming shit up. Once he figured out the principles of his speaker, he went to two of the biggest speaker companies of the day with his design, only to be dismissed out of hand. Altec Lansing told him if his idea was any good, their corps of elite engineers would have already thought of it. Bozak said what he'd proposed "was impossible."

Villchur went back to tinkering and teaching and met Henry Kloss, a student in the NYU night class, who understood the merits of his design immediately. Excited by Villchur's talk of his proto-type, the two men drove up from New York City to Villchur's home in Woodstock (already a haven for eccentrics, artists, and bohemians) to listen to it. Villchur is said to have played an LP of organ music by E. Power Biggs, and its playback reached such massive low pedal tones, Kloss went nuts. Kloss already had a shop in Cambridge where he made speaker cabinets. And Villchur had a magical design. On a shoestring, the two men started their own company, Acoustic Research, and began to produce sealed-box loudspeakers out of Kloss's shop. Right away, the speakers set the hi-fi world on fire, and the revolution in speaker design took off from there, influencing the entire home-stereo industry.

Yet even after amazing technological and commercial successes, Villchur wasn't done. He started thinking about another weak link in loudspeaker design—the tweeter that produces the high frequencies. In 1958, he came up with a new one for his AR-3 speaker.

Villchur explained the challenge in Birchall's article: "The secret of high-frequency dispersion can be stated in two words: small size. Shrinking a cone tweeter to a small size doesn't help because the voice coil becomes too small to handle any appreciable amount of power. I placed my voice coil at the large diameter of the diaphragm [instead of within its circumference]. When you do that, the shape of the diaphragm emerges almost naturally as a dome. That has the further advantage of making unnecessary a second suspension . . ." In Villchur's new concept, tweeter size shrank considerably, going from the 8-inch Western Electric 755A in the AR-1 to the much smaller 2-inch and 1⅜-inch Villchur-designed tweeters in the AR-3 and AR-3a speakers.

Yet Villchur's AR-1 and his subsequent models (AR-3a notably among them) were still fairly largish affairs, bigger than an orange crate, nearly the size of a twelve-gallon trash can, taking up a fat lamp table's worth of space, and needed at least 60 to 100 watts of amplification power to function right—about ten times more power than the older, more efficient speakers required. As new, higher-powered solid-state amplifiers were being introduced, this wasn't a commercially significant problem. But the sound the new combination made got away from the traditional, liquidinous warmth of the vacuum-tube amplifier.

A few years after Villchur's contributions had been established, engineer Victor Brociner closed his own audio speaker and electronics company in New York City and came to work at University Loudspeakers in White Plains. He had a knack for cramming and miniaturization, having already figured out how to shrink down electronic gear to fit into some of the first stereo consoles—the furniture-like, lifestyle products that combined electronics, speak-

ers, and a turntable in one long and lavish buffet-sized unit. These had become all the rage as signs of prestige and cultured leisure in the American home. At University, Brociner designed an even more compact speaker than the AR-1, trimming the size of Villchur's twelve-gallon enclosure to four-tenths of a cubic foot (less than the volume of a quart of milk), using a more efficient bass driver (requiring fewer watts

to power it), and matching it with other tech Villchur had created in the meantime (the dome tweeter and dome midrange driver) to make for an ultra-compact, sealed box loudspeaker that would sing like a mini-Caruso. Brociner's Mini-Flex speaker was an easy fit into modest-sized living rooms like ours, needed much less power than an AR speaker to run it right (a tube amp of 20 watts could do it), and looked cool too. But its main feature remained the stiff acoustic air spring of Villchur's provenance, creating superior bass down to 40 Hz (human hearing goes to about 20 Hz).

My father powered his University Mini-Flex speakers on the mere 35 watts per channel his home-built Dynaco Stereo 70 amplifier put out. And the sound it made tutored my ears, the S-70's EL34 vacuum tubes providing the current to the Mini-Flex speakers that produced the music of warmth and lavishness in our living room, upholding sensuousness over raw power, its sound the wellspring of my passion for hi-fi and my entry into the worlds it let me travel.

The Good, the Bad and the Ugly

One day in 1966, my father came back from the movies and stopped at a record store on his way home. He'd just seen a matinee of the Sergio Leone spaghetti Western *The Good, the Bad and the Ugly* starring Clint Eastwood and, though I don't think he recognized it was derived from Akira Kurosawa's *Yojimbo*, about a dispossessed samurai (as a sequel to *A Fistful of Dollars*), he reacted as though the Italian film had sprung from a legacy rediscovered. He bought a recording of the soundtrack. And he put it on his stereo right away, snicking the cellophane along the edge of the jacket, sliding the record in its paper sleeve out, then extracting the shiny black disc so he could place it on his turntable, the stereo already fired up and ready to go. He dropped the needle into the lead-in groove, and the theme song came on.

A simple, childlike drumbeat started the track, reverb heavy, then an ocarina (a folk ceramic flute) resonated a trilling two-pitch melody, answered by a harmonica (I think). Another flute, fluttering mournfully, repeated the trilling figure, a half-octave down. A deep set of drums, perhaps the timpani, got hit and resonated in soft

circles of sound, and then someone whistled the trill, now a theme. Chimes got struck, then a toy piccolo was blown softly, a guitar got lightly strummed, a male chorus hooted forcefully. Then an electric guitar, heavily reverbed, plucked out a bold, quaverous melody, while the male chorus intoned nonsense vocalizations. *Ho-hee-ho, ho-hee-ho, hah!* Symphonic drums rumbled, a drum kit morphed a march rhythm into a galloping one, and the full orchestra crescendoed into full tutti. Then it all repeated into an allegro movement in a kind of concerto for electric guitar, choir, and movie orchestra with spotlit brass, and folk drums, flutes, and whistler sprinkled in the mix. It was a brand-new mash of musical strategies that hopscotched across pop, classical, and rock genres.

Of course, this was Ennio Morricone's famous theme performed by Unione Musicisti di Roma (an orchestra) and I Cantori Moderni di Alessandroni (a choir), but what I heard more than any catchiness or intricacy of orchestration was the fact that the folk drums were left, the ocarina was right, and the harmonica squarely in the middle. When the whistler came in—most definitely a virtuoso—he was deep left. The electric guitar's image took up the back, just right of center. Then the harmonica moved to the deep right and rose like a turkey vulture over everything. It was sixties stereo sound at its boldest and most sophisticated, the instruments in specific placements in a soundstage between my father's two speakers. The choir sang from depth, mostly left of everything, and when the strings and brass came on, they were layered somewhere in the middle. The music *moved* and yet the images of the instruments shimmered in place. The stereo shoved the walls of the room out to the Spanish desert where the movie was filmed and into the sound studio where the music was recorded. It brought Morricone's thrilling orchestra into the drab living room of our house, where the surflike sounds of the 405 and 110 freeways came leaking through the vent over our stove, where dark residues from the flameout stacks of Union Carbide out by 190th and Carson Boulevard settled on the shake shingles of our roof and on the withered trumpet flowers, indigo-blue, falling from the bending jacaranda tree over the parking strip, patchy with zoysia grass, by the curb.

Part Five

It's My Life

The Muse Café on Shirakawa-Dori

Through correspondence in an elegantly penned, almost calligraphic script, the Beat poet Gary Snyder had said to go see Cid Corman in Kyōtō once I got there. It was the fall of 1973, after I'd graduated from college, and I'd arrived during the height of *momiji* season—the autumn leaves turning brilliant colors of russet, yellow, and gold all over the ancient city. On the street near Ryōan-ji Temple, I saw a bus pull to a stop under a red bough of maple leaves jutting out from a grand tree—a dragon's crown of clustered stars scraping over its forehead of ideograms announcing its destination. Everywhere trees burst with a kind of glory I'd never seen before, having been acquainted only with the dingy palms of Los Angeles and the bending funiculars of coconut trees in Hawai'i. The world was magical, alive with beauty and a kind of natural sentience emanating from all the leaves, each gesture of wind setting them atremble over gray streets, murky streams, and the flower-print umbrellas decorous ladies unfurled and held above their heads in the soft rains.

As instructed, I trekked across the grid of the city to its far eastern edge, to a street called Shirakawa-dori, "The Avenue of the Silver Stream," where I'd find alongside it, about halfway up a short asphalt pathway, a chalet-like building and a shop sign that said MUSE CAFÉ—my destination. Snyder had written that the poet Cid Corman would be there every Tuesday, available for anyone who came along to chat.

Cid was legendary, the founder and editor of an important literary magazine called *Origin,* which had, in its day, from the fifties through the seventies, published groundbreaking work by American poets Charles Olson, Denise Levertov, Robert Creeley, Louis Zukofsky, Philip Whalen, and Gary Snyder himself. It also published lively translations from Chinese and Japanese literature and letters between Corman and these poets on topics ranging from

their new aesthetics to politics, music, and classical literature. Many of the poems I'd read in anthologies while I was an undergraduate had been first published in Corman's *Origin*. Cid's own poems, terse, written to an esoteric yet discernible syllabic and rhythmic music, also occasionally appeared in its pages.

> *Do you really want*
> *to know what lies ahead for*
> *you? Isn't it the*
> *not knowing—not having to—*
> *makes it all possible?*

A kind of contemporary American master of a haiku-like verse in English, Corman was a legend to me, a sage to seek out for wisdom and tales, an elder who, if I showed proper respect and sufficient intellect, might introduce me to the arcana of his world and life's work.

The Muse was a second-floor establishment, and once I reached its entrance, I hiked up a narrow set of wooden stairs to its space. I remember the café itself was dimly lit, with wood-paneled walls and only tiny windows that looked out over the barren lane of Shirakawa-dori, the stream and its greenish waters, and cherry trees stripped of their leaves. There were no *momiji,* and the outside seemed a gray drained of all colors. Inside, there were numerous dark brown wooden booths with black leather seats and a few long tables with benches. I took a seat in a booth facing the stairs so I could spot whoever might enter. I ordered a cup of coffee when the waiter—a tall Japanese guy dressed in black slacks, white shirt, and a bow tie—came around.

The coffee turned out to be so bitter that I had to add two of the white cubes of sugar served beside it. Tastefully curated chamber music was piped in overhead through a sound system with some fidelity. The speakers were spread too far apart, so there was no depth or spaciousness, but at least I could tell what instruments were playing. I thought most of what I heard was piano trio music, a few quartets, and maybe a baroque ensemble here and there. Japanese coffeehouses and listening bars were (and are still) differentiated by

the music they played. The Muse, a bit highbrow, featured Western chamber music. I glanced over to where the turntable was behind the bar and noted it was a Garrard, a British-made piece highly prized by audio buffs like my father.

After about a half hour killing time and listening to this music I only nominally understood, I noticed a stout man in thick glasses and a plaid mackinaw bustle in, shaking out his umbrella, leaning a beat-up leather messenger bag against the clothes tree near the top of the stairs. He pulled a black beret off his head, which was mostly bald with skin the color of cream, except for dark brown hair cropped closely at his temples. He looked more rumpled than distinguished.

Once free of his coat, the man gave me a longish glance, sizing me up, and came over, gesturing with his open palm to ask if he could sit and join me in the booth. I was mildly startled, as I'd never had an elder nor anyone of eminence ask to sit with me anywhere. I think I gave a hurried nod and Cid sat down across from me, smiling as he settled himself. He was a fairly burly man, not quite six feet, and about fifty-something I guessed, clean-shaven and pale with the mien and aura of a shopkeeper about him—modest. Through his thick glasses, he squinted and seemed to grin or grit his teeth as he spoke.

"So you must be Gary's friend?" he said. "He wrote a while back that I should be expecting you."

I think it must have been the Beaux Arts Trio that was playing in the background. It was one of Beethoven's piano trios, a sprightly piece with lots of trilling from the piano, an echo of its repeated figures from the violin, and a sonorous, dark line of accompaniment from the cello. The recording was on vinyl LP, and I'd heard the needle drop onto the lead-in on the edge of the record, that soft, ticking burst of midrange bloom that signaled the groove had been engaged just a moment before the music got started in a repeated descending line on the piano, tailgated by the violin. It lent a kind of stately ambience to the whole space of the café, filled with light talk mostly in Japanese, the clink of porcelain cups, saucers, and teapots, the soft ring of a metal spoon stirring against china. The tune was a minuet of sorts, its theme pitching and receding back

and forth in its dancelike rhythms, inspiring a kind of serene call-and-response in our conversation, meting out a calm and politesse between us, easing us into quietly joyous talk.

The waiter came over again and Cid ordered tea. We spoke of Gary a bit, the famed dharma bum of Jack Kerouac's novel, the scholar-poet who'd studied Zen so long at the nearby Daitoku-ji Temple, who'd met me when he'd visited my college and who'd generously offered up introductions once I told him I'd be traveling to Japan. But I had numerous pent-up questions for Corman that I wanted answers to. A bit of a grump, Cid fended off my enthusiasm with frequent demurrals but spoke to some of what I wanted. Not that day, but eventually, I got him to talk about his correspondence on "projective verse" prosody with Charles Olson, that gigantic intellect who'd served as rector at Black Mountain College, which had been the proving ground for artists like dancer Merce Cunningham, painters Josef Albers and Cy Twombly, and architect Buckminster Fuller. I got him to tell me about the great poets of Modernism, about Amiri Baraka when he lived in Greenwich Village and was still known as LeRoi Jones, and about the seventeenth-century haiku poet Matsuo Bashō. Cid praised all the learning and lyricism of Bashō's prose diaries, having translated *oku-no-hosomichi*, the greatest of them, as *Back Roads to Far Towns* in his version. And he told me about the young Gary Snyder, whose first poems and translations from the ancient Chinese Cid had published in *Origin*. Snyder had given the debut reading of his first book at Kyōto's Yamada Art Gallery—the publisher of that book.

"Gary was shaking, he was so nervous," Cid said, fixing me with his myopic eyes.

Gradually, meeting by meeting, he sketched out the lore regarding an elder generation. He told me of visiting Ezra Pound at St. Elizabeth's Hospital, where he'd run into Allen Ginsberg and William Carlos Williams, who also came to visit Pound, the great Modernist poet who'd been captivated by Mussolini's Fascist politics. He told me Pound would sit up in his hospital bed and be alternately rambling, incoherent, remorseful, and rageful against "usury" and capitalism in his intermittent remarks. He told me Williams, another great Modernist poet who was a pediatrician and

Pound's college classmate at Penn, would eye the pretty nurses as they walked in and out of the room.

"Doc Williams always liked the ladies." Corman winked.

Though seemingly inconsequential, details like these enlivened my sense of these poets, pretty much only academic up to this point. I got him to talk about the parade of American literati who'd come through Kyōto in the past—a translator of César Vallejo, the Peruvian Surrealist who'd died in Paris on a rainy day; a composer of minimalist music who'd written incomprehensible epics in free verse; a freewheeling fullback of an American Buddhist who wrote of love affairs, heavy drinking, and drug-induced Beatnik hijinks in San Francisco's North Beach. They all came to see him, a kind of ex-pat magus of literature.

But not all the tales came at once. I continued to meet Cid just about every week throughout that fall and into early winter that year, my appreciation for chamber music building surreptitiously the whole time, my ear developing not just for its instrumentation and clarities of score, but for the folk tunes and rhythms that lay underneath much of it, their transformations from the rustic into sophistication, from folkloric lustiness and simplicity into disciplined and complementary instrumental parts that combined into a kind of elegant intricacy, which, in turn, had a way of ordering thought into a renewable and consistent pacing, mysteriously chantlike in its effects, intellectual and yet meditative. Hearing chamber music felt a bit like the way *Kind of Blue* had affected me back in college—a calming balm to the tumult of my anxieties, the difficulty of my literary studies, the frustration of unexpressed emotions. Snyder may have been uncharacteristically fraught with nerves at his Yamada Gallery debut, but I was a manic bundle of unrealized potential, inadequacy, and youthful enthusiasm then. Music came to be a kind of meditation and process of *pharmakos*—a drawing out of the admixture of pain, ignorance, and frustration that welled within me when I was twenty-one and the world was opaque and pearlescent in the face of my hopes.

Meetings with Cid would alternate between his stately and calming talk and the bright, foamy effervescence that I usually contributed. But I can remember one occasion when we both fell silent,

and it was not over poetry. At the Muse one winter afternoon, big flakes of snow like inch-wide pancakes fluttering down in the stilled air outside the café, a piece of music came on that caught our ears and demanded we cease talking. I glanced over toward the bar, where the turntable and electronics were, and saw a barista shove an album sleeve into a display rack mounted on the wall behind him. It was the Heifetz-Piatigorsky recording of Felix Mendelssohn's Octet for Strings in E-flat, op. 20, an unusually ebullient and complex string piece for the normally sedate taste of the Muse's regular DJs. Most of the cover was a murk of black except for a startlingly luminous trapezoid in the lower right corner, which showed what seemed to be floating saucers of amber lights shining down upon the distinguished ensemble dressed in white tuxedos, seated with their instruments on a darkened stage.

What we heard, after the murmurous stirrings from the string section, were captivating notes from a sweet and nimble violin stating the composition's theme—a sprightly melody, repeated, abbreviated, quoted, then played with a silvery gusto as it built to a crescendo in the *Allegro con fuoco* (allegro with fire) of the first movement. Cellos punctuated the rhythm with thick pizzicatti that evolved into a sonorous accompaniment in their lower registers. Violins and violas added richness and harmonies. Virtuoso Jascha Heifetz's violin sounded to me like a soul singer—a reason the music caught me, bathed as I was through my adolescence in R&B tunes. But this was different—richer, fugue-like, with cascading pieces of the music's theme emanating in thrilling alternation from violin to violas and cellos and then back to Heifetz, the magnificent soloist. And yet, in the main, it remained a joyous and romantic dance, springlike in its refreshing character, bouncy and insouciant as the bells of daffodils dandled by a cool wind. Every note seemed painted in vibrant colors, as though a Fauvist like Raoul Dufy had dipped each bow of the ensemble in rich oils and saturated every one of its thirty-six strings so they reverberated shining hearts and arrows into the air, scattering blooms of primary colors around the sedate quarters of the café.

The movement ended and we took up our conversation again, Cid stirring sugar into his tea that had gone cold. He might have

told me a story about the first time he'd heard the octet, perhaps in Boston where he'd gone to college, or in a church in Italy where he'd lived for a couple of years during the fifties. In these stories, the events were always special. Cid was forever involved in a shadow game of upsmanship, competing with some ghost he'd felt hovering over his life, and I'd no notion of its incarnate identity. It was as though unknown souls swirled over our talk sometimes, a generation of being and gossip barely emanant to me but intimate to his own consciousness. It was these ghostly things I wanted him to impart, the life of poetry before me, the vibrancy of speech that was once exchanged among both the gigantic and the obscure, being uttered to me over the stirrings of a spoon in a porcelain cup, over the soft vibrato of a cello extending out a melodic line thickened with romance and soul.

Wild Horses

It's hard for me to explain, but early Dylan, *Hot Rocks: 1964–1971* by the Rolling Stones, and the Bartók string quartets mean the winter I lived with my American girlfriend in a small converted teahouse just outside the inner walls of Zuishun-in, a sub-temple of Shōkoku-ji, one of the oldest Zen monasteries in Japan. I'd taken a room as a lay student there, allowed to practice without having to take vows or shave my head like the Japanese *unsui* (monks). Instead, I was asked to pay rent for my little three-mat sleeping room and took part in only a portion of the rituals, duties, and meditation sessions. That suited me fine, as I was something of a tourist in the country of Buddhism, curious about it certainly, but not ready to make any sort of commitment. I'd do some light housekeeping, chant with the monks in the mornings, meditate with them when I could, but was not obliged to maintain a full schedule of study. *Zen Lite*. Perfect for me. Then, I got into a clusterfuck of contradictions and ecstasies. It was my first year out of college, I was on a fellowship to write poetry, and I ran into a vivacious architecture student from Berkeley, whom I got to move in with me after a night of dancing in the clubs along Kawaramachi-Shijō, the entertainment district of Kyōtō.

I'd fallen in love with her from the start, just as she stepped out of the shower at a guesthouse where I'd been invited to have Thanksgiving dinner with an American couple I'd just met—a poet and his wife, a distant cousin from Hawai'i. I was sitting around a *kotatsu,* a low warming table the Japanese use in winter, my hands tucked under the quilt splayed out between the tabletop and frame, when a young woman came out of the shower adjacent to the room where we were. The door opened and she walked out of it, bending quickly from her waist and throwing a cascade of honey-blond hair tumbling forward over her head and face. She was wearing jeans and a deep red leotard and shook her head violently as she rubbed her hair with a towel. Then she wrapped the towel in a turban around her head, threw it all back in a whipping motion, looked up, and smiled at the gathering. There were sequins across the breast of her leotard, iridescent circles of glittering foil pressed tightly against the Danskin. I was a goner.

Her name was Nadja, a name she'd given herself, she told me, and she was from Colorado, a graduate of the University of Chicago. She was in Kyōtō to study temple architecture for her master's thesis at Berkeley, could talk about anything—Spinoza to seashells—and could read signs, maps, and menus in Japanese, though not speak

it. About my height, she was a compact beauty—appealingly wide in the hips, slender at the waist, and a complete knockout going up from there. Her face was a narrow oval and the complexion even except around her cheeks, which had a slight orange glow from the prior summer. We became quick, easy friends and then lovers after one chilly night when I'd taken her dancing. We stayed in a cheap inn nearby and watched the first snowfall of the season come down at first light the next morning. After that, there were so many nights when I sneaked back late into the temple after *mongen* (curfew), the priest's wife came up with the gentle proposal to rent the both of us a small, four-and-a-half-mat teahouse (called a *hojō*) outside the inner gates of Zuishun-in, cognizant that I wasn't, after all, suited to monastery life and also aware that she needed the rent money to sustain the temple's more important affairs.

I was still naïve about sex, knowing little beyond my limited college experience, which had been almost exclusively with just one lover throughout. Nadja was something else—more insouciant, expressive, and lighthearted with it—even as our entanglement was easily becoming the most profound experience of my life. But I'd kept that from her, knowing her attitude was different, having been told she'd go off, once summer came, to join her chemist boyfriend in Sweden on a Fulbright there. I tried to play it as cool as she was, but I was doing a bad job. I fell big-time and was happy rolling with her in her ice-blue Gerry sleeping bag stuffed with goose down, meeting her for *soba* lunches, joining in *saké*-drinking parties with her fellow architecture students, and rolling with her in my own claret North Face bag. We'd spend evenings reading and rolling, tittivating like baby sages while lying upon the *tatami* floor, exchanging stories of our childhoods, promising each other completely nothing. When we could afford it, we went out to clubs with live music and I'd drink "black and tans," mixing Guinness stout with light Japanese beers, while she sipped tea and read *The Dream Songs* of John Berryman. A lot of other guys would come around and try to meet her, but when they'd ask what she was up to in Kyōtō she'd point to me, say "Fucking him," and chase them off. She'd stung them so hard, the air seethed with the odor of ozone.

Most evenings, though, we'd stay inside the teahouse, read-

ing, listening to music. I had tapes we'd play on her Wollensak recorder, a basic box with drive, speaker, and a row of plastic, push-down controls. These were cassettes my father had made for me, recorded from LPs I'd left with my parents when I'd moved out of my dorm room and headed for Japan earlier that summer. Patiently, he'd recorded from analog playbacks on his own stereo in the living room. He'd sent *Hot Rocks* by the Rolling Stones, *Rock of Ages* by the Band, *Live/Dead* by the Grateful Dead, *At Fillmore East* by the Allman Brothers, Bartók's quartets, and Bob Dylan's *Greatest Hits*. They'd come by mail, of course, probably after I'd asked him for them, wrapped in heavy brown paper tied with string expertly knotted, and covered in stamps.

By the time the tape got to "Wild Horses," the last track on the Stones double album, we'd have spent almost all night long with each other, drenched in a slick science, our souls keeping the same rhythmic time, pledging our bodies against the infinite and separate futures we both affirmed we'd have.

The song opened with Mick Taylor's chords on an acoustic tuned Nashville-style, the three low strings tuned an octave up, then Keith Richards strummed a twelve-string, and they joined to make an intricate tapestry with an electric slide (Richards overdub-bing, I'd guess) chiming notes of a repeated figure, launching the slow dirge of this ballad, like a funeral march for a love gone wrong. Taylor switched over from chords to mostly harmonics. And then the drums, bass, and chorus came in, driving the tune, starting a heavy stomp that attenuated back into another one of Mick Jagger's verses, sung with a plaintive twang, the drums and bass dropping off, the arrangement suddenly gone sparse as sagebrush in a desert wash. But on one of the next verses, they'd added a tack piano played by a session man not part of the regular group. It lent lovely fills and a honky-tonk feel, contrasting with the picked guitars, thickening the fine roux of the choruses that came after.

Nadja and I would quietly warble along sometimes, *Wild horses . . . ,* meeting each other's eyes in the glints of music that spun around us in the cold *cha-shitsu* room that winter in Kyōtō, trying to hit the same notes as we felt for each other's hands and fingers, hips and hair, dragging ourselves into a welter of unbuttoned

jeans and shirts, odors of tea on our tongues, the unsaid arrowing the splinter of shadow in the shrinking space between us into a brilliant spear . . . *Couldn't drag me away* . . .

Kissaten

In Kyōtō the fall and winter I lived with Nadja in Zuishun-in, when I wasn't studying, meditating, or sweeping the temple grounds, I'd hang out a lot at *kissaten,* the tiny coffee shops specializing in one kind of music or another—jazz, rock, classical, or even just blues or pop. It got that specific sometimes. Japanese houses and apartments would not have the room necessary for a home hi-fi system, and, like flats in Paris for Hemingway, heated quarters were a luxury. So, *kissaten,* literally "tea-drinking shop," filled this need for folks to gather, relax, listen to music, and enjoy each other's company in a cozy, heated room.

I had two "go-to" places. One was the Muse Café on Shirakawa-dori, where I'd meet Cid Corman to hear his stories over those cups of burnt coffee with too much sugar. There, I'd hear mostly Beethoven quartets and Schubert piano trios. At Honyaradō on Imadegawa, near the Imperial Palace and Doshisha University, over yet more cups of burnt coffee, I'd sit for hours and listen to records by the Band, the Flying Burrito Brothers, and Taj Mahal, along with lots and lots of Dylan, as one of the café's founders, Yuzuru Katagiri, was Dylan's Japanese translator. He'd also translated Allen Ginsberg and Jack Kerouac, so the place had a bit of the feel of the San Francisco Renaissance, North Beach, and the Caffe Trieste.

For some reason, though, Nadja and I once wandered into a place that played only Beatles records. The décor it had was a forerunner to Yayoi Kusama—everything pink, white, and black with polka dots, as I recall. The tables were of prefab white plastic, walls pink with white and black polka dots, chairs black and white. It was as though you'd stepped into the nightmare scene from Disney's *Fantasia* or into the Kolova Milk Bar from the Kubrick movie of Anthony Burgess's novel *A Clockwork Orange.* But we sat down anyway, perhaps needing a break from the cold winter air, ordered our coffee, and maybe Nadja pulled a book out of a backpack to read

and I got out a notebook and pen to write something. We weren't there for the music, I know. In those days, you could divide our generation into two types—those for the Beatles and those for the Rolling Stones. Nadja and I were for the Stones—staunch, blues-based rockers that they were. The Beatles were mainstream—too cleaned-up, often treacly in their sentiments, and ubiquitous on the airwaves. Yet we both stopped, hearing something new come on the stereo of the *kissaten*. It was the first few, simple chords from John Lennon's "Imagine," and then the ex-Beatle began to sing the haunting notes of the song's start: *Imagine there's no heaven. / It's easy if you try* . . . The sweet innocence of the lyrics and the humility of their sentiment were as penetrating as the weight of the piano's chording that thudded against the soft pulsings in our blood, chiming with its nearly silent flow. The high strings in accompaniment made a sweeping drone, a kind of gentle keening that seemed part of the air until it dropped away and we were surrounded by a brief silence, quickly filled by Lennon's repeated figure on the piano, then a deft drumroll and an electric bass locking under it.

Nadja reached her hand out across the scuffed white plastic of the café table and my fingers twined briefly with hers. We stayed like that through the song, then let go when it ended, giving each other the curious looks folks do when, suddenly, they've shared the same ether of emotion, dispersed as quickly as it had arrived. Music can be like that, a faint gauze of certainty gone as quickly as it was surely present.

There were baby spotlights in that café, maybe rows of track lights pointing down at us and the tables, on throughout the song, but I don't remember them except the moment "Imagine" stopped playing and I felt the timed lights winking on and off in the curt silence that seemed as ephemeral as alternating moons of hope or pain.

What's Going On, Robert Hayden?

"What's Going On?" was something different than I'd expected from Motown and Marvin Gaye, the soul singer already famous for his catchy cover of "I Heard It Through the Grapevine" back in

1968. Black women in the dorms at my college played it constantly on the stereo during most of my freshman year. I guess it was a way to draw lines, re-create Black space, and feel more at home at our predominantly white school. But the new record was nothing like it. The jacket was a headshot of Gaye, taken in three-quarter profile, his face bearded, slick black raincoat collar up, and surrounded by a green background, lush and out of focus.

The title track was the first on the A-side and started with a falsetto trill, like a vocal guiro with a rising top, then broke into a collage of male voices greeting and murmuring, *Hey, what's happenin'?* You had the feeling you were entering an apartment where a party was going on. Then came the launch of a wailing figure from an alto sax, accompanied by big, tuneful pulses from an electric bass. This was James Jamerson, part of the group that came to be known around Motown as the Funk Brothers rhythm section, plucking out a walking line with catchy accents and syncopations. A set of congas hit a chopping beat, while someone slapped lightly on a box drum. Then Gaye sang the opening lyrics, softly crooning, rising to falsetto whoops and wails at the end of each verse: *Mother, mother . . .* I heard a xylophone playing fills, then fading into the scrum of lush background sounds—backing male voices *oohing* and *ahhing,* Gaye's second lead vocal tailgating in falsetto and stretching out the first lead. His voice swirled around the melody and sounded like a higher-pitched echo of himself, doubling on the line *To bring some lovin' here today,* fixing the thought of the verse and emphasizing its pure sentiment. There was a Mellotron in there too, backup vocals switching to *Sister, sister* hard on the downbeat, fingers snapping sharply. Then an ensemble of strings joined in, swelling and lengthening the vocal line, evolving into sweeping twirls of figures that underscored the overarching sweetness and romanticism of the song. The male chorus took up chants of *Brother, brother* backing Gaye and his plaintive lyrics, then Gaye hooted out the song's insistent question—*What's going on?*—no longer the casual, throwaway greeting but addressed now to love's lack in the world, to the futility of the Vietnam War, to soldiers losing their lives there (Gaye's brother got sent to fight), to demonstrations and pickets, to police brutality and all the hurt of everyday life among Black people and

a whole generation of young in America standing up in protest. Gaye punctuated the sentiments by scatting in falsetto, beguiling and mournful, as the instrumentation wrapped him in a braided flag of sounds, the symphonic sweep of strings, he and his backup boys calling *Right on,* in response.

I put the 1971 album on all during the summer of '75 as I was trying to make sense of the disparate pieces of my life. I was twenty-three, studying Japanese literature at the University of Michigan, loving Black music and Anglo-American poetry, hearing Whitman to the tune of Coltrane's tenor sax, reading William Wordsworth interwoven with Weather Report's jazz fusion albums blaring from my stereo. Afternoons, I stretched out on the cool love seat downstairs in the house in Ann Arbor where I rented a room, closed the blinds to the light outside, and thought back over the short tracks of my life.

I was only a year past my temple studies in Japan, going to readings by poet elders whenever I could, and trying to sort out the warring cultural influences and social forces cresting against each other in my blood—my upbringing in a Japanese American family, my immersion in Black urban culture during high school, my academic studies in college and grad school, my white and multi-ethnic friends. I felt deeply that I was ready for nothing as a young writer. Not only had I no wisdom, but I'd no prospect to stand upon to take in the vast, unarticulated sweep of all that I'd seen in my twenty-three years on the planet. Every admonition I'd heard, every view that claimed to be comprehensive and organizing required that I reject whole swaths of life I'd witnessed—the cruelty of three generations of plantation life for us as children of Japanese immigrants was now instantly to be canceled by our move to the suburbs in L.A., the inclusion by my Black classmates in high school was belied by my privileged education in college, the devotional purity (albeit wobbly) of my single year living in a Japanese temple was countered by my current erotic life with my Chinese American girlfriend, an economics student, whose warm brown skin bore the scent of candlenuts.

Only music seemed to bring it all together in a secluded space of respite and approval. Even beyond the sentiments expressed in

the lyrics of Gaye's song, it was his conviction that lovingness and not hate could lead that reached me, along with the serenity in his verses, the great swelling beauty in his music, his voice like that from a tribe that didn't yet exist, sacred and longing as any *oli* chanted at the island shore by my childhood neighbors. I read in *Rolling Stone* that Gaye wrote the songs for the album after a great time away from the music business, during which he was grieving the loss of Tammi Terrell, his duet partner on many hits; he'd quit touring, refused to perform love songs, even decided that he wanted to play wide receiver for the Detroit Lions, a few from the team having befriended him. He'd been moved by letters he'd gotten from his brother in Vietnam, and by witnessing what happened after that brother returned home, a dispirited veteran shunned by the young in his own community. Gaye's smooth and jazzy song was not only a new sound, but a new message completely different from the teen love lyrics that dominated the growing archive of hit tunes from Motown. And it was also completely unlike the militant declarations I'd been hearing from Black poets at the podium—for just cause, mind you, decrying racism, insisting that anger and a macho defiance were the only proper responses to a system run by *the Man*. "What's Going On?" was something else completely—a contemporary spiritual—and matched a wish I wasn't yet brave enough to have. It put social hope to music, gave it a beat, and amped the volume to the ghostly anthem of the otherwise frail lovingness in our hearts.

Robert Hayden was a teacher of mine that year, carrying on the quiet discipline of his life as a poet and scholar despite being vilified by younger writers for being an "Uncle Tom," as well as by the prominently militant poet Amiri Baraka, and being ostracized, even (what now is called) "canceled" by Black students. But I'd no idea. Along with asserting ethnic identity and insisting upon the acknowledgment of suppressed histories, the movements that empowered people of color too often carried within their messages strong nationalistic imperatives that indeed canceled what was per-

ceived as the culture of the master. For Hayden, well schooled in formal, canonical verse and open about his affection for it, this was a terrible burden, as African American poets of his own generation and younger then condemned his personal style and his poetry as "like the white man's," a fact I was ignorant of until he told me one day in the poetry room at Michigan.

Along with maintaining an appreciation for traditional verse, Hayden had written numerous poems celebrating Black life and

historical figures like Harriet Tubman and even contemporary ones like Malcolm X. He wrote about the Middle Passage, lynchings, the flight of freed slaves, about urban riots and student demonstrations in the sixties, and he wrote about his difficult childhood growing up in Detroit—a ghetto ironically called Paradise Valley, where heavyweight champion Joe Louis was also from. He'd won honors, distinguished himself, and taught for many years at Michigan and Fisk University, a historically Black college. He had been a student of the English poet W. H. Auden and professed admiration for canonical white poets while at the same time remaining loyal to Black experience. Hayden wore thick, Cokebottle eyeglasses and was always dressed formally whenever I saw him. And he was tall as a rising sunflower in a bow tie.

I'd invited him to read with a slate of "ethnic poets" I'd organized, together with two African American students, and gotten money for it from several organizations on campus—the "Third World Alliance," we called ourselves. The other poets were to be African American Etheridge Knight, Native American Leslie Marmon Silko, Japanese American Lawson Fusao Inada, and Chinese American Mei-mei Berssenbrugge. The venue was the Trotter

House, a lovely manse of many rooms and a posh meeting space that housed the African American fraternity at the University of Michigan. I reveled in the hope that this would be a big event, bringing multi-ethnic voices together, that my fellow students would talk about it for a long time.

But Hayden surprised me by declining outright as soon as I asked him, saying, "While I am most flattered by your kind invitation, I'm afraid I must decline as you would not receive the audience you'd hope to were I to appear at the podium alongside your other guests."

I was shocked. I asked what he meant.

Hayden said, "Have you ever heard the phrase *Uncle Tom,* son? I'm afraid I am persona non grata at the Trotter House. I am seen as a somewhat corny Negro and my appearance on the bill would only diminish your attendance and not accomplish the accord you likely hope for. Therefore, with regret, I feel I must decline."

I was devastated. Not that he would not join in, but that he'd been ostracized by other Black writers and students. I was naïve, of course. Calling two of my old teachers back in Claremont on the telephone—the Black poet Stanley Crouch and the Jewish poet Bert Meyers—I asked them about it. They both decried the shunning and explained. There had been a public censuring of Hayden at a conference of Black writers at Fisk University back in 1966. While onstage together, fellow poet Melvin Tolson, author of *Harlem Gallery,* had denounced him. The attitude spread after that. The Chicago poet and a leader of the Black Arts Movement Don L. Lee (later Haki R. Madhubuti) denounced him. This pained me deeply. When I pressed Professor Hayden, he offered to attend but insisted he would not read.

The night of the event in Ann Arbor, I saved two seats in the second row (readers sat in the first row) for Hayden and his wife, composer and pianist Erma Morris Hayden. When they came down the aisle, filing to their seats in the middle, Etheridge Knight (once a student by correspondence, while in prison, of Gwendolyn Brooks), stood up and applauded, saying something like, "I want y'all to know that the great Robert Hayden has graced us with his presence." One by one, the other poets all stood and applauded. Then

the mostly student audience did too, some sixty or seventy people, slowly, wondrously.

I don't often think of this, and had not in many years. Mainly I recall the moment when Hayden refused to read—a story I have told to myself many times over. But the whole scene comes alive for me, when I let it, and I can see everyone in my mind's eye—Etheridge in a midnight-blue Hawaiian shirt with tiny yellow quarter moons scattered on it, Lawson Inada and Leslie Silko rising to their feet, the two Black co-sponsors of the event rising. The memory of it makes me weep.

Earlier that fall, when I'd signed up to study with Hayden, it was to be a one-on-one tutorial, and naïvely, I asked to read "the Black poets," African American writing of the sixties, but he simply said, "We shall read Keats." I'd not studied English Romantic poet John Keats very closely at all and thought it a lack, so I agreed. I'd actually have agreed to study penguins if he'd proposed it. He met with me every Monday afternoon on his porch, at tea, no less. His wife would bring out a tray full of cookies, cups and saucers, and a teapot, once in a cozy, I recall, the first time I had seen such a thing. He had me read aloud and tutored me to slow my pace, enunciate the words, give them feeling, letting the emotions "emerge organically" as I read, with savor, patience, and sentience, appreciating not only the words but the cadences, phrasing, startling insights, and rhetorical complexity and density. It was a lesson in acting, although not in the theatrical sense, but in terms of inhabiting not just the meaning of the poetry, but its shadings into emotion, its rhetorical flavor and flourishes, and its "essential modesty," I think Hayden's phrase was. He meant not to overdo it. We didn't talk themes or details or history much, just read the poems aloud and admired them. He asked me what I thought about a passage here and there, and suggested lines of inquiry I might take, but also remarked that I'd "return to ponder" much in the poems in the years ahead if they'd had any effect on me. He was right.

Why Hayden compels me, continues to, is because he was the first and only Black poetic voice I'd heard from back then who, while recognizing the history, the culture, the oppression, urged lovingness rather than defiance and fury. He'd ushered me into the

entryway to two of the loveliest and most fiercely beautiful poets in the language—himself and Keats. It is like hearing Haydn and Hindemith in music. You feel the sweep, not only of their gorgeous phrasings, but of how a tragic loveliness persists in the culture and finds its poets. Aside from Marvin Gaye, no other prominent Black artist was saying it quite like that. Everything was about revolution, attacking the Man, performative defiance, castigation, exhortation—so one-dimensional, I found, and self-damaging to read. Hayden's poems were crafted, emotional, and loving. He was a beacon of compassion and self-respect. And as I felt I knew him, it hurt me that other Black voices had denounced him.

The Harder They Come

That summer of 1975, as I was finishing out a year in Ann Arbor as a graduate student in the Japanese Literature program, I was enrolled in language classes, but mostly focused on playing center field for the department's softball team and dating Susanna, a Chinese American woman from Princeton, New Jersey. We'd met at a mutual friend's birthday party where we ended up dancing together on every record. She could switch styles no matter what came on, and I, having been schooled on Motown and Stax-Volt records by Black kids in high school, could almost match her, step for step, funky chicken to skate, the twine and cool jerk. We were dancing to the Average White Band, a Scottish soul group that had fashioned their style after classic American rhythm and blues. Then something different came on. It was the soundtrack to *The Harder They Come,* a movie I hadn't yet seen, starring Jamaican singer Jimmy Cliff.

We stopped a moment, listening for the beat and its intervals, hearing a hard *cha-cha* but with a twist, a powerful syncopation that was new to both of us. Organ chords seemed just a half-beat off a simple *one-two-THREE* from the bass, and drums chattered and high-hatted in a loose skiffle style. Cliff's high voice wound itself in rhythmic inflections that emphasized the third beat, the organ tailgating him throughout, and we fell into the herky floppiness of the sound, improvising off a step adapted from the temptation walk.

This was the first either of us had heard reggae, and its sound as much perplexed as captivated us while we danced, heads bobbling like souvenir dolls at a Detroit Tigers game. As soon as the tune ended, we rushed over to the record player and asked to be shown its cover.

A single 33⅓ rpm LP, it was nevertheless given a gatefold sleeve by the record company. We opened it and read about a completely new world—groups like the Melodians, the Slickers, Desmond Dekker and the Aces, and Toots and the Maytals. Though it had crept into the underground scene in New York, reggae had not quite made it to the radio waves across the States just yet, so it was all a revelation—not only a new sound, but a new perspective on the relationship between colonialism and oppression and a sweet resolve inspired by both the Christian faith and a thing I later found out was called *rasta,* a spiritualism that claimed Ethiopia's emperor Haile Selassie was the incarnation of Christ in our time. It all seemed simultaneously mystic and radically provocative. But what moved me especially were the hymns of struggle and survival on the album, which I got on my own in a shop along State Street in Ann Arbor just as soon as I could. When I heard "Many Rivers to Cross," Cliff's piercingly high voice sailed over a churchy organ. And like so many songs in the Black gospel tradition, it was about struggle, the singer confessing himself to be at the brink of giving in to rage or temptation, ultimately roused by faith into renewed determination. I felt a powerful identification with the sentiments of the song, feeling my own generational losses, my ambitions to live in a life dedicated to poetry, as crazy a thought as the visions of any mystic from Jamaica or Hawai'i. I wanted to see the movie.

When I did, I went with a crowd of friends and we took up the middle sections of two rows in the theater where it played somewhere in Ann Arbor or Ypsilanti. The movie has an off-kilter, bildungsroman-like plot about its hero Ivan, played by Cliff, a singer who moves from the country to Kingston, cuts a record, then falls into drug-dealing, thuggery, and misadventure, only to find that his song goes wide throughout Jamaica while he falls in a hail of police bullets. Its dialogue was completely in Jamaican patois, the creole language of its people, and if not for the subtitles in standard

English, almost no one in the audience could understand what was being said. Yet once I caught on to its rhythms and inflections, I found it was almost as understandable to me as Hawaiian pidgin, the language locals spoke in the islands where I grew up. The patois was remarkable. When a song came on, cutting into the narrative, it was as though a "calabash" cousin were singing to me across time and the waters of deep mystery.

Susanna and I almost lasted through the summer, going to Chicago and its blues clubs one weekend, taking in Sydney Pollack's *The Yakuza* starring Robert Mitchum and Ken Takakura (which I'd glimpsed being filmed in Kyōtō's Gion one night the year before), writing linked-verse *renga* together, and going to lakeside picnics with friends. But our connection dwindled once we both realized I'd not be staying, that I wanted out of Ann Arbor and back to the West Coast. I wanted people who spoke my language of urban streets, coastal screwpines that lilted and wove like waves pulled by a mystic moon.

The Bridge

I returned to Gardena and rented a one-bedroom apartment on a loop of a street near 164th and Normandie Avenue. The neighborhood was lined with jacaranda trees that littered the patchy grass of its parkways with purple blossoms that dried slowly in the scorching heat of Los Angeles in early fall. My place was a downstairs "efficiency" unit in a complex that featured twin bungalows on either side of a narrow courtyard. It was the kind of thing you'd see described in a Raymond Chandler novel—hot and airless, the patter of frying meat in an oiled skillet audible through every screened back door in the alleyway as you walked by, batting flies and brushing by the gaudy heads of Lily of the Nile sprouting beside each doorstep of painted concrete. I had a tiny living room, a dusty carpet on the floor, a love seat with stiff springs, a lamp on a small table, and an easy chair that worked better as a backrest when I sat on the floor in front of a balsawood coffee table and struggled to write on weekends and weekday evenings after work. I tried to write poetry, but the words were hard to come by, my life

taken up with so many unanswered questions. *How was I to make my way?*

I was trying to write something cynical about American commercialism and how it dominated popular consciousness. I hadn't any education in cultural critique or even parodic surrealism then, and so was only capable of creating images of folks in L.A. chasing dreams instigated by TV shows, movies, and commercials—descriptions of shoppers at a mall, someone snapping their chewing gum in a movie line, nondescript folks in a carpool discussing a sitcom while driving to work. Behaviors I found revolting. I tried to contrast these with an image of a girl in hot pants and a leotard roller-skating down the boardwalk by the beach in Redondo. It didn't go anywhere, and the poem failed. I was trying the same approach I'd used a couple of years before to describe street scenes in Kyōtō—a kind of imagism under a sybaritic lacquer of accidie. A Buddhahead Baudelaire.

I'd quit graduate school, a dropout, and taken up my old summer job again as a meter reader for the City of L.A.'s Water and Power division. I spent every day hiking through neighborhoods much like mine, dodging dogs in a rage that I'd violated their turf, asking homebodies to open their back doors so I could read their electric meters, hopping fences and brick walls, pausing under billboards, jumping rusted train tracks, now and then catching a glimpse of the Pacific, a rippling blue rectangle as though a window blind had been drawn down between two buildings. I was making money but only biding my time until my next move, maybe going north to join my college girlfriend up in Seattle, where she was studying musicology. I needed to bank a few thousand to keep me going if I moved up there without a job. I had a vague idea to try my hand at playwriting next.

Meanwhile, though, my free time was my own and I filled it reading the Chilean poet Pablo Neruda, L.A. novels by Chandler and Nathanael West, and expanding my listening repertoire with more jazz. I'd been going to the Lighthouse, a waterfront dive by the pier in nearby Hermosa Beach, and had heard a variety of local and international musicians, but despite patient tutoring by Bill Taylor in college, the music hadn't yet come together for me in any

kind of coherent way. I really didn't know cool from bebop, swing from standards, the Art Ensemble of Chicago from Art Blakey. I went to a few record stores and, almost at random, picked up a few things—albums by saxophonists Ben Webster, Coleman Hawkins, and Sonny Rollins. The sounds of Webster and Hawkins, older players than Rollins, were familiar to me, their improvisations affecting and easy to follow, sticking close to the melodies they played, giving them bounce and sometimes an explosive kind of power. But the music of Sonny Rollins enthralled and perplexed me both.

Completely by chance, the Rollins album I had was *The Bridge,* a milestone in the history of contemporary jazz released about a dozen years before, in 1962. A reissue pressing on RCA, it came in a sleeve with a light bluish cast over a black background, the image of Rollins playing his sax, his puffed cheek superimposed next to a night shot of a single white tower and the looping suspension cables of New York's Williamsburg Bridge. I might have read the backstory about the album—that Rollins had taken a long sabbatical from touring at the height of his stardom, "woodshedding" for three years, using long, nighttime walks along the pedestrian passages of the bridge as his practice sessions, playing his tenor sax all the while, trying to improve his technique and come up with something new, something that was more original and heartfelt than the hip, somewhat angular music that had made him famous. It sounded like what I thought I had to do as a poet—to get introspective and search *within.*

I'd written poems in college, in Japan, and in graduate school at Michigan, winning a few prizes along the way. This had gratified me, but I felt I was a bit of a fraud, what poets derisively call a "magpie," imitating the styles of others without one of my own. I'd found it easy to "sound" like another poet, whether it be someone terse and fairly direct like Gary Snyder, the outdoorsy Beat poet who'd studied Buddhism in Japan, or someone lavish with metaphors and rhetorical figures like Pablo Neruda. One day I could write a slow, quiet poem full of lyric sentience like those of Austrian Georg Trakl; on another, I could write a raucous, barbaric yawp of a piece, irreverent and comical as Allen Ginsberg. But none of it was my own. None of it came from my own experience or a deep meditation on that experience. As a poet, I was like a cover band, a

television fake like Bobby Sherman in a hair-sprayed pompadour, singing tunes of the famous in styles copped from their hits. *What was my song?* I thought. *Had Rollins figured out his while walking the bridge?*

So I listened carefully to the album, night after night. I took in the lyric balladry of "Where Are You," a captivatingly languid tune that sounded like a waltz taken achingly slowly, Rollins alternating solos with the guitarist Jim Hall, who played deft chords and brilliantly articulated runs in both accompaniment and solo. I studied the stumbling, thrilling post-bop style of "John S," full of staccato tenor runs, chattering snare and high-hat work, and a speedy, walking bass line, Hall comping in ringing, then quickly damped chords in accompaniment. I heard the beats, the breakneck improvisations, the fine timbres of Rollins's tone and expert embouchure punching and slurring notes, the rich harmonics of Hall's sinuous electric guitar. The title tune was itself an even more provocative lesson in speed and post-bop expertise, Rollins taking the band through intrepid time changes, alternating rushed passages and decelerando interludes, hitting his notes hard, sustaining a note, then rapidly slurring through chains of them. It was heady, intellectually energetic music, a kind of jazz Modernism, full of retards, then explosive bursts of notes in intervals unfamiliar to me, sounding vaguely Asian and pentatonic.

But it was "God Bless the Child," Billie Holiday's great ballad, that established a center for all I heard on the Rollins album. Hall and Rollins took the melody and played with it, throwing in odd hesitations, lyric liltings, and long-held notes full of conviction that ended when Rollins simply blew audible, noteless air through his horn. Then, in a solo near the middle of the take, he hit sweet, smooth, echoing notes, bending them in bluesy fortissimos. Hall, too, caressed the song on his guitar, Rollins then blowing softly in accompaniment. They ended with Rollins repeating the melody in a burnished style, finishing it with honking, then fluttering flourishes. The song was arranged in progressive solos like a suite of lessons in the liberties and poignant lyricisms possible in jazz. I came away realizing that I was less than an ephebe, and that, if I were to approach any artistry at all, I'd a long career of "woodshed-

ding" of my own to accomplish before I could get close to the array of knowledge, expertise, and feeling Rollins and Hall threw into Holiday's song. *God bless the child that's got his own,* yes—but I had yet to earn mine.

When I took the needle off the record, picked it up from the platter after it stopped, put it back into its paper sleeve, perhaps absently rotating it, and slid it into the album cover, pushing it gently through the sandwich of the cardboard, I was still locked in meditation, spellbound with the music spinning unheard and silent in my imagination, which had just found a new order, a polestar of challenge and a bridge to what might come next.

Kokoro Ga Odoru

Wakako Yamauchi (1924–2018) was a Nisei, a second-generation Japanese American, who became the teacher of my heart, *kokoro* in Japanese, a word for spirit and mind together. I started visiting her in midsummer of 1975, after I dropped out of graduate school and was back in Gardena, working for Water and Power. She was about fifty; I was twenty-four. I had a bungalow apartment only a few blocks from her home, a contemporary with a swimming pool in the back instead of a yard, and I was always over there for conversation and the meals she'd feed me if I showed up at the right time. Her house was decorated with her own paintings, impressionistic landscapes and cityscapes that demonstrated a fine eye for muted shades rather than splashy colors. She'd also hung the paintings of her friends and her brilliant teacher Matsumi Kanemitsu, who had studied with Léger and Kuniyoshi, and whose work was as strikingly bold as hers was subtle.

Wakako was also a writer of short stories and published steadily for over thirty years in a Japanese American newspaper, starting soon after having been released from Poston Relocation Center in Arizona, where she'd spent most of World War II with her family as a teenage internee—one of 120,000 persons of Japanese ancestry our government had imprisoned in such camps. She'd begun writing for the camp newspaper then, apprenticing herself to Hisaye Yamamoto, also a teenager only a year or two older, who was then

taking a correspondence course in poetry from Janet Lewis and her husband, Yvor Winters, both noted poets at Stanford. I saw mimeographs of the lessons once—a list of "100 Best Poems in English" compiled by Winters for his M.A. students along with devilish assignments in various meters, rhyme schemes, and forms. To Wakako, though, formal lyric poetry was tangential, as her interest was in storytelling and capturing the lives of Japanese American tenant farmers and migrant workers she'd witnessed in California before and after the war.

When I met her, she'd just started writing *And the Soul Shall Dance,* her first full-length play, set in the Great Depression. She was dwelling in remembrance, recalling the wonder she felt as a child at the moment she first encountered an artist, denizen of a different world, an *odori* dancer who, through misfortune, had been married to a dirt farmer. The play was a kind of spell, invoking not only a lost history and testifying to its sufferings, but rising in its images to conjure a madwoman, a magnificent ghost, dressed in elegant kimono silk and brocades, dancing sorrowfully across a windswept desert landscape.

Through the years, when we got together, as we did often when I returned to Gardena to visit her and my parents, Wakako would sit on her sofa, light up a slim cigarette, take a casual pose below one of her paintings on the wall behind her, and tell me stories about the Nisei past. They were about an itinerant farmworker who could play Bach and Vivaldi on the violin, about an infant brother who drowned in a washtub by the pump on their family farm, or an esteemed teacher of art and calligraphy before the war who'd become addicted to gambling (so broken and shamed was he by the internment, he lost all but the clothes he wore shuffling along the streets of L.A.'s Japantown, asking passersby for money). She spoke about the withering effects of love gone unexpressed between couples long married, about sexual and emotional thirst, about an old woman's nausea at her own wretchedness. She was, for me, a kind of Colette and Chekhov at once, an elder who shared with me the insights and passions of her life and those she'd witnessed.

And she painted—abstracts, still lifes, and, I remember, an urban scene of picnickers in a public park with patches of sunlight

floating down from a canopy of eucalyptus trees above them. Her colors were light, specular and just slightly washed out, and her human figures abstract outlines, elongated and thin like Modigliani's. Race seemed as absent from them as it was front and center in her plays, stories, and conversation. One evening, she brought out an old compact tape recorder and put a cassette in it, saying her brother had made it and given it to her, that the songs on it brought back the past—images from her childhood, wraithlike in memory but instantly transformed by the music into vivid bodies stooped over a stove or hoeing at a row of cabbages. She heard voices, not only of the recorded singers, but of people who sang the very same songs, tending a flower garden amidst the dusty winds and desolation at the internment camp at Poston, sitting on a bench with other teenagers at a Friday-night dance in Barracks 12-1-A, pouring miso into cheap porcelain bowls in the kitchen of a migrant hotel.

At first, she played the tape, listening and inviting me as well, the gray smoke from her cigarette curling before her face in the still air of her home, tears gently forming in her eyes, then falling like petals from a withered flower. She sang along, her voice chesty and resonant, the tone irredeemably plaintive, the stories inevitably about loneliness and abandonment. These were *ryūkōka*, Japanese popular songs of the early twentieth century that themselves grew out of the older *min'yo* folk tradition of festival dances, narrative balladry, and teahouse love songs of the floating, impermanent world. *Ryūkōka* maintained the Asian pentatonic scale but added Western instruments and styles, creating the Japanese pop of the twenties and thirties, with an occasional American tune mixed in amongst them, but sticking mostly to Japanese melodies. It was this prewar Japanese culture that Wakako's immigrant ancestors brought over on Victrola records. They invoked a separate past and came from a wellspring of aestheticized melancholy almost completely unavailable in literature written exclusively in English. There are, in these invocations, a special, almost religious quality, being emotional evocations in slowly revealed, deliberate lyric cadences quite unlike anything I know of in American letters. They seem closer to the spirit of Appalachian gospel tunes but have the resolute quality of African American spirituals. Finally, they are perhaps closest to the

tragic arias of Italian opera—dramatic monologues, testimonies to love, laments for love lost, and confessional soliloquies before irredeemable acts, all in lyric registers of grief, amorousness, or even vengeance. When Wakako called them forth, I felt, deep in my bones and in the strange, nearly imperceptible rushings of my own blood, the banishment of the emotional gulf and historical distance between myself and those first-generation immigrants who might have been my own ancestors. The Japanese term *sabi* is a word for the indescribable, plangent sadness we feel at the poverty of being itself, a mode that invokes pity and the cast of mind where real compassion is born. This was Wakako's way—as a teacher and a writer. She gave me gateways into the heart, *kokoro:*

> *Akai kuchibiru*
> *Kuppu ni yosete.*
> *Aoi sake nomya . . .*
> *Kokoro ga odoru.*

> *

> As I press these red lips
> Against a glass and drink
> This green absinthe,
> It's my soul that begins to dance.
>
> (translation mine)

She sang a teahouse love song, a lovelorn lament in a solemn alto. The English is smooth, direct, almost a flat statement, but it does no justice to the Japanese lyrics or to the wrenching measures of the song. It's sung haltingly, in elongating vowels that climb in register, pirouette a moment, and then drop precipitously, taking up a sliding syncopation in the middle of the second line, before the cadence smooths again in the third, only to gather a mid-line lilt in the fourth. The notes swoop up and then down in each line, shimmer in an almost tortured vibrato at the mid-line, then lilt or get a slight hitch in their cadence as the stanza's measures proceed. The effect

is haunting, even pitiful, pitched to a delicate and ephemeral recollection and a stylized solemnity.

It was the song that had given her the play *And the Soul Shall Dance* and figures prominently in its story. It tells of a *maiko*, an apprentice *geisha*, who becomes a kind of fallen woman not unlike Verdi's Violetta in *La Traviata*, someone who renounces her artistic family and the life of gaiety and succumbs to singular passion, only to get dropped, forsaken by her lover. Wakako's madwoman teaches this song to a ten-year-old girl, daughter of the neighboring farmers, reminisces about her former life in Japan as an artist and entertainer in the teahouses, and begins to tutor the girl in the life of imagination, artifice, and splendor. It's a tease for her mind to travel away from the careworn lives of drudgery and necessity around her, a lesson in fanciful escape and the pain of having tried for it. At the play's climax, the madwoman, dressed in her finest *nihonbuyō* robe, sash, and dangling sleeves, runs away to the desert, singing the song as wind whips around her, as though through a tumbleweed.

> *Kurai yami no yume*
> *Setsunasa yo*
> *Aoi sake nomya*
> *Yume mo odoru*

> *

> Dreams through the infinite dark
> Scatter like leaves in the light of dawn.
> But the absinthe I drink
> Makes dreams that dance even for the lonely.
>
> (translation mine)

After a while, during sessions like these, Wakako would shut off the tape and just sing the songs to me herself. How she sang them, the spirit I felt rising from within her thin body and shimmering in the slow, mournful warble of her quaverous voice, taught me the patient cadences of emotion, the way a story can inhabit just a few

words, how a melody can carry an entire lifetime in its measured unfoldings.

She told me that, when she was first released from camp in 1945, still a teenager, she got a work permit sponsored by a company in Chicago. Because of her talent for painting, she was given a job decorating shower curtains and men's ties, working from early in the morning to evening, making a few cents an hour. She'd get her coffee in the morning, take it to the steps of the Art Institute of Chicago, stand there below the sculpted lions on each railing, and dream she might someday take classes there, meet real master artists, and apprentice herself to the passion she felt for rendering the bleak world she came from in images that might be bettered by more tutored understandings of Manet, Bonnard, Dufy, and Cézanne. She was seventeen.

That was the start of a lifelong ambition to make art out of obscure and dispossessed lives. As we sat together, Wakako shared not only Japanese folk ballads and *enka,* but tunes from Duke Ellington and the big band era. She told me about a swing dance one Friday night in Poston, all the Nisei teens gathering in the commissary. She ended by singing "Sophisticated Lady" in a breathy alto voice.

When I remember Wakako, I can't help thinking of Mimì in *La Bohème* too. The first time I heard the opera, at La Scala many years later, when I heard Mimì sing her first aria to Rodolfo, the poet

she has just met, she declares *Mi chiamano Mimì* in beautiful soprano notes bathed in a steady, modest pathos. A moment later, when she sang, "I embroider silk with false lilies and replicas of roses," I thought of Wakako, and a sob of recognition caught in my throat.

"The ties were ten cents apiece," Wakako had said. "And every morning before work, I'd dream I could be in there, over a canvas at the Institute, learning how to paint."

"Each day, alone in my whitewashed room, I make the same lunch for myself and look upon the roofs of heaven," Mimì sings. "A rose blooms, and petal by petal, I watch its slow birth upon the artificial earth." She is eighteen, too, I think, and like my teacher, she saw things differently than the plain, workaday eye could reveal.

In My Solitude

In the late seventies, I was living in Seattle and a manic mess, directing a community theater group of Asian American players and other theater artists (costumer, set designer, etc.), raising funds through grant applications and whatever the take was on a given performance night. I'd moved there to be with my college girlfriend but had split up with her, then tried taking up in rapid succession with an actress, a poet, and a salesclerk at a women's boutique, and even proposed marriage to a playwright from Los Angeles a decade older than me. I was a loser in romance, and my love life was as much a disaster as my attempts to write poetry and plays. I spent nights in intense rehearsals, midnights to early mornings banging nails on the sets, then slept until noon, writing grant applications and making notes on scripts in the afternoons.

I just could not sit still long enough to do the deep dives into daydreaming that real writing took, so I wrote superficially, glibly, and produced a comedic script about returning World War II veterans, Nisei from Hawai'i who'd exiled themselves in a Northside Chicago bar, ashamed to show themselves at home for having no medals, no careers, and no wives or families. A memoirist I once told about the script said the play was about "castration," and it shocked me, but I knew it was true. There were generations of Japanese Americans who were denied a whole truth about their

bodies in our culture, denied full social and sexual empowerment, the Asian face and torso constantly denigrated, mocked, or exoticized by movies and TV. I'd meant it as a kind of tragicomedy, a testimony to the sere sadness I'd witnessed among the generation of my father and uncles. But what worked for the actors was playing my script for laughs, and they were damn good at it. The play got noticed.

I was flown down to L.A. to work for ABC Studios as a writer for a sitcom starring comedian Pat Morita, who'd felt trapped in Oriental schtick, cast in the show as a foreign-born Japanese and frustrated with the cascade of Chink and Jap jokes that were all his veteran team of writers, themselves refugees from *I Love Lucy,* could come up with for him. He'd been given jokes about buck teeth, slanted eyes, and inane, supposedly Asiatic malapropisms. My charge was to write "genuine Asian American comedy," work something hip and culturally current into the identity of the show. But I too got frustrated, unable to break through the stubborn resistance of the old-school writers, who were completely against a punk kid getting more lines in the script than they did. They shouted me down in writing conference each afternoon, declaring *You don't know funny, kid!* Not yet tough enough to contend with them, I slunk away early and sat on a curb outside the meeting room most every day for the two or three weeks that I tried to stick with it.

Evenings I'd drift over to the Hollywood apartment of the playwright I was trying to romance. She drank highballs of amber bourbon every evening and had an original script in development rehearsals in an Equity-waiver theater. She had another script being filmed for television, spent days on set there and early evenings at the waiver house. She was busy.

But after hours, a few members of each cast would drift over to her place for talk and conviviality, swapping stories, and trying to outperform each other in wit and repartee or just plain hilarity. I was fascinated by these actors, all of them older than me, most by an entire generation. They'd played in musicals like Rodgers and Hammerstein's *Flower Drum Song* and Stephen Sondheim's *Pacific Overtures* on Broadway, appeared on variety shows hosted by Red Skelton, Dinah Shore, and Frank Sinatra, acted in movies like *Aun-*

tie Mame and *The Teahouse of the August Moon*. But it wasn't only fascination I felt, because, in so many of their roles, they'd played stereotypes and supernumeraries who were caricatures of our Asian identities—bowing and scraping onstage and on camera, slurring their speech, singing Broadway paeans to chop suey and chopsticks. They'd endured levels of humiliation I could not imagine, performed racially degrading roles I could neither respect nor wish to remember, except with enmity and anger. I felt their ideas of emotion were stagey, running like quicksilver across their faces and through their bodies, and masking the undercurrent of bitterness and profound notes of sorrow that, if only for a nanosecond, crept into their voices when they spoke of the past. I was unsympathetic at first, of another generation and cultural experience, arrogant and unbowed, ferocious in defiance. I told myself I would be different, yet I remained attentive in their presence, observing their after-hours moments of candor. The two I paid particular attention to had, throughout their careers, regularly refused to play fake Asians, had always insisted on their American identities.

I remember the comedian Jack Soo, recently cast as Detective Nick Yemana in the sitcom *Barney Miller*, drifted in one evening when the summer light was as material a presence as the off-white apartment curtains it shone through, a kind of luminous dust settling on our skin and hair. And Soo's thick hair rose higher than most, his skin agleam with television fame. There was a legend, apocryphal but typical in its inventive conflations, that went around about Soo in those days—that, born Goro Suzuki on a ship traveling from California to Japan in 1917, he'd grown up in Oakland, excelled at sports as a teenager, and, in order to avoid the evacuation and internment during World War II, changed his name to sound more Chinese and hid out in San Francisco's Chinatown during the whole war, working as a singer and bandleader in the nightclubs. The truth was he'd been sent to Topaz War Relocation Center in Utah, where he entertained everyone in *shibai*, Japanese for "revues," where he sang, told jokes, and served as emcee. He'd developed a kind of All-American persona, his accent almost New Yorkerish, perhaps fashioned after cinema gangsters, and his postures, whether sitting or standing, were languorous and slouchy like

an old hat or well-worn sofa. He was most comfortable, it seemed, in an easy chair in the corner of the apartment next to an open window, curtains periodically billowing behind him like the gigantic gills of a translucent carp. He'd light up a filter-tip cigarette, poise it like a corkscrew between his index and middle fingers, accept a sweating highball from the playwright, and hold forth with a story, modulating his baritone voice in gentle tones, confident everyone was listening.

Schlumpy as a tall bag of hotel laundry, he must have been handsome when younger. He had the supple moves of an ex-athlete (which he was), hooded eyes, dyed eyebrows and hair, and an incredibly relaxed manner of speech that reminded me of Dean Martin, his vowels plosive, the syntax of his long sentences sashaying like a trombone's, his head cocked to one side in dismay or disbelief at what you might say, or held ready for a double take. People laughed easily at his jibes, kept his drink filled as though he were at a casino table in Vegas, and asked him to tell stories about his days playing Chinese nightclubs in the Midwest. The tales were always about hijinks with the dancers and their costumes, a night when he dressed in drag and came onstage doing the can-can, or tried and failed to hit the high notes in "Begin the Beguine." Soo croaked like a frog when he told that story, standing up from his chair and striking an Olympian pose, his neck wattling, his eyes clamped shut. He mocked himself and made everyone love him. I could see that the role of Sammy Fong, the hip Chinese American nightclub owner he played in *Flower Drum Song,* wasn't at all a stretch.

But Soo never talked about that role that I could see. I sensed a tension about it between him and Pat Suzuki, the Japanese American jazz singer who was in that show too, playing Linda Low, one of the female leads on Broadway. Her showstopper was the tune "I Enjoy Being a Girl," but few recall it as hers because she was replaced in the film version by the non-singing though curvaceous Nancy Kwan. The lyrics are artifacts of a seriously different, unliberated time, and the point of the number was that this Oriental beauty was completely *American*—assimilated into the "healthy, red-blooded" gender values of the day. Today, the song seems parodic and is mainly performed by drag queens to hoots and squeals from audi-

ences geared to camp and gender-bending. It likely can't evolve much further than that.

But Suzuki herself, a middle-aged woman when I met her—my boss Pat Morita called her, cruelly I thought, "a washed-up chick singer"—was something of a chameleon. Short and petite when young and performing to big-time audiences and on hit television shows hosted by Lawrence Welk, Jack Paar, and Sinatra, she had thickened some, become matronly of figure, yet still possessed the pixieish face and quick, catlike moves of a veteran stage performer. She could curl up like a Manx in the corner of a sofa, bat her eyes, and accept a cold drink, staying silent while others told stories. But she sprang to her feet when reminded of a scene she'd performed, doing a quick dance routine all in a kind of deft pantomime, without saying a word, holding us spellbound with her movements, precise, showy, and done to the rhythm of a silent tune in her head. It was a kind of *t'ai chi* in invisible top hat and tails, as I recall, imaginary cane in her hand, a soft-shoe routine she executed, feet darting between the coffee table and a love seat, her hands like fluttering doves. The folks who'd gathered would call for her to sing then, and, after a few modest demurrals, she'd oblige, falling silent for a moment, finding her emotional focus, gathering energy for the rendition of the tune she'd chosen. One time, it was "In My Solitude," the stately ballad by Duke Ellington, taken slow, sensitive to the hidden drama only traced by its lyrics: "In my solitude, you haunt me . . ."

She sang it completely unlike the brassy show tunes she'd been known for, emphasizing the loneliness of its theme, stretching the cooing, long third vowel of "solitude," drawing it out with a light, glassy vibrato cool as sheet ice. Her elocution reminded me of Ivy Anderson, maybe Ellington's finest female singer. There was a theatricality to it, a powerful aura of poise and sophistication even as she slimmed a note down to a breathless whisper at the end of a phrase. And, when she sang *There's no one could be so sad* in a following verse, she gave the briefly elongated *a* in the last word a subtle, almost metallic sheen as though from a muted trumpet. When she sang *I'm craaazy,* the *a* had a slight, initially emphatic jump to it, then lilted off in a long, tailing decay as though cigarette smoke

from a detective in a film noir were being put to rest in a satin-lined casket. There was so much control of timbre and phrasing in the way she sang, an unmistakable mastery of style that testified to years of discipline and a pure love for the art of singing that Suzuki had practiced as a child in California's farm country, as a teen interned in Granada War Relocation Center (yet another government prison camp) in Colorado, and as a young woman performing thousands of consecutive nightclub gigs at the Colony Club in Seattle. When a low note came, it rose quaverously from within her throat, its sound sculpted as it passed through her mouth and then blossomed slowly like the bell of a morning glory in the stilled, summer air of the playwright's parlor. She had transformed our session of shallow frivolity into something supernatural—a spell that reached across years of solitude and devotion and sparkled briefly like evaporating dew on a flower.

A German Requiem

It was hard for my mother to accept what I'd chosen to do with my young life. She'd wanted a doctor, a lawyer, or an engineer, and was angry for the longest time that I'd chosen to study liberal arts and poetry and spend days on end listening to music, trying to catch the inspiration to write a poem. But when I was twenty-five, she relented—I think in a moment of tenderness that she had repressed with anger so much of the time—and bought me the gift of a Sony CF-2500 four-speaker cassette player capable of stereo and mono playback, depth of sound, and a fineness of resolution beyond the simple boom boxes of the time. It was a sleek but heavy unit, weighing just under twenty pounds, a kind of black ocean liner of a thing packed with dense electronics. It had lattice speaker grills; a superstructure of brushed and polished aluminum with push-button controls; three tiny top-hat, die-cast metal dials for bass, treble, and volume control; a telescoping antenna for catching AM/FM radio; and a thick, swing-away metal-and-plastic bar for a handle.

I knew where she'd gotten it too—at the narrow storefront of an electronics shop in Little Tōkyō, near L.A.'s City Hall, where

she worked. The place was basically a fold-down wooden counter in front of a long storeroom of Japanese import electronics in boxes stacked in racks and cubbies all the way back to the alleyway door. She'd taken me there once or twice when I was a teenager and had bought me my first transistor radio, a hand-sized thing in a black leatherette case. The man who owned it ran back and forth to Japan all the time, identifying and then ordering up good gear and hawking it to his regulars at deep discounts. My mother was one of those.

She sent it up to Seattle as a birthday present, and it became my musical companion during the itinerancy of my twenties, when I never lived anywhere for more than a year. In those days, I mostly played a lot of rock and was just beginning to explore jazz, but it was classical and chamber music that captivated me at times, the Sony cassette player somehow well suited to the tones of piano trios, string quartets, and, miraculously, even a few orchestral pieces. It was *sonorous,* its tiny speakers producing a wondrous midrange capable of carrying the acoustic nuances of vibrating strings, fine tonal changes of the human voice, and rendering, albeit in miniature, the scale and depth of an orchestra.

Nights alone in my studio apartment, I'd sit at the fold-up dining table trying to write while playing something that would afford me companionship without distraction. This would usually be the solo piano music of jazz virtuoso Art Tatum, full of astonishing off-rhythm improvisations that would curl back to the beat, but only after several measures of stretching out. Or I'd play John Coltrane's *Ballads* or the Budapest String Quartet performing Beethoven's Razumovsky quartets, two LP favorites from college that I managed to re-acquire in stereo cassette. As fine as all these were, there was one recording that created an unmistakable aura that both stilled my mind and stirred the deepest rumblings from my soul, cloudbursts of music that made soft rains of sound, filling me with a lapping ocean of peace that subsumed all the disquiet of the day. It was Johannes Brahms's *Ein Deutsches Requiem,* a recording by the Chicago Symphony Chorus and Orchestra, directed by Georg Solti.

Brahms composed his *Requiem* after the death of his mother in 1865, and that was about all I took away from the scanty liner

notes that came tucked inside the cassette. There was no information about the structure of the music, but mentioned were the various and intermittent dates of its composition. Brahms I knew mainly from the grand melodies of his Symphony No. 1 and Concerto for Violin and Cello. Aside from having attended a handful of concerts, I hadn't yet any education in Western art music, and what I took away from having heard its great compositions was merely bits of melody from Beethoven and Brahms and a loving, untutored response to its mystifying stirrings and provocations.

But the *Deutsches Requiem* struck me as something ceremonial and magical, a sequence of elegiac bible verses, emanating from a small, singing cloud that came alive on the little stage of the table in front of the Sony player. The voices of the chorus furled and billowed in air like the stately blossomings of a great tree of flowers, and I would find myself attending to every note as though each one marked not only the ephemeral, passing articulations of life, but the pure instances of their *becoming*—the swellings of a tremulous drop of dew, the sinuous skate of an albatross over an oncoming shoreline wave, the small cascades of grief as they shuddered through your body when mourning the loss of a beloved.

> *For all flesh is as grass*
> *And all the glory of man*
> *As the flower of grass.*
> *The grass withers*
> *And the flower falls away.*
>
> (Peter 1:24)

If I were trying to write a line of verse while Brahms's great music made a hearth of sound before me on my little dining table, if I had paused a moment trying to think of an image for the lone shrub of a stunted pine stuck on the craggy side of a trail in the Sierra Nevadas, or if I were simply getting up to boil a pot of water on the two-burner hotplate in the kitchen so I could make yet another cup of instant coffee for myself, choruscades of song and a pricking accent of harp strings would swirl over me, lifting whatever emotions I might have been struggling to describe or dispel into another

empyrean of feelings beyond my own imagining, notes from an oboe like the topmost curls of a wave cresting over me, tenors and sopranos achant with sorrow, mournful organ a sonorous undertow of pedal tones, and the smother of violins soothing the air of my small apartment with translucent strings of a showering light.

> *Blessed are they who mourn . . .*
> (Matthew 5:4)

It's My Life

In the fall of 1978, I'd just moved back to SoCal from Seattle where I'd failed three times over—as a novice playwright, as a graduate-student scholar of Japanese literature, and as a steady man for any of the three women I thought I was in love with. I ran from them all and started the M.F.A. program in poetry at UC Irvine, where I'd been admitted and given a sweet, non-teaching fellowship for the first of two years.

Since elementary school, I'd always written poetry, and thought privately that a poet is what I wanted to be, but could never commit to being disciplined about it, could not reach beyond thinking of it as a kind of foolish, emotional indulgence, and an inward passion and wistful avocation. But failing three times over by the age of twenty-seven had thrown me back into solitude and renewed my passion for poetry. With the little money I had, I'd bought a *Complete Poems of César Vallejo, Memorias* by Pablo Neruda, translations of Swedish and Spanish poets, and the poetry of Charles Wright, a contemporary American poet from Tennessee.

I read all of these as closely as my father might have studied his electrical circuits, trying to feel the emotional sense of them and the new worlds they introduced me to—faraway worlds that obstructed lyricism with political brutalities, sparse inner landscapes rife with a sere and constant wishing for transcendence, and, in the case of Wright, the rich and florid landscape of Tennessee as a backdrop for a sweet grieving for the evanescence of things, spidery hints of ascension in the glinting dews of morning. The works all moved me, grooved their feelings into a mental rhythm being born within

poems by CHARLES WRIGHT

Bloodlines

my own dim imaginings, and I was inspired to apply to study how to write my own poetry under Wright's tutelage down in Irvine, only about an hour's drive from where my parents lived in Gardena.

What I read in Wright's poetry was a style taken from the T'ang dynasty Chinese, scholar-poets gathering images from the outer world in nature, then pivoting to the heartache within, a sudden flash of emotion about the loss of home, children, friends, or a life of peace in exchange for wandering. These ancient Chinese poets had inspired me early on, ever since my studies as an undergraduate, and I'd written some early poems in imitation of English translations by poets like Kenneth Rexroth, Witter Bynner, and Ezra Pound. What Wright did in his poems was, in more condensed and elegant diction than theirs, what I thought I wanted to do. He understood the Chinese and joined it to his own background as a Southerner brought up within a different landscape, with spiritual worries from his own faith as a Christian, with a precise hold on reticence, decorum, and lyric rapture. His poems were beautiful—

sometimes lavish, sometimes perplexingly cryptic—and I needed to learn from him.

At the same time, Wakako had become my *kumu mele*—my teacher of song and spirit. Staying in the small cabana house adjacent to her pool, I'd spend evenings listening to her stories and put questions to her about the first generation, what it felt like to be suddenly evacuated from her home as a child and thrust into a relocation center guarded by soldiers, what it was like in camp, how hard it was to come back to freedom and try to retake her life after the war. She told me *everything*—about the privation and isolation of prewar farm life, about the hidden artistry among some of the farmhands (one guy could play the violin, another the piano, both charming her and inspiring her written stories), and her own pre-teen wonderment staring out across farmlands at cathedrals of thunderheads rising in the distance. She revealed to me the social history of Japanese Americans and unlocked a trove of emotional lore that had been denied me through strict cultural and familial silences. The Japanese American community wanted to forget and move on. And the prevailing culture around us wanted nothing more than to erase that history too. But Wakako remembered it all. She remembered sitting among other teenagers working on the camp yearbook. She said someone had a record player with a stack of records that they'd sing along with. In a quaverous voice, she'd sing something like "I Remember You" or "Don't Sit Under the Apple Tree (with Anyone Else but Me)," then recall a *saké*-drinking song in Japanese, infusing her notes and the Japanese lyrics with a mordant sorrow that made me weep. She sang to me and told me who we *were*.

During that first month of my M.F.A. study in poetry, I'd drive from Wakako's house in Gardena to Irvine twice a week. I had an old Toyota Corolla painted taxi yellow that I'd bought a few summers before with money I'd won for a poetry prize in Michigan. While I was in Seattle, my father had used the car commuting to work to save on gas. He gave it back to me, and it became the beater I used to make my own commute. It was basic—stick shift, no A/C, AM-radio only, spare in the trunk, the body stubby and chassis light for mileage. To get to Irvine in the L.A. heat, I'd drive

the 405 freeway with the windows down and the radio turned way high to a Top 40s station.

Billy Joel's hit "My Life" came on a lot back then, and it was like a springboard to my own insouciance and lack of care about my past failures. His song had an infectious jump to its rhythm and an insolence to its lyrics (*I don't care what you say anymore, / This is my life . . .*) that freed me and gave me a puckish courage that carried me through the first awful poems I wrote in Wright's workshop. I wrote about R&B howler Wilson Pickett as a surf instructor in Waikīkī. I wrote about evening floodlights on Wakako's swimming pool penetrating its blued waters as though in a painting by David Hockney. I wrote about a Japanese stand-up comedian insulting his mainstream audience with slurs. I wrote from a mixture of cowardice and resentment, and affected a stance of cultural defiance that added up to little more than fakery.

But I told myself I didn't care, that my life was renewing itself, that I might soon belong with the poets I so admired, Wright and the Chinese among them, and discover what it was I had to say and how I had to say it. Billy Joel had the right idea. Sling it and sass it, *Whooo . . . yeah.* The bouncing left-hand notes of Joel's piano doubling with a driving electric bass set the pace of the tune, a chattering high-hat, then a drum roll and snare shot, and then Joel's five brisk chords and their catchy repeat, followed by his tinkly, taunting right-hand riff and the soft, rhythmic cooing that introduces the tune. The song told the story of a guy who split on his old life and took off for another, a kind of misogynist spitting on someone he'd left behind, telling her to get lost, that he's got his own life now and it was headed somewhere undeclared and likely alone, but nonetheless without her and the life that used to be. It was about not only a breakup, but a guy taking a chance on himself, on a new path, wanting something different.

I didn't really know this then, but I'd taken this song to heart, turning the cheap car radio up almost to full blast every time I heard the bouncing notes of his intro come on, whatever DJ nonsense was layered over the music, and singing along with its lyrics only half-grasped, scatting and mumbling notes to the words I couldn't make out. Windows down, the monoxide-hazed air of the

405 blasting through and rippling over the chest of my T-shirt, I sang it as though it were my own song, an anthem for myself as a young artist full of fear and the half-thought I was a phony. *Leave me alone*, Joel sang, and I said it too, spitting on my failed past and turning the yellow Toyota compact into the southerly corona of my new life, owning nothing but its own sheer sass.

Part Six

Wandering
Rocks, 1

Through the years, a chronicle of perfect sounds:

November 2006

The earth-shuddering notes of "White Room" jumped from the speakers and I could tell things were going to rock and rock hard. Eric Clapton's overdubbed guitars wailed like banshees over Ginger Baker's ominous tomming and Jack Bruce's under-drone of bass. When the trio swung out of the intro and into the main part of the tune, there was a sticky slap and satisfying crunch to Baker's stickwork on the snare, rapid-fire and deeply resonant impacts to his drumming. Clapton's wah-wah lead squawked and screamed like he was hooked up to a Marshall amp sitting in my living room, exciting and shoving invisible waves in the air. Baker's kick drum boomed and stomped like a Clydesdale stuck in its stall, itching to rumble. And Bruce's bass propelled them both along from underneath, locked to the rhythm of Baker's kicks. When Clapton quivered some chords with his tremolo bar, it felt like he reached out and grasped the interlocking bones of my skull and spine from the inside, shaking me from the marrow out. Bruce's baritone voice drove the beat and Clapton's falsetto sailed like a light schooner over his own liquidinous chording.

I'd just hooked an old solid-state amp I'd been given onto speakers I'd bought the year before. Those speakers had never worked well for me. They were Tyler Acoustics Taylo 7U towers, three-way floor-standers that I'd bought from my old friend Peter down in Laguna Beach. About a decade old but still handsome in oil-stained cherrywood cabinets, these were his own, personal pair he'd bought from a maker in Kentucky. He'd offered them to me for what new bookshelf speakers from the same manufacturer would have cost. The bookshelf units were much smaller and,

I'd figured, less capable of creating a large, satisfying sound, so I'd jumped at the deal. And I'd bought what I thought would be a great match for them—the gleaming Electrocompaniet ECI-3 integrated solid-state amplifier. But, after weeks of trying, the rig just hadn't worked at all.

A piece of audio equipment, no matter how good it is intrinsically, needs to be used in a way that its characteristics demand, respecting its operating parameters, and in conjunction with other gear appropriate to its technical features and limitations. But to a novice, it's hardly ever obvious how to accomplish that, especially when operating by chance or attempting to emulate carefully cultivated advisors, throwing money down for gear you're clueless about incorporating into an evolving system. Too many times, you just don't know how. Speakers have impedance curves and ohm loads to consider, an amp needs to have its input impedance matched to a preamp or source (say, a CD player) that it takes signal from—facts little understood by a stumbling beginner such as I was. An audiophile will inevitably indulge in various follies, upgrading and tearing up his system many times in his odyssey for the perfect sound. This happened to me a lot.

Impedance, measured in ohms, is here the speaker's electrical resistance to incoming current, measured in wattage, carrying the audio signal. This impedance varies depending upon the frequency of the incoming signal, presenting more resistance (and a moving electrical target) to different ranges of the audio band at different times. In a complex signal with a lot of treble information such as operatic or choral voices, the speakers I bought from Peter proved very difficult to drive for the ECI-3, an amp with an output far lower than the speakers needed. On a CD of Renaissance choral music sung by Chanticleer, my new audio combo sounded like shit spit by a raven. Though small group jazz sounded okay, smooth and supple, rock was seriously wimpy and orchestral music just crapped out. Violins sounded scratchy and sour, horn fanfares grating and edgy.

Discouraged, I put the speakers and that amp aside and started over, assembling a completely new system with a tube amplifier and the speakers built in Italy. I put the new system in my study, where

I listened and worked every day, and left the failed one in my living room, where it got played hardly at all.

I'd gone through scores of different components and at least a dozen distinctly different systems before I thought I'd found my Ithaca in audio, as it were. I'd certainly spent good money for equipment inappropriate for what I was trying to match it with— either other gear or else a kind of perfect sound that was in my head. These were audio versions of Nausicaä, Circe, and Calypso, each of them bewitching in their ways. But meanwhile, I'd gotten smashed between whirlpools and wandering rocks of gear, time and again, fleeing audio failure, a cave-dwelling Cyclops who tossed boulders at me as I withdrew back to my own ocean of sound.

About a year after I'd abandoned Peter's speakers and left them idle in my living room, my friend in Carmel sent me, as a gift, an impressive beast of an amplifier. It was a solid-state Aragon 8008-ST, then about ten years old (*forty* in audio years), a beat-up black cube of a thing with a row of ventilation ribs and heat sinks at the top, making it look like it had metal gills. And it had the design feature of a V-notch along its top right edge, giving it a frontal profile as if it had black hair, sharply parted and styled with product. My generous friend had it lying around the house unused, so he just got a notion one day, packed it up, and sent it to me in a big UPS box stuffed with Styrofoam peanuts. Weighing nearly seventy pounds, the amp was missing some screws on its top plate, had scratches all over the anodized finish, and needed a fuse. Faithfully, I sized and replaced the screws, polished it up as best I could, and put a fuse in it.

What I didn't know was that it was blessed with some very special Japanese output transistors then famed for their clarity and power, a rare smoothness, and a warmth of sound that was still uncharacteristic of solid-state amps when it was released in 1996. The Aragon's output was 200 watts per side (nearly three times the power of the ECI-3) at 8 ohms, doubling to 400 watts at 4 ohms, which was exactly the impedance load of Peter's old speakers—that thing I was barely conscious of at the time. I simply did not understand these technical matters. I was just "chasing," as is said among audiophiles, headlong in the race but distracted by desire like Atalanta by the golden apples of Hippomenes.

Almost immediately, I got the idea to try the Aragon with the Taylo 7U speakers. I moved the amp to the living room, hooked it to my chain of electronics, picked out a CD, and fired up the system. I'd chosen the classic acid rock album *Wheels of Fire* by Cream, the sixties supergroup of Clapton, Bruce, and Baker. Their sound was psychedelic. I starting laughing in delight and relief, dancing around the living room like a teenager. The formerly failed speakers just rocked the house, finally being fed signal by an amp they could really use. To no one, I kept shouting over the music, *This is how it was supposed to be all along! Like this! This fucking rocks!* And, without companions, sonic wheels of fire bursting in air around me, I slapped five with an invisible friend.

System: Tyler Acoustics Taylo 7U speakers; Aragon 8008-ST stereo amp; Rotel RC-995 linestage; California Audio Labs CL-15 CD player; Audience Au24 speaker cables and interconnects; PS Audio power cords; Pottery Barn console used as an audio rack.

January/July 2007

Going to an audio show can be something like visiting a museum with a splendid collection of paintings, each with its attractions and emphases for the eye's pleasure and stimulation. Occasionally, wandering the galleries, you come upon something that changes your way of seeing—it expands an aspect of perception in a way you hadn't before thought possible. Picasso's paintings had this effect when they were first unveiled—a cartoon eye weeping bullets of tears out of a caricature of Dora Maar's long profile, hair like a black waterfall, a sausage nose, mouth like the keys of a toy piano. Or maybe a luminous landscape catches your eye—the peculiarly angled way an intensely yellow wash of sunlight strikes a humble row of haystacks in a newly mown field in a painting by Van Gogh. You look and you look and you look. Somehow life has been rearranged. Grief has a face. And joy its portrait. In Las Vegas in January 2007 for a pair of audio shows, I came upon speakers that affected me that way.

It happened when I was running between the Consumer Elec-

tronics Show at the glitzy Venetian Hotel and Casino, the industry's annual showcase event, and T.H.E. Show, a companion exhibition for smaller manufacturers at the St. Tropez Hotel. Excited to be there, I'd been assigned to write a show report for *Vacuum Tube Valley*. I must have visited a hundred rooms in three days, giving myself blisters, taking notes on every sound system that I heard, bumping into acquaintances, eating on the run, and feeling dehydrated all the days long (water fountains were few and inadequate for all the heat thrown off by amplifiers in every room).

In a small room at the St. Tropez, I came upon a fairly modest demonstration put together by Randy Bankert of O.S. Services in Southern California. His Sonist Concerto 2 was a new speaker then, designed after French and Italian monitors that could be powered by the low watts of single-ended triode amps. After years of acting as distributor for European speaker lines, Bankert had decided to make his own speaker, using characteristics and specs he'd long been suggesting to manufacturers.

"I took the best from what I heard—the emotion of the Reynaud, the transient response and dynamics of the Loth-X, and the creamy tonality of the Zingali—and tried to combine them," Bankert said, mentioning perhaps the leading French, English, and Italian makers of speakers sensitive enough to use with low-watt tube gear.

I was listening to his Sonist Concerto 2 speakers, a pair of two-way monitors, then driven by an Italian integrated amp that used 300b output tubes. Their sound was a rare blend of sweet and articulate extension into the extremes, but with a warm midrange.

In late summer, Bankert sent me a pair of these speakers to review (I'd since become a writer for *Ultra Audio,* an online audio magazine), and I was loving their warm sound on blues by

Albert King, rock by Stevie Winwood, and vocal albums by Sarah Vaughan, Kiri Te Kanawa, and Kathleen Battle. But it was Mingus's *Changes One* that really knocked me out—the album by the great jazz bassist, bandleader, and composer. In his compositions, he blends Ellington-like lushness with field hollers, political sloganeering, and the rhythmic cacophonies of Dada and free jazz. But Mingus never forgets the beat, never fails to reach for the gorgeous in harmonies, the sublime in melodic line.

Changes One had been one of my favorite albums ever since I heard Mingus and the band he called the Jazz Workshop play tunes from it at the Bank, a short-lived but essential underground dive in Seattle's Pioneer Square district, early in the spring of 1978. On three successive nights, his sextet knocked the bricks out of those basement walls. I've owned the LP and the CD, and often ran an imaginary DVD of those performances in my mind. It's music I've relived time and again, taking images from those three nights and relishing the blend of jazz and sonic theatrics, George Adams preaching, howling, and striking sonic poses on his tenor like a kabuki actor strutting down a runway through the audience.

On the CD, the band is a quintet, and the tunes run juxtaposed into each other—beautifully romantic melodies, such as "Sue's Changes" and "Duke Ellington's Sound of Love," alternating with "Remember Rockefeller at Attica" (a reference to the deadly quelling of the prison riot there in 1971), and the rawboned "Devil's Blues." With Mingus's bass thumping and humming along the bottom of every run and solo, it's music that is at once gorgeous and savage, like the novels of Herman Melville or Tuscan cuisine—jellied marrow served in a crystal flute alongside a beefsteak bathed in a platter full of its own warm juices.

His band played "Duke Ellington's Sound of Love" as if the All-American ghosts of jazz were having their musical reunion inside the walls of my study qua listening room—Mingus pegging his bass through my indigo Chinese rug and into the floorboards below, Dannie Richmond setting up his kit in the corner by my art books, Don Pullen comping and filling in front of the bookcases crammed with poetry. And the best—George Adams on tenor and Jack Walrath on trumpet harmonizing the theme written in a style

borrowed from Ellington, the two horns coming from center stage over my desk (which sat between the speakers), the long line of the tune billowing out like a sail of solitude to my heart hungry for the soul of this song. When the entire quintet kicked back in, Mingus plucking out the slow waltz of the tune like a nightingale on bass, the horns drew a lazy river of trailing harmonies across the music's melodic surface.

Did I think of the speaker's ribbon tweeters? The big Cyclopean eyes of the Concerto 2's paper-cone woofers? The cherry finish on its poplar baffle? Do I think of vocabulary and symbolism when tattooed Queequeg hoists a harpoon and sights on the arching back of a breeching sperm whale as I turn the pages of *Moby-Dick*?

System: 300B SE integrated amp; Marantz SA-11 S1 SACD player; Sonist Concerto 2 monitor speakers; Audience Au24 RCA interconnects, speaker cables, and powerChords; Finite Elemente Signature Pagode five-shelf rack; FIM amp stand; Aliente speaker stands.

October 2007: An Analog Demo

I'd been visiting my friend George Radulesk every few months since we'd first met, when he invited me to hear an exotic Japanese amplifier he hoped to sell. He is a man my age from Cleveland, Ohio, who had moved to Portland, Oregon, to work as a statistician with the regional power company. We'd gotten to emailing each other, at first intermittently, every few weeks or so, then almost every day as we discovered the depth of our mutual passion for all things audio. George knew so much more than I did and could advise on most anything I asked about, from where to get small screwdrivers and Allen wrenches to how to control the little vibrations emanating from my electronic equipment that, I discovered, smeared the audio signal and put a soft haze of obscurity over what might otherwise be pristine, sharp, and more thrilling to hear. He'd told me about little polymer feet, carbon fiber shelves, ceramic cones, and myrtlewood blocks that I could place under the CD player or preamp, "cleaning" the system of the tiny vibrations that took away audio clarity. In actuality, I found out later, each of

these things represented slightly different approaches, but what was important was that George had a major repertoire of strategies and familiarity with equipment far beyond mine and was generous in sharing what he knew. That we listened to different music, George mainly to jazz and me to classical, wasn't as crucial as our daily exchanges about the hobby.

One time, likely my second or third visit to his Portland home (a three-story manse in the hilly, tree-lined suburbs), he treated me to a session with LP playback that took my breath away. His analog chain started with a Swiss moving-coil phono cartridge with a relatively low output voltage. Then came a tonearm and turntable made in England, its motor an outboard and heavy thing a few inches away from the turntable's plinth made of black phenolic material. The platter was massive and weighed almost twenty pounds all on its own. To start the whole thing, you had to give the platter a gentle push, as the motor was designed to maintain a steady speed rather than have the torque to overcome inertia and start the platter spinning. It was an eccentricity that helped create an aura of specialty. Cables ran from the back of the tonearm and connected to a black rectangular box with a chromed faceplate on it that read EAR for "Esoteric Audio Research," it, too, made in England. This was the phono preamp, and it took the tiny voltage feed from the cartridge through the tonearm's phono cables and amplified it so another preamp could amplify it further and send it along to the power amps in George's system. The power amps were two separate and gorgeous pieces of tube equipment, one monoblock for each stereo channel, each a small boat of audio beauty emanating its own amber and magenta lights.

From his collection, sequestered behind the sliding doors of a sixties-style console at the side of his room, George slid out an old favorite of mine from the mid-fifties, *Bill Evans at Town Hall*, and walked with it across to his audio rack. He slid the record out and placed it on the turntable, giving the platter a soft push to get it spinning, and then dropped the stylus of the cartridge into the outer groove of the record. Accustomed as I was to digital sound, what I heard then astonished me. I'd had the CD for a number of years and had played it in regular circulation for just about forever. But this

was sound of a completely different order—it was so organic, warm, luscious, bountiful in harmonics, and achingly slow in its trailing tails of decays, palpable in air. I felt the softly chiming brass of cymbals brushed and stroked as though with metallic thistles by drummer Arnold Wise. The resonant bowings of Chuck Israels's bass vibrated as though it was in the room in front of me. Evans's piano trilled and tinkled with sprightly touches, and complex chordings played crisply, yet holding notes for thrilling milliseconds in the soundstage before us. And when the tune "Make Someone Happy" came on, his piano felt so real I thought I could just rise from my seat, duck under the soundboard of the piano, a grand sonic image before me, and dance among the hammers and strings with joyous, sashaying feet. After that, I had to have a rig of my own and resolved to set about assembling it.

Within a few weeks, I made an online deal with a man retiring from the Merchant Marine in California's East Bay. He sold me the same English turntable and arm as George's and threw in a good Japanese stereo cartridge too. Then George offered me his EAR phono stage and I jumped at it. I was all set.

George's system: Benz Micro Glider SL MC phono cartridge; Nottingham Spacedeck turntable with nine-inch Spacearm tonearm; EAR 834P phono stage; Joule-Electra LA-100 Mk III preamp; Joule-Electra Heaven's Gate monoblock power amplifiers; Verity Parsifal Encore speakers; Cardas Golden Cross RCA interconnects; Signal power cord to phono: MIT Z-Chord II power cable to preamp; AbsoluteCables speaker cables; Synergistic Research Power Three power cables for amps; Zoethecus five-shelf audio rack; Zoethecus amp stands.

June 2008

I was down underneath the floor of my study, squirming around alternately on my hips, ass, shoulders, and belly in a narrow crawl space, trying to jack up the crossbeams of the suspension to stiffen things up so the floor wouldn't vibrate anymore. I'd gotten five twelve-ton hydraulic bottle jacks from Harbor Freight, a half-dozen concrete pier blocks from Home Depot, and another half-dozen

squarish wooden blocks rough-cut from a two-by-four to do all this. The heavy pier blocks got shoved, rolled, and shimmied into place, barking my knuckles, each positioned under a separate crossbeam, and the bottle jacks martialed on top of each one of those. The blocks of wood were to spread the pressure under the beams from each of the jack heads as I pumped them up. I was told to do this by the guy in Memphis who made my custom speaker wires and exotic interconnects from copper foil.

I'd told him about my current audio problems—that I couldn't play Mahler's Second Symphony on my new turntable because something was making my speaker woofers pump like crazy and the stylus jump from the record groove, sending the tonearm skittering across the surface of the prized Decca LP of Georg Solti's recording with the London Symphony Orchestra.

A few days before, anticipating sublimity and great gravitas from my new analog system—an English-made Nottingham Spacedeck turntable and tonearm, an EAR 834P phono preamp (also English), and the Shelter 501 Mk II moving-coil cartridge (Japanese, vaunted but affordable), all carefully chosen—I had run into quick disaster instead. No sooner had I dropped the needle onto the groove of the record than something mysterious and malicious started making the audio rack shimmy like a clapboard house in an earthquake. I had barely heard the LSO's double basses ripping through the opening bars of the magisterial theme to Mahler's piece, dubbed *The Resurrection,* when all hell broke loose. The bass woofers of my speakers pumped rapidly in and out, the ass-end of the tonearm jittered, and the needle jumped the record track like a subway cheat leaping a turnstile. When the first bass drum *thwack* capped the first orchestral

crescendo, the energy coursing through the tonearm to the stylus needle made the whole rig bounce completely overboard. This
happened not only once, but every time I started the needle in the
groove on the edge of the LP. I'd changed the stylus's tracking force
from 2 grams to 1.8 grams to 1.5 grams, a gradual diminuendo of
downforce pressure every time. Same result.

Technically speaking, though I didn't know it right away, what
I had was a feedback loop caused by a tremendous amount of subsonic noise trapped and recirculating within the audio system. This
likely came from a slight rumble in the turntable's bearing as it spun,
which then got picked up in the electronic amplification chain,
sent through to the suspended floor as vibration, and then picked
up and re-amplified until the woofers got involved, pumping rapidly, themselves adding subsonic vibrations, until the whole chain
produced so much energy that it charged through the cartridge
stylus in a flood of silent attack sending the needle leaping out of
the record groove. It's a phenomenon not all that rare in an analog
system (i.e., one based on records rather than CDs) that isn't put
together with appropriate care for what's called *isolation,* meaning
the proper insulation or draining of extraneous noise and vibrations
from components. You have to either isolate or redirect vibrations
so that they don't build up and get amplified within the system in
the kind of feedback loop that I had going. You can do this either
by using compliant footers under each component (isolation) or else
by channeling the energy through to a "ground" not unlike the
one needed for electricity itself. What I was trying to do by jacking
up my floor was stiffen it so that it would be a better conduit for
vibrations to exit through its support Y-beams to the earth below.
Instead of flexing and bouncing energy back up through the metal
footers of the speakers and component rack, I wanted my floor to
drain its pent-up audio energy literally "to ground."

All was explained to me by my distant guru in Memphis, a
former sales manager for a nationally known speaker company.
I'd gravitated to him quickly, corresponding frequently after I'd
ordered my cables, and he'd willingly shared all of his experience,
telling me about cartridges, speakers, and little technical things in
audio. He said I could stop the build-up by trying any of three

methods. I could get an electronics gizmo called a "rumble filter" that would take away the initial subsonic signal, treating the problem at its root. I tried this and not only didn't it work, but it made the system sound lousy—which he'd predicted. Or, I could get a wall-mounted plinth for my turntable, bolting it to a stud in the wall of my study so the subsonic energy might drain directly through the structure of the house. But I couldn't do this, as I didn't have the space to give up to such a platform. My study was already jam-packed full of desks, bookshelves, and audio gear. So, I went for the third method—reinforcing the suspended floor from underneath.

The day I went spelunking in the crawl space under my house, there were little clods of dirt that busted up beside me as I twisted my torso so I could lift my shoulder into a better position to pump the first bottle jack. I heard the long crossbeam, likely cut from Douglas fir, creak a little—giving out a small, wooden grunt as I jacked it up. I went from beam to beam this way, dragging a little battery-powered lantern alongside me, shifting on my elbows and scrunching my shoulders flat or to the side, blinking away the tiny motes of dust that fell onto my eyeballs as I worked. I thought of Michelangelo hanging from a sling below the ceiling of the Sistine Chapel, his guts shoved to his chest, his paints spattering his face, his boiling rage at the pain damped, his art of minutely incremental planning so finely, patiently, relentlessly executed. No, not Michelangelo. That was too much for me to claim. I'd have to give it a think.

Once back up in my listening room, I fired up the system, feeling apprehensive still, not knowing whether all my ministrations underneath the floor would have done any good. I pushed the turntable's heavy platter to give it a softly spinning start. I picked up the tonearm and dropped the cartridge needle into the lead-in groove. The familiar *shussshh* came on and then I heard the stirring sound of violas humming like a swarm of bees. The ripping bow-strokes of the double basses followed, and Mahler's *Resurrection* symphony came alive in monstrous swells of sound. I let out an exclamation of pure delight. The bottle jacks had worked, killing the feedback loop. The woofers behaved. And the needle stayed in the groove.

That night, I had the flattering picture in my mind that I was

a kind of virtuous aquanaut toiling under the floating island of my own Venice, tending to the underwater pillars that were the foundation of my humble, ranch-style villa above. I described what I was doing in emails to a few friends, calling the adventure with practical mechanics and home improvement "my own private Serenissima." They wrote back indulgently, knowing I had become, in my late middle age, a geekish and genial slave to my audio hobby.

System: Nottingham Spacedeck turntable with nine-inch Spacearm and Shelter 501 Mk II MC cartridge; EAR 834P phono preamp; Thor Audio TA-2 Mk III preamp; Air Tight ATM-2 stereo amp; Analysis Plus Solo Crystal Oval, Verbatim, and Atlas Questor interconnects; Verbatim speaker cables; Fusion Audio's Venom and Predator, Cardas Golden Reference, and Harmonix xDC Studio Master power cords; Sonus faber Grand Piano Home speakers; Finite Elemente Pagode Signature five-shelf audio rack; sound panels from Acoustic Sciences Corporation.

August 2008

One reason to enjoy the wonderful paintings of the French primitivist Henri Rousseau (1844–1910) is that they present a world as if one were suddenly just given the gift of sight. There are boldnesses of outline, saturations of color, and a magical quality in which each flower and leaf in the painted landscape stands out from every other thing as one gazes at the whole, though they make of their imagistic collective a grand gestalt of details deepening into splendor. Rousseau painted numerous fanciful jungle scenes—dream tableaux of a lion nuzzling a sleeping traveler in the desert, a dusky flutist charming a boa amidst surroundings of Jurassic foliage, a Parisian woman in her Sunday dress strolling through a Hawai'i-like rainforest of exotic plants. Though we may have seen such in a flower shop, never are their colors or phantasmagoric auras so sensual as when we see them depicted and placed as elements in a landscape by Rousseau.

I think of the best listening in this same way: when I hear a musical passage, it can be as if for the first time, my previous experience of listening to it renewed, then suddenly displaced by a recur-

rence that makes my memory of having heard it before evanesce as
the more beautiful notes of this present listening trail out over the
air, striking new chromatic registers. The harmonies are deeper and
more complex each time, and the melodies are as new footprints on
a beach that, just a moment before, had been washed smooth by a
wave.

I'd been saving the *Romantic Warrior* LP for the moment when
I thought a new phono stage I'd gotten—a Herron VTPH-2—had
been run in sufficiently to give it a fair listening test. I anticipated it
would sound fast, well timed, and tonally and rhythmically coherent
with such technically precise electronic music. Back at the height of
the jazz-fusion movement of the seventies, two supergroups were at
the top of the heap: Weather Report, led by Wayne Shorter and Joe
Zawinul, veterans, respectively, of the Miles Davis and Cannonball
Adderley quintets; and Return to Forever, led by keyboardist Chick
Corea, a veteran of the group Davis had assembled (and which also
included Shorter and Zawinul) for *Bitches Brew,* the 1970 landmark
album that broke fusion jazz/rock into the mainstream. Corea made
five albums with various lineups of side musicians before he came
up with the group that sealed his reputation: classically trained funk
bassist Stanley Clarke, acoustic and electronic drummer Lenny

White, and flamenco refugee and jazz-conservatory-trained guitar-ist Al Di Meola. *Romantic Warrior,* released by Columbia Records in 1976, pretty much set the standard for this music.

Once I dropped the needle in the groove of the record, at the start of the first track, "Medieval Overture," Clarke's propulsive bass line launched the entire ensemble forward into the score. It was as fully articulate and chest-thumping a slap bass as I've heard—a combination of Funkadelic and *Bitches Brew.* It met the challenges of pace, rhythm, and timing—White's soft-thumping kick drum, splash cymbal strokes, and rapid passes on the floor toms all rooted to the repeated figure of Clarke's bass line. Further into the piece, Corea's Mini-Moog ostinato was a revelation of what seriously good audio equipment can do, as notes whirled from channel to channel as if on a carousel, his arpeggios as liquid as guitar runs. And after Di Meola's guitar stated the major theme, backed by Clarke's pre-cise, contrapuntal bass, both flew off into dueling solos—the notes rippled furiously through my listening room like the animation in the "Sorcerer's Apprentice" segment of Disney's *Fantasia.* After they dropped away, White's drum solo was full of depth, multiple sur-face strike tones, and just plain, raw speed.

About two thirds into the track, a deep, painterly, unfret-ted moan rises from Clarke's special Alembic bass (equipped with Instant Flanger)—a dark, shadowlike presence that flickers through the music's current like a Balrog's malicious roar. Through the Her-ron phono stage, it was at first almost subsonic, then rose swiftly from the infernal depths to create an immense backdrop for the entire soundstage. These electric-bass notes are subtle, refined, almost light in character in comparison with those of a plucked or bowed string bass, but sublime and a touch dreadful for their immensity across the soundstage. In general, the imaging was larger than life—instruments outsized for their amplifiers, but tight, well balanced, and precisely placed in relation to each other. To me, this was electronic chamber music, vigorously played, both a spiritual anagoge in itself and a musical forerunner to the Kronos Quartet playing acid rock on classical instruments. My new phono stage was up to every bit of it, on time all the time, generous in its tonal presentation without being overripe or fat, rapid with entries and

exits, as precise as Corea's New Age compositions themselves. The phono stage returned my listening to a primitive passion, allowing me to sit back in my seat as a gentle Rousseauvian jaguar padded soundlessly out of the dark woods of my discontent, bringing to me its sublime kiss amidst a bountiful dreamscape full of the most glorious things in music.

System: Herron VTPH-2 phono stage; Nottingham Spacedeck turntable with nine-inch Spacearm and Shelter 501 Mk II MC cartridge; Thor Audio TA-1000 Mk II preamp; Air Tight ATM-2 stereo amp; Analysis Plus Solo Crystal Oval, Verbatim, and Atlas Questor interconnects; Verbatim speaker cables; Fusion Audio's Venom and Predator, Cardas Golden Reference, and Harmonix xDC Studio Master power cords; Sonus Faber Grand Piano Home speakers; Finite Elemente Pagode Signature five-shelf audio rack; sound panels from Acoustic Sciences Corporation.

September–October 2008

Kara Chaffee leaned over the exposed interior of the chassis of one of her KE50A amps, full of varicolored and carefully routed wires, and pointed the tip of her soldering iron over the frail, bare wire end of a resistor, a tiny piece in the amp's circuitry, this one as transparent as a jellyfish and nearly invisible. Then she pulled the tip back, touched it to the end of a coil of solder lying on my kitchen table beside the amp, and pointed it back in again, touching it to the other wire end of the resistor. She'd pulled a vintage carbon composition resistor out of the circuit and was replacing it with one of the new foil resistors from Texas Instruments. This, we hoped, would adjust the sound, pull it away from the fuzzy, thickened tone I'd heard in the midrange, particularly on violins and voices, and opening things up so both would sound clearer and without that barely discernible but opaque scrim I'd heard as a smearing of resolution, obscuring not only individual notes, but the "travel" from note to note, the trail of movement between them. I was locked into the sound I wanted, and so Kara hoped to incorporate it into the capabilities of her new amplifier design.

We were "beta-testing" various small, final changes to the amp she'd dreamed up based on the 1954 Fisher 50A amplifier, which had once powered a console stereo system meant for the living rooms of upper-middle-class American consumers. It was a mono-block, meaning there was an amp for each side of a two-channel stereo system, with discrete power supplies and individual casings, each driving only one speaker in a stereo pair. She'd come up with the idea after restoring a beat-up, vintage pair for a friend, reading around first in old audio catalogues and then in a handbook of published circuits until she had a grasp of its qualities and approach, and eventually seeing the design as one that might produce exactly what I was after in my quest for the perfect sound. And we were near the end of our many ministrations, having already settled on the final tube complement (two Gold Lion KT88s, one Amperex 12AU7 for the first gain stage, two RCA black plate 6CG7s, two General Electric 6CL6s, and two National 6CK3 rectification tubes) along with the capacitors that sounded the best (cheaper Jensens over exotic Mundahls), and we were now into the finer, inner delicacies of the design.

I'd become friends with Kara after asking for a listening session, which she granted, at her home in Vancouver, Washington, one late fall afternoon in 2007. I'd first met her a year earlier at an audio show in Las Vegas, visiting her exhibit room and finding a memorably rich, yet sparklingly detailed sound that came from a pair of her monoblocks designed around the Shuguang 845 tube, a contemporary copy of a huge triode invented by RCA in 1931 (said one of my tube mags) for AM radio transmitters. The tube made for a powerful, room-filling stereo sound that was clear, rhythmic, full of funky impact, with treble extension that was liquid and quite supple on the solo guitar music Kara was playing. I took note, and another year later, discovered she lived barely over a two-hour drive from my home in Eugene. We arranged for a demo of her stereo version of the amps that I loved. Then, driving her way south to California to visit her sister that Christmas, she stopped in Eugene and dropped the amp off for me to try in my own system.

The amp sounded glorious on combo jazz music, creating a flowing sound that was like water in an onrushing tide spilling over the outer reaches of a reef that sheltered a lagoon, filling my listening room with a lusciously romantic sound that reminded me of my father's system. I reveled in the romance of its presentational style—a slow sinuosity in the music so long as I didn't play anything taxing. It sounded superb with recordings by Miles Davis and John Coltrane, their instrumental solos burnished with an attractive sheen. It sounded great with Sarah Vaughan, rendering her way with Ellington tunes, rapturous and aching. But on orchestral music like the First Symphony of Johannes Brahms, it lost punch and resolution in orchestral tutti, my fairly conventional, three-way speakers wheezing just as the music tried to rise to a grand crescendo. Listening to opera made it even more evident that, appealing as the amp was, it just wasn't powerful enough, with its output of only 25 watts or so, for the speakers that I had, the top notes congealing, crackling, and then breaking up just as they had with my first audio system a couple of years before.

About a week later, Kara took the amp back on her way home to Vancouver, and, regretfully, I let it go, wishing there was a way to capture the sound I searched for along with the raw power I

needed for the music I wanted to play. Kara said that what I liked about the amp was its big, old-time triode qualities, a kind of fulsomeness like cream, but that what I *really* wanted was a quality she called "transparency"—a lack of coloration and an openness that got out of the way of the music so that its genuine qualities could shine through. I wanted nothing between me and the music.

Sometime over the next year, a friend asked Kara to restore to operating condition a pair of original Fisher 50A mono amplifiers. She took them apart and, after getting completely familiar with their schematics and design, put them back together, this time with a bit more capacitance in the power supply. She came to see that the design was not precisely vintage, if *vintage* meant an overripe midrange, high coloration, and woolly bass. The 50A's stock features—dual-tube rectification, choke-input power supply, interstage transformer drive, tube-regulated bias supply, and 1614 pentode tubes operating in triode output mode—all shouted quality and translated into excellent sound by contemporary standards.

"The old Fisher 50As were transparent as hell, yet full sounding," Kara said. They produced excellent detail without being pushy, fatiguing, or forward.

After shipping the restored amps off to her friend, Kara couldn't get the sound of them out of her head. She sat down at her drafting board (she still used a physical drafting board and mechanical pencils), began to draw parts, schematics, and values, and soon got the idea that she could manufacture this amp herself—upgrade its parts, clean up its looks, and recapture its sound but make it better suited to contemporary tastes. She built a pair of her own prototypes in her shop—the two-car garage attached to her home where she kept workbenches and a CNC machine she used to cut the metal chassis for her amps and preamps. She created a milled aluminum chassis, a machined top plate, and transformer end caps—and beefed up the circuit. She adapted it with features of her own—little lamps to control and ameliorate voltages, modern resistors and new old-stock Russian and American paper-and-oil capacitors in the signal path instead of the paper-only caps Fisher had used, and installed a Lundahl interstage transformer to produce better bandwidth than the original. She simplified the amp's front

end and further increased its transparency of sound (no scrim of artificiality covering up the pure signal). Finally, she replaced the original 1614 output tubes ("which no one loved," she said) with KT88s, the kinkless tetrodes designed by Marconi and Osram in England back in the fifties—the same tubes my Air Tight stereo amp used. But instead of running them in the classic ultralinear manner, pushing them to their full output power as the Air Tight did, the Fisher design ran the KT88s in Class A triode mode, lowering the power output some but achieving that transparency of sound I craved. For all of the Air Tight's great qualities of extension into the highest treble range, easily rendering the top notes of any operatic singer, it had some irregularities in frequency balance that, eventually, I'd begun to discern and feel disappointed by. It receded in the midrange timbres and seemed to short bass power in its presentation, orchestras sounding all too often light in the cellos and double basses, which robbed symphonies and concertos of their natural richness and amplitude. Though it was indeed powerful with operatic voices, its presentation was "mail-slotted" as Kara put it, emphasizing the upper half of the full register of hearing. The KE50A's output power was only 40 watts—about half of what the Air Tight produced—but it was still more than enough, she theorized, to drive speakers like mine and output an even balance

through the frequencies. Her version of the 50As came out sounding better than even she had expected, she told me, driving her open-baffle speakers easily and providing enough power to handle the one CD of orchestral-choral music she possessed—Herbert von Karajan's version, with the Berlin Philharmonic and Vienna Singverein, of Mozart's *Requiem*. She called me up, the grand voices and orchestra playing in the background, and said she thought she'd built just the amps I needed.

When she brought them down to Eugene, I saw their look was certainly "old-school"—lots of tubes, with two transformer cans sticking up from the amps' semi-perforated, raw aluminum case. Fairly smallish, each amp was about nine inches wide and a foot and a half long, and the transformer casings stuck up along one side, measuring another nine inches from its base. Its weight was just over forty pounds—half what the Air Tight weighed.

"You gotta remember that the original Fishers had to fit into one of those fancy wooden consoles that looked like furniture," Kara said. "So they're fairly compact, and nobody was worried about looks since they'd be hidden anyway." It was all about sound and a small footprint.

The pair of Kara's updated amps, all gleaming aluminum and shiny glass tubes exposed, ended up fitting side by side on the bottom shelf of my audio rack. Ultra-retro in appearance, the KE50As gave my system a touch of throwback suavity and radiated an aura of pipe-smoking dads, stockinged feet stuffed in bedroom slippers, back in the heyday of hi-fi.

As a woman, Kara is a rarity in audio circles, and she is a transgender one, having made the change in her early thirties some three decades ago. "Being a woman is a mixed bag," she said to me once, "but it's livable. The other way was not." She'd had the thought of transitioning all through adolescence and her college years at UC Berkeley. She'd grown up in the Bay Area, observing that, while her older sister was treated gently in the family, she was treated quite roughly, not only in terms of the standard expectations of male achievement (athletics and academic prowess, career goals), but also coming in for more abuse. Kara was the child of alcoholics who had visited her young life with much woe. Gender difference

was always on her mind, and within her soul she felt decidedly feminine. She made the transition with the hope, even though her parents had passed away, that the world might treat her differently, that she herself would be transformed not only physically and sexually, but in terms of psyche and an inner identity as well. She embraced being a woman.

A few years after she finished college, but before her transition, Kara found work in the East Bay at a machine shop that fashioned precision parts, and apprenticed herself to the craft of working with metal. There, she was exposed to a wide variety of fabrication techniques and had access to a full toolroom—this allowed her to gather basic experience on machine tools such as a milling machine, lathe, and surface grinder. She learned all about the cutting, bending, shaping, drilling, and carving of metal sheets and blocks, worked with soldering irons and welding torches, punch and stamp presses, pneumatic drills and CNC machines, yet she never quite reached the coveted status of a journeyman in the trade. She'd had the stereo hobby since high school, however, and got more and more interested in audio electronics. She subscribed to the magazines, studied the DIY handbooks, and eventually found herself holed up in a farmhouse in Santa Rosa, gifted with a pile of old audio equipment—spare parts, remnants, and broken-down gear. Over the course of a decade, she went through everything, researching designs from published schematics, familiarizing herself with circuits and their operation, figuring out how to get that pile of old audio gear into working order. She took transformers from one amp and put them together in circuits designed for another. She pulled resistors and tubes from an Eico and placed them into a Scott or a Harman Kardon. She scrounged for caps and chokes, swapped Fisher parts for Marantz. She figured out how everything worked and saw into the hopes and ingenuities of scores of audio engineers from the fifties and sixties. She decided to make electronics of her own. In 1998, with a partner she later bought out, she founded the deHavilland Electric Amplifier Company.

Kara had proposed she finish the final iteration of what she called her "KE50As" with me as the critical pair of ears to fine-tune their sound. "I've got the main points covered," she said, "but there

are a couple of things we can still fiddle with." She mentioned input tubes, resistors, and silver-and-oil versus paper-and-oil capacitors. We'd run the amp in my own system using different things, do A-to-B comparisons, and I'd tell her what I heard and liked better. It would be a kind of collaboration.

"You got something we can use for test music?" she asked.

"Yeah," I said. "I think so."

Transparency wasn't what I thought of when I first heard the amps in my listening room. Instead, I thought of tonal density and weight. The KE50As created an audio tapestry—an overall lavishness and permeable luxuriance open at times to stunning accents and sparkle, anchored by the gorgeous weight of its own fabric. Yet nothing seemed out of balance or over-the-top—no artificial "detailedness" bordering on brashness, no thickened and overemphasized midrange masking a missing top end or bass, no too-forward treble or bass heaviness at all. These characteristics were only somewhat akin to the signature deHavilland sound I'd come to know from Kara's big triode monoblocks that used transmitting tubes from the forties—her Aries 845 and GM-70 amps. She'd described her KE50As as "easy-goin' and relaxed," but to me they sounded more refined and distinctive than that. I'd call their sound "romantic" except for their precision, the wealth of detail they provided, their sonic palette that ranged fairly evenly through the audio frequencies. The deHavilland KE50As could capture the glorious weight of symphonic music, as well as the captivating airiness of a full choir's voices abloom in air.

We were listening to a CD of the famed Russian conductor Valery Gergiev leading the Kirov Orchestra in Stravinsky's *Le Sacre du Printemps*. It was a stirring piece of music, of course, and I had chosen it to test the sonic palette of my system with Kara's KE50A amplifiers now powering it. Through Part One's "Introduction" and "The Augurs of Spring," Stravinsky's modernist ballet presented a series of exquisite miniatures, as the bassoon, flute, and oboe took brief solos. Then the dance began with a series of sharp,

pulsing, bowed strokes from the strings that were full of eccentric, thrilling accents and stirring fortes. After the first fanfare and initial powerful blast of bass-drum strokes, a spinning drama built among the horns, strings, and woodwinds, which shared and alternated in taking up the theme. The "Mock Abduction" culminated in assaultive, elephantine crescendos followed by a rapid series of thunderous strokes on the bass drum that could have been the death of an ordinary amplifier—if not in tonal accuracy, then in the congestion of timbres; if not in a shallow, shrunken soundstage, then in the faintness of bass slam and speed. In my experience, I'd heard solid-state amps achieve such slam, but not without sacrificing tonal color and treble clarity. And I've heard tube amps deliver the sparkle, but without such impressive slam. Kara's 50As reproduced *all* of Stravinsky's magisterial complexity without shying away, with no diminishment of all the orchestral flavors, and with thundering blasts of bass drum and timpani. While we were listening, and only at fairly moderate levels, mind, I pulled out my handheld RadioShack sound pressure meter and measured the peak output. Its digital readout jumped to 95 decibels on the bass-drum strokes—loud as a motorcycle or a power mower.

When the music ended, Kara was sitting quietly beside me in a folding chair I'd set up next to my listening seat, positioned dead center between my two speakers and about ten feet across the room. She angled her head, locks of her dark blond hair sweeping her jawline, her face impassive except for the trace of a smirk. *Waddaya think?* she seemed to be saying. My mouth gaped. I was already saying *Wow!*

I'd yet another test CD readied in the queue for the 50As, however—*Utopia Triumphans: The Great Polyphony of the Renaissance,* a recording by the Huelgas Ensemble, led by Paul Van Nevel. On it, a forty-three-voice choir sang a motet entitled *Spem in Alium* by English Renaissance composer Thomas Tallis. It is as sublime a piece of music as I have ever heard and a real "system crusher" for audio reproduction. The massed voices of the ensemble put the power, resolution, and transparency of a system to perhaps the most difficult test I could think of. Tallis's composition calls for eleven sopranos, six altos, fifteen tenors, and eleven basses. On a system

with otherwise adequate amplification (like my first few), the voices will cave in on themselves in a weltering mosh pit of audio hash without any differentiation among the choral parts. That sound is not only hard and edgy, it's glassy, full of time smear and distortion. I didn't dare play it first, but, having heard Kara's amps on the Stravinsky ballet, I broke down and spun the difficult choral piece on my player. With the KE50As driving my speakers, the worshipful voices were clear, the sections distinct, and the polyphonic vocal lines pulsed sinuously, as called for by Tallis's magnificent music. The presentation was clean and airy, the voices blooming in space, the notes given wraithlike embodiment and long, impressionistic decays, the full choir sometimes hanging cloudlike amidst a wide soundstage spread across my audio room. I was in an apse of ascension until Kara spoke again.

"You wanna give that opera stuff a try?" she asked.

I pulled a CD by Renée Fleming out of the carousel rack sitting on my writing desk between the speakers. After dithering a moment, I chose "Endless pleasure, endless love," from Handel's *Semele,* on the recital disc *Handel,* a recording that features Fleming's distinctively pure, creamy soprano. The aria is taken allegro, Fleming accompanied by the Orchestra of the Age of Enlightenment, a chamber orchestra that performs these intricate baroque scores under the precise direction of Harry Bicket. Fleming sounded agile and sprightly at the top of her range and chesty and pleasantly darker in her lower notes, which made for contrasts not only in register but in texture and timbre. At some point in the aria, Fleming struck a crazy coloratura note, ornamented and vibrant, lyric and sweetly piercing, testing the upper reach of the amps' extension. The KE50As nailed it—no spike, no glare, no hole in the voice, and no ornaments of melisma or vibrato disappearing and breaking up Fleming's supple rendering of the aria's most dramatic moment. Throughout, the sound field was beautifully interwoven with voice and instruments, including the occasional light touches of a harp. I could also feel the viola da gamba gently pressurizing the room as Fleming trilled and thrilled gorgeously, her voice shimmering in shallow sonic waves across the room as though a pale, translucent scarf were trailing behind her as she sang. It was glorious and there

seemed nothing between me and the music. Certainly not an audio system.

And so it went, voice after voice, orchestra after orchestra, Kara's amp taking on all comers, undefeated, it seemed, by any requirement. On LP, I played Luciano Pavarotti singing "Nessun dorma" from Puccini's *Turandot,* and heard that famed tenor voice powerful enough to fill a stadium, sailing out open and expressive throughout his wide ranges of pitch and dynamics. The man from Modena's vocal power didn't faze the KE50As one bit. The amps remained unaffected as they delivered the requisite steadiness of sustained current to my speakers. Just three years before, when I began my audio journey, Pavarotti's recordings were the most difficult for each of my rudimentary and mismatched systems to reproduce—so many of my amps, tubed or solid state, clipped and distorted, the speakers spitting and breaking up under the burly Italian's immense vocal power. At last, not so here—all the heroism and bravura of Pavarotti's silvery notes came through with a fine, dramatic purity.

System: Nottingham Spacedeck turntable with nine-inch Spacearm and Shelter 501 Mk II MC cartridge; Herron VTPH-2 phono preamp; Cary 303/300 CD player; deHavilland Mercury 3 preamplifier; deHavilland KE50A monoblock amplifiers; Von Schweikert Audio VR5 HSE loudspeakers; Cardas Golden Reference and Verbatim interconnects (RCA); Verbatim speaker cables with jumpers; Cardas Golden Reference power cords; Finite Elemente Pagode Signature five-shelf audio rack.

Part Seven

The First
Amplifiers

Herbie's Mouth

I might've been five when my cousin Herbie Shigemitsu did a memorable and astonishing thing out near the Japanese grave-yard by the piggery on a sandy promontory in our home village of Kahuku on the island of Oʻahu in Hawaiʻi. Herbie was my elder by several years—he was eleven or twelve then—and was impressive to me in every way. For one thing, like an adult man, he wore large aloha shirts and khaki pants rather than the T-shirt and shorts that I did. His clothes rippled in the trade winds as he smiled and reached into a shirt pocket and pulled out a pack of chewing gum, a speckled cowrie shell he'd found on the beach, or a cellophane pack of dried cuttlefish he'd share with me and his other young cousins. For another, he thought up our games and adventures, leading us in raiding the piggery of a few new piglets he'd grab from their pen and stuff under his shirt as we trotted behind through the sandy roads that wound through our village, eager to witness what he'd do with them. Would he pet them or eat them? Would he let them go in the canefields? But he only wanted us to hold them, to see us squeal and giggle as the tiny animals, smaller than cats, wriggled in our hands, eyes closed, nudging their wet noses against our flesh. Then he gathered them all up, dropping them back inside his billowing shirt like it was a large, marsupial pouch, and raced back to the pen, where he straddled its sides like a colossus and lowered each piglet back onto its bed of straw, while its huge mother grunted and squealed in anger below his legs. To me, cousin Herbie was a Titan who brought fire to my life.

One afternoon, I think we younger kids were playing on the carpet of poky temple grass amidst the stones and wooden grave markers in the graveyard where two generations of our ancestors had been buried. It was on a sandspit that jutted out into the Pacific, and the winds whipped around our bodies in whirling blasts of cool

air that seemed like ghosts coiling around us in the otherwise constant subtropical heat of our island. Herbie was over by the wooden
fences of the piggery, about a baseball field away, and I saw him
bring his hands up to his face and make a kind of cave of his fingers
and palms, surrounding his mouth. And then I heard him.

"Garrett! Tommy! All you guys! Quick you come! Come now!"

He waved then, knowing he'd caught our attention. And
we gathered up our things—marbles, slingshots, paper targets,
whatever—and ran over to him as he peeled away, headed toward
the main part of the village, where there was a gas station owned
by one of our aunts, a grocery store that sold us our treats, and our
great-grandmother's shack with her garden abundant with papayas,
squash, and melons. We obeyed his beckoning without question,
his call for us to gather and follow him. And the astonishing thing
was, we'd *heard* him through all that distance and buffets of wind.
His voice carried from the piggery across sand dunes covered in sea
grapes and the emerald hillocks of the graveyard to our ears. *How
did he do that?* I thought. *How did he throw his voice so far?*

I can say now that what Herbie made out of his cupped hands
was a small acoustic amplification device. It altered the relation of
his voice to the air so that the compression of the sound waves he
made by shouting was increased. At the same time, the volume of
air it immediately had to push was made smaller. His cupped hands
had restricted the wide field of air from the infinite universe around
him to just the amount inside the cone of his hands. He'd changed
the ratios of sound wave compression versus the resistance of the air
volume it had to pass through. Simply put, his hands made a megaphone around his mouth. He had made an air-bomb of his voice
and a launchpad with his hands. In the audio world, this is known
as a *compression horn*.

Of course, I'd no understanding of any of this at the time. It
was plain magic to me, and when I asked him how he did that, he
simply said he yelled into his hands "ladat."

"Ladat?" I asked. *Just like that?* And I tried it, holding my hands
in a basket around my mouth. The sound just rebounded back at
my face.

"No," he said. "No make one basket in front your face. Jus'

make one tunnel ladat, open at dah mout' and at dah oddah en'."
He showed me again, launching his voice through his cupped hands
into the swimming air, backing up farther and farther away from
me. "Garrett . . . Gaa–rrett!" he softly cried, letting the horn of his
hands amplify his kind voice.

"Now you try," Herbie said.

In the days that followed, we younger kids practiced shouting
at each other through the tunnels of our hands, from opposite sides
of the graveyard, from one tip of the little cove on the side of the
promontory to the other.

"Hey, Tommy! You can hear me now or what?"

"E, Garrett, hear you? More worse I can smell your dirty mout'!"

We teased each other with insults, gradually lengthening the
distance between us, bouncing from sandspit to canefield, standing
beside a row of ironwood trees and shouting down the dirt road that
led to the dump where we once found a cache of army K–rations,
saying fouler and fouler things until an adult or teenager would hear
and start scolding us and we ran away, laughing, marveling at our
new powers, our ability to send our voices across the landscape like
the curling winds scouring the sands and making scallops on the
waters of the ocean that surrounded us.

Megaphone

What my cousin Herbie had shown me was probably the most
primitive if not indeed the first megaphone, originating in prehistory, perhaps around the same era that fire itself had entered the life
of the species. Like fire, then, amplification may be an invented tool
of the late Paleolithic, an ancient part of human culture and deserv-
ing of respect, even a measure of worship.

In a YouTube video, Paul McGowan, the audio engineer who
presides over PS Audio, a contemporary maker of fine stereo equipment located in Colorado, says, "A megaphone directs and focuses
sound the way a lens might an image." But, more than that, it *amplifies* sound in the specific way its bell or tunnel compresses air around
the speaker's or singer's mouth, and then flares out at the end where
it contacts the open air, thus expanding the pressure wave from an

initial small source to a larger one. It takes a small connection with the air and, through a process of mechanical coupling, transforms it into a larger one, making the sound bigger too.

The megaphone as an instrument perhaps made its first appearance in ancient Greece around the fifth century BC, as there are ceramics, bas-reliefs, and sculpture that depict actors who wore masks with cones protruding from their mouths. But the written record of the megaphone begins with Samuel Morland in 1655, a man who experimented with a variety of acoustic horns, his largest made of a copper tube more than twenty feet in length. He claimed it could project the human voice over a mile and a half. A contemporary, Athanasius Kircher, did him a little better, constructing a wall-mounted "cochleate" or spiral device that could overhear voices outside a house and also project a voice from inside. Kircher installed his only slightly more compact megaphone in the side of a house with the narrow end inside and the flared end out. In 1878, the American inventor Thomas Alva Edison topped them both with a speaking/hearing trumpet for the deaf and hard of hearing. It was made of three parallel cones deployed in a single row, two outer for hearing and one inner for speaking. The two outer cones were six feet eight inches long, made of paper, and connected to the listener through a tube inserted in each ear. The middle cone was similar to Morland's trumpet but had a larger mouthpiece. It could throw a low whisper over a thousand feet and a normal voice about two miles away—longer than the range of a high-powered hunting rifle. It could also pick up voices from over a thousand feet away, but was a bit cumbersome to carry around, adding, despite its auditory and acoustic prowess, what was equivalent to the bulk of three NBA power forwards onto a normal human body.

Much later, during the early twentieth century, Herr Senger, a Swiss opera singer, invented a papier-mâché megaphone with an elaborate mouthpiece that fit around a speaker's face, making for a device with much better clarity, fuller vocal tone, and more pleasing resonance than the usual megaphone of the time, which had been made with a sheet of brass shaped into a cone. Its mouthpiece covered the lower part of his face and nostrils too, making for great vocal projection as well, finding use in a production of poet Edith

Sitwell's *Façade: An Entertainment* (from poems written in 1918 but first performed in 1923), a kind of drawing-room opera made of a series of her poems written in rhythms styled after dances like the waltz, foxtrot, and polka. Its music was written by William Walton, now best remembered for his Violin Concerto (1939), commissioned by Jascha Heifetz, and Cello Concerto (1956), recorded by Gregor Piatigorsky, Heifetz's frequent chamber partner. The Sengerphone was used to project a voice from behind a large drop curtain with a hole cut into it for the mouth of the device, so the audience would respond only to the voice and music and not the personality of the singer—a completely uncanny and wacky Modernist strategy of performance, at once making for a more organic and natural sound and yet allowing it to emerge from a kind of alienating, mechanical presence.

Also early in the twentieth century, movie directors in Hollywood used the megaphone to direct "casts of thousands" on location and on the immense studio sets of the time. Cecil B. DeMille used megaphones to direct the extras who played Egyptian slaves in *The Ten Commandments* (1923) and the Israelites and Roman soldiers in *The King of Kings* (1927). Charlie Chaplin used one too, demonstrating the political neutrality of the instrument, serving the needs of both conservative and leftist filmmakers. It's ironic that an amplifier would be an important tool used to direct so many silent films.

But there's a big downside to the megaphone. The human voice (or a parrot's or chimpanzee's, for that matter) gets amplified, but its frequency balance is thrown off, the bass sound dropping away and the mid- to higher frequencies getting a boost, thus producing the characteristic nasal sound. Everything below about 90 Hz (high bass) gets attenuated and below about 50 Hz (low bass) gets cut out completely. Concomitantly, the upper middle frequencies (between 2 kHz and 5 kHz—the higher registers of the human voice) are emphasized, resulting in the megaphone's distorting squawk. And if the megaphone is made of brass, as most were in the early twentieth century, you get that "tinny" sound of the later Gramophones, with their iconic flower-bell amplifying horns as a bonus. This is evident in the vocal style of the early crooners of the twentieth century like Rudy Vallee, a Yalie whose natural voice was thin

and reedy compared to the operatic belters of earlier recordings like the Italian tenor Enrico Caruso. To compensate for the weakness of his voice during live performances, Vallee sang through a megaphone so he wouldn't be drowned out by the musicians in his band, the Connecticut Yankees. When they broadcast their shows on the *Fleischmann's Yeast Hour,* bandleader Vallee still sang through a megaphone as that was the voice his public was accustomed to and it had become a signature sound. When they recorded tunes like "Makin' Whoopee" (Velvet Tone, 1929) and "The One in the World" (Victor, also 1929), he sang into the microphone through a megaphone too. Later, to add that antique, old-time American music hall flavor to some of his tunes, Beatle Paul McCartney imitated this sound using modern electronic techniques in songs such as "Lady Madonna" (Parlophone/Capitol, 1968) from *Hey Jude* (Apple, 1970) and "Honey Pie" from *The White Album* (Apple, 1968). So did John Carter, the lead singer of the New Vaudeville Band, on their recording "Winchester Cathedral," a Grammy Award–winning hit from 1966 (Fontana Records). In fact, in order to get that particular sound during the recording session, Carter is said to have sung the tune through his hands.

Acoustic Amplification

The megaphone is just one item in the history of acoustic amplification. Humankind has been trying to amplify sound for millennia, using tools, architecture, spatial designs, and natural land formations in order to project sound across varied distances. The Greek amphitheater at Epidaurus, built in the fourth century BC, is so ingeniously constructed that a whisper or a coin dropped on its stage can be heard from its rear seats, fifty-five rows away. The poet Etheridge Knight said that, when he was in prison, inmates would sing rhymes and vulgar toasts into the chamber of the cellblock, letting their voices echo and bounce, entertaining and annoying each other, playing cutting games, competing to be the funniest, the loudest, the most profane. Maurice White, lead singer and founder of the funk group Earth, Wind & Fire, in a song from *Open Our Eyes* (Columbia, 1974), their fifth studio album, praised the African

kalimba, a kind of piano that sings its sweet notes from a wooden resonator board fitted with metal tines plucked by a player's thumbs. June Kuramoto of the fusion group Hiroshima plays the koto, a Japanese lute over a thousand years old in design (derived from the even older Chinese guzheng) that uses thirteen taut strings strung over thirteen movable bridges and a resonating board made of paulownia wood, carefully planed by traditional craftsmen. Its ethereal song is said to emanate from the head, horns, fiery tongue, and stomach of a dragon. And when I was ten, my maternal grandmother, Tsuruko Kubota, who worked as a maid in Beverly Hills, took me to an outdoor concert at the Pilgrimage Theatre in Los Angeles to hear violinist Jascha Heifetz and cellist Gregor Piatigorsky, with a chamber group of distinguished others, play Brahms's String Sextet no. 2 in G Major and Mendelssohn's Octet in E-flat Major from a stage surrounded by rock stairs and a background *skene* of stone pillars and cubicles that spun the sound of their incomparable instruments, vibrating wood and strings into the night air over my head in a way I've never forgotten, rich and sinuous, ample as time.

Bandshell

My cousin Tommy, a year or two older than me and a few inches taller, stood facing the concave hollow of earth cut by the sea and wind into the cliffs below the Japanese graveyard near our village. The wind blew and whipped around us, flapping the cloth of our swim shorts against our thighs, streaming our hair like black banners of seaweed into the currents of air. He was a skinny kid, but he'd wrapped a bedroom quilt around his shoulders and torso, making him look bigger and oddly transformed into something like a statue or a store manikin. The quilt was a thing made of colorful triangles of leftover scraps scavenged from the sewing of women's dresses and *muumuus*, throw pillows, and the *zabuton* adults used to sit on the floor while they played poker and *hanafuda,* a rummy game of cards with flower suits. He was intoning something I did not understand, though it was in English. It was something from a film or television, I gathered, as he spoke like a film star with an English accent, deepening his voice, throwing it out from the shal-

low cave of earth to the stones shining with sea spray on the beach, to us other kids standing stupefied as he declaimed:

> *Four score and seven years ago,*
> *Our fathers brought forth*
> *On diss Hawai'i one new nation*
> *Conceive in liberty and dedicated*
> *To dah sugar plantation*
> *Where all men go be paid equal.*

His voice was rich and resonant—not like his normal speaking voice—and it rose above the light chatter of stones as we, barefooted, circled awkwardly around him, gradually falling under the spell of his melodious speech as he continued:

> *Now is dah winter of our diss kine tent*
> *Made glorious summer by diss ton of pork;*
> *And all dah clouds made lava on us guys' house*
> *In dah deep bosom of dah ocean buried.*

Tommy was taking speeches he'd learned from a book and twisting them around, talking like an actor, but making up meanings that only he understood. We younger kids meanwhile felt only the weight of his words, the stentorian tones of his speech—how it fell against our ears in stately rhythms—the amplification of his voice as he intoned against the cliffside that rose up like an earthen flame cupped around his robed figure.

Tommy had found a natural bandshell and occupied its space like it was a reverberant stage where he could launch his voice over us and the stones of the beach toward the pitching sea, the only audience that could confirm his grandeur.

Sound Waves

On music and sound the ancients were a lot like my parents—at times fanciful, at others stern, and in the end prone to mistakes, yet not without a sweeping charm that excused all error for their

illuminations of the invisible world singing around us. Pythagoras, Greek mathematician of the sixth century BC, applied his considerable genius to a description of the universe by making an analogy to music, naming the relations of celestial bodies as "the music of the spheres," keeping all of its ten thousand things in balance and accord with an invisible, harmonious force. Two centuries later, in the fourth century BC, the philosopher Aristotle may have been the first to describe sound as like the action of waves, stating (erroneously) that air moved continuously along the direction that sound traveled, that the notes he heard from something like a lyre were the result of its strings transferring movement to the surrounding air. Current scholarship, sparing Aristotle from blame, now attributes this thought to a lesser thinker named Strato or one of his contemporaries. Yet it is certain that it was the Stoic philosopher Chrysippus in 240 BC who emphasized the wave analogy, observing a body of water (I like to think he sat beside an old pond like the haiku poet Bashō, perhaps watching a frog jump in) and speculating that sound itself was like a water wave. Plutarch, the biographer of noted Greeks and Romans and, for thirty years, the priest at Delphi, wrote that "the air is not composed of small fragments, but is a continued body . . . and being struck with the breath, it is infinitely moved in waves and in right circles, until it fill that air which invests it; as we see in a fishpool which we smite by a falling stone cast upon it . . ." And there it is—the clearest declaration in antiquity of sound as a transverse wave. Finally, Anicius Manlius Severinus Boethius (480–524 AD), author of the great treatise *The Consolation of Philosophy,* an early medieval philosophic text later favored by the likes of both Dante Alighieri and Geoffrey Chaucer, repeated the comparison of sound to the ripples of waves when a rock is thrown into a pool of water.

But the most extensive description of sound as a wave in early times comes from the Roman engineer Marcus Vitruvius Pollio (c. 80–70 BC–after c. 15 BC) in his discussion on the design and acoustics of the Greek amphitheater in *De architectura.* Vitruvius describes the movement of sound as a voice that flows and ascends through air, stair-stepping up the rows of a well-constructed amphitheater of Greek antiquity:

*Voice is a flowing breath of air, perceptible to the hearing by
contact. It moves in an endless number of circular rounds, like
the innumerably increasing circular waves which appear when a
stone is thrown into smooth water, and which keep on spreading
indefinitely from the centre. . . . In the same manner the voice
executes its movements in concentric circles . . .*

But from modern science, we know that waves in water are
transverse waves and sound is not. Though the ancients were right
in comparing sound to a wave, their choice of a wave on water for
comparison wasn't quite right. Sound waves are in fact *longitudinal*.
A transverse wave, like those of a ripple caused by a stone tossed
into a quiet pond, moves perpendicularly, up and down, in the
direction away from its source. In other words, the molecules of
its medium excite themselves and pass along energy in a manner
that causes them to bob up and down, yet return to place, while
the action travels away from whatever has been its source—either a
pebble tossed, an earthquake, or a kid's cannonball in a swimming
pool. The wave at an American sports stadium is something like
this—the spectators stand up and lift their arms, we see the wave
traveling, but everyone still sits down where they were before they
stood up and waved their arms. And yet the stadium wave travels,
as we see it on jumbotrons and television sets and from seats across
the ball field. That's a transverse action.

A sound wave behaves differently. The currently favored analogy is to a Slinky toy stretched out and anchored at both ends. If
you were to gather some of its coils from near one end, compressing
a small bunch of them together, and then let go, the compression
would move along the length of the Slinky to the other end, the
coils passing the movement along to the opposite end of the toy,
but the gathered coils you initially grabbed together rebounding to
their original, non-compressed resting point. This action is a longitudinal wave, moving away from the direction of its source but
in the direction of travel rather than up and down. You can think
of air as being composed of longitudinal layers that are like these
coils, with a sound wave constituted as layers of air that push and
pull at one another much like the compression moving down the

Slinky. But again, the molecules of air don't quite move, but are agitated back and forth along the path of the wave's travel, causing alternating compression and rarefaction, areas of high pressure and low pressure, passing along their energy, but rebounding to their initial positions. So, yes, sound is a wave, but like the longitudinal, compression wave of a Slinky and not one moving through water.

Nonetheless, Vitruvius's notion of sound being able to wash up like a light tsunami over the ascending staircase of an ancient amphitheater was a very poetic and not altogether false analogy. He recommended that all the seats in a properly designed amphitheater should be built so that a straight line could be drawn from the lowest to the topmost rows, touching the front angles of each of the seats. He warned against obstructions too (he wouldn't like the exposed vertical girders in present-day Wrigley Field, I guess), declaring there should be nothing that could alter the flow of the human voice in its upward and outward flow from the stage to the audience. The architects of antiquity, Vitruvius said, regulated the flow of the human voice in "a true ascent of steps in a theater and contrived, by musical proportions and mathematical rules . . . to make it fall on the ears of the audience in a clear and agreeable manner." Furthermore, along with the semicircular bowl of the theater and the wooden floor and *skene* (background of the stage), the stone seats themselves were a means to lift the sound of the human voice along to the topmost rows of an arena:

> . . . *as in the case of the waves formed in the water, so it is in the case of the voice: the first wave, when there is no obstruction to interrupt it, does not break up the second or the following waves, but they all reach the ears of the lowest and highest spectators without an echo. Hence the ancient architects, following in the footsteps of nature, perfected the ascending rows of seats in theatres from their investigations of the ascending voice, and, by means of the canonical theory of the mathematicians and that of the musicians, endeavoured to make every voice uttered on the stage come with greater clearness and sweetness to the ears of the audience. For just as musical instruments are brought to perfection of clearness in the sound of their strings by means of bronze plates*

or horn so the ancients devised methods of increasing the power of
the voice in theatres through the application of harmonics.

Vitruvius was talking not only about geometries of theater construction here, but also about the mystery of resonance control. He alluded to *echea,* brass and ceramic urns placed at critical junctures, determined by mathematical formulae, along the seats and stairs of the amphitheater, supposedly to drain away unwanted sound (like the murmurings of the audience) and to emphasize the voices of the actors. This means drawing away bass sounds and boosting higher frequencies like the human voice. Vitruvius wanders far afield in his writings, touching on architecture, acoustics, and hinting at audiology as he goes. And it's true—similar urns have been discovered embedded in the excavated ruins of a few such amphitheaters in Greece and Crete, confirming Vitruvius's observations and giving rise to current folkloric thoughts of serious sonic mumbo jumbo possessed by the ancients. My contemporary friend and audio pal George Radulesk in Portland, Oregon, collects tiny resonators made out of brass, silver, and gold, modeled after Tibetan prayer bowls said to alter acoustics, and places them at specific nodes all around his elaborately tricked-out listening room. He swears they improve the sound of his stereo system—little esoteric cups of sonic gladness.

And Vitruvius wasn't all wet, as it turns out. Modern audiologists have made studies of his notions at the amphitheater at Epidaurus. Designed by Polykleitos the Younger and considered his masterwork, the theater initially featured thirty-four semicircular rows of limestone seats (later expanded to fifty-five by the Romans, fourteen thousand seats in all) and was discovered on the Peloponnese peninsula under a layer of earth in 1881. It was said that a performer standing on the open-air stage, at the center of the orchestra circle at the bottom of the theater, could be heard in its back row over sixty meters away. In 2006, as reported in an article in *Live Science* by Tom Chao, scientists from the Georgia Institute of Technology determined that the limestone material of the seats possessed a profound filtering effect, suppressing low frequencies, minimizing crowd and background noises, and reflecting

high-frequency sounds back to the audience up to its last row. Both Nico Declercq and Cindy Dekeyser, mechanical engineers on the team, initially suspected that the slope of the theater, as suggested by Vitruvius, was an influence, that surface waves of sound climbed its rows with almost no damping. But ultrasonic wave experiments and some minimal computer modeling soon showed that frequencies up to 500 Hz were lowered and those higher than 500 Hz were left unchanged. It appeared that the corrugated seat surfaces of the semicircular theater acted as natural acoustic traps, filtering out the bass range (wind and crowd murmurs) and passing along sounds of the onstage human voices unaffected. And, through a trick of mind called *virtual pitch,* a phenomenon of the human brain, listeners would reconstruct the missing frequencies of the sound spectrum so that what they heard would seem normal—as we do when listening to a voice on the telephone that's missing its bottom end or using small studio monitors or bookshelf speakers with no bass woofer.

Previously, in another experiment while at Ghent University, Declercq had shown that the stepped surface of Kukulcán, the Mayan ziggurat at Chichen Itza near El Castillo in the Yucatán, had the ability to transform the sound of handclaps at its base so that they would be heard at the summit as the chirpings of the sacred quetzal bird or as the soft peal of rain being poured into a bucket. The architectural design of the temple's steps, with high risers and short treads, created this astonishing, even magical phenomenon, most certainly a tool used in ceremonies by Mayan priests. I'd love to go there, record it, and put it on iTunes as a sound grab for pop songs or samplings backed by a hip-hop beat. But at Epidaurus, Declercq found that the seats just filtered out bass background noise and improved the ability to hear the performers' voices, rich in high frequencies, across the considerable distance of the stepped rows of the amphitheater. Unlike the clever ventriloquisms practiced by Mayan priests, the only motive for the Greeks was the achievement of clarity in boosting the acoustic travel of lines from Aristophanes, Aeschylus, or Sophocles.

On YouTube, Rick Steves, the genial travel writer, posted an episode on *Epidavros,* as he calls it, entitling it "Epidavros, Greece: Perfect Acoustics." In it, he stands like a classical actor, taking up

a position, hands outspread, at the center of the orchestra circle in a telephoto shot that's taken from the top of the stairs an indoor sprinter's dash distance away. Steves utters a bowdlerized version of Mark Antony's funeral speech from *Julius Caesar,* substituting a theme of international travel and understanding. For the betterment of humankind and the ruin of the bard's elegant strophes, you can hear every word Steves is speaking.

I would like to think that if, on a quiet, late spring day of intermittent, westerly breezes (the theater is sited to face west), someone were to take my hand and lead me up within the semicircle of the fifty-five rows of limestone seats to the top edge of Epidaurus, then command me to turn and listen to a match being struck by a tour guide at center stage nearly three hundred feet below, I might try to blow it out. There would be a sound wave launched by the quick scratch of the match head, then a flare and combustion of its sulfuric compounds—a kind of soft hashing of high frequencies in a finely articulated radiance of noise that might seem so intimate and nearby to me, I would be taken in by the trick of it.

None of the many such amphitheaters I've occupied at different times have been as grand or ancient as Epidaurus. And yet I feel like I've taken from them my own rightful inheritance of this piece of human culture. Not only did I marvel at my cousin Tommy's speech from the concave formation in the wall of a beachside cliff, but I also sang joyously under the stucco ceiling and beside the locker-bay walls of my junior high in Gardena, California, three classmates crooning along with me the lyrics to "In the Still of the Night," our voices bouncing and reflecting and amplifying like a lavish, finely granulated cloud of celestial sound that encircled us. At the Hollywood Bowl one summer evening after I graduated from college, I heard the thundering timpani and gorgeous strings of the *Un poco sostenuto/Allegro* movement of Brahms's First Symphony as performed by the Los Angeles Philharmonic under the baton of the incomparable Zubin Mehta. I treasure the tiled bathroom in a house in Woodstock, New York, that the roots rock group the Band used in the late sixties to fill out their sound on *Music from Big Pink,* and the bathroom in another house in the Hollywood Hills they used on the tune "Jawbone" from their eponymous second album. From

a bandshell in Honolulu, I can still hear Israel Kamakawiwo'ole, big and childlike, strumming his 'ukulele and singing, in his pure, high falsetto voice, a lyric from "Kaho'olawe," a 1986 song about a racing canoe named in commemoration of the 1976 occupation, by a brave band of young kānaka 'ōiwi activists, of the sacred island off the coast of Mau'i that was at the time used by the U.S. Navy as a practice range for carrier jets on their bombing runs. I like to think it was the combination of the earnest acts of the protesters together with Bruddah Iz's heavenly voice—pulling down the music of the spheres and reverberating outward through the universe from the Waikīkī Shell (after all, an open-air amphitheater)—that stopped the bombs.

Resonators and String Sound

My friend the poet and novelist Nicholas Christopher was walking up the ancient stone steps of the temple at Delphi in Greece early one fall evening in 1973, when he felt overcome with a mild panic at the loud chorus of cicadas chirping and whirring from seemingly everywhere around him. He felt himself wanting to flee as though Erinyes were pursuing him, feasting on his eardrums.

"I was stalking up these steep steps crunching under my boots and it was like noise from the fiercest howling wind coming at me through the rocks and trees. It was as though the gods were aghast at the punk way I dressed—leather boots, leather jacket, cigarettes tucked into my rolled-up T-shirt sleeve—and making the point I shouldn't approach sacred antiquity!"

Nicholas had just graduated from Harvard and was taking a triumphant tour through Europe on a motorcycle, reaching his ancestral homeland a few months into his wanderjahr. Of course he had to visit Delphi, the seat of prophetic wisdom, the omphalos of prophecy since pre-classical times.

The myth, though, is that the voice of the cicada is not that of a god, but a human one, shrunk to size. The story is that a group of revelers, after having been enchanted by the Muses, sang and danced so long that they stopped eating and died without noticing life had slipped away (I've nearly done this myself—on a hideaway

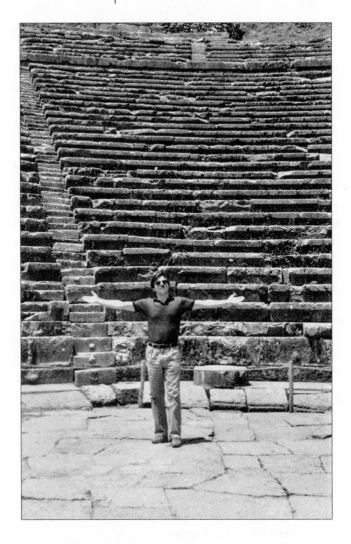

tryst near New Smyrna Beach, Florida, one fall). For their devotion to passion, the Muses rewarded them by arranging that they'd
never need food or sleep, but assigned them the task of singing from
birth to death—as the male cicada practically does, once he emerges
from his life underground and shrugs off the nymphal exoskeleton.
Furthermore, as spies for the gods, cicadas were to report on who
honored the Muses and who did not.

Cicadas are celebrated in classical poetry from Hesiod to Homer
and achieve honorable mention in Plato's *Phaedrus* and Pliny the

Elder's *Natural History*. There is also a beautiful strophe in Virgil's *Georgics* about their summer song:

> But when joyful summer, at the west-wind's call,
> sends sheep and goats to the pastures and the glades,
> let's run to the cool fields while Lucifer is setting,
> while the day is new, while the grass is still white,
> and the dew on the tender blades is sweetest to the flocks.
> Then when day's fourth hour has brought thirst on,
> and the plaintive cicadas trouble the trees with their noise,
> I'll order the flocks to drink the running water
> from oak troughs, at the side of wells or deep pools . . .
>
> (Virgil, *Georgics*, Book III, translated by A. S. Kline)

Aristotle has a note about cicadas' "foreparts" in his *History of the Animals*, and they appear as items of charm and sensuousness in Asian literature, too. A heroine of the fourteenth-century Chinese prose text *Romance of the Three Kingdoms* takes her name from cicada-shaped adornments on the hats of court officials. And Genji, the Shining Prince from the eleventh-century Japanese romance *Tale of Genji*, likens one of his lovers to a cicada for the way she sheds her brocade robe from her shoulders before lying down with him. Elsewhere in Japanese, there are numerous haiku and *waka*, decorous poems full of praise and whimsy about the cicada's song. Yet the Greeks characterized it as shrill, dinning, and tiring to hear. George Seferis, a contemporary Greek poet, mentions in an essay what a relief it is that the incessant song of cicadas has mercifully quelled while he hikes down from Delphi one late summer evening. I guess the gods favored him over his American cousin that time.

But how do these insects make such a racket? Not like the cricket that rubs its hind legs together, creating noise by what biologists call stridulation. Instead, the cicada uses tymbals, resinous structures inside its abdomen that it bends with its muscles, making a small clicking sound, then pops back out in another muscular effort, making a slightly louder click, all in rapid alternation. But these noises would make only the faintest sound if not for another feature—

acoustic amplification. The abdomen of the male cicada, made up of large chambers derived from the tracheae, is actually hollow and acts as a sound box for the madly vibrating tymbals, able to combine their minute ticks into the oscillating battery of continuous notes that plagued my friend Nicholas that night at Delphi. The cicada's belly is a powerful resonance chamber that amplifies each click up to 120 decibels, among the loudest of all insect sounds, according to scientists. If a cicada were to sing just outside your ear canal, for example, the sound could cause you permanent hearing loss.

The cicada's call comes from a resonator—a biological adaptation arriving naturally out of evolution. Perhaps taking a cue from nature, human culture has developed resonators of its own, affixing soundboards and sound boxes to diverse acoustic instruments, string and percussive, in order to amplify their inherently pleasing yet ephemeral and somewhat fragile sounds. The lyre is one such instrument. So is the harp, the zither, the African mbira, the Japanese koto, the lute, and the guitar.

The ancient lyre, a U-shaped harp instrument, has been chronicled in literature since biblical times. In order to soothe his soul after the spirit of the Lord has left him, King Saul asks for young David, Jesse's son from Bethlehem, to come and be his armor bearer and play the lyre for him:

> And it came to pass, when the evil spirit from God was upon
>   Saul,
> that David took an harp, and played with his hand: so Saul was
> refreshed, and was well, and the evil spirit departed from him.
>
> (1 Samuel 16:23)

Lyres such as David's have been around at least since ancient Mesopotamia (2500 BC) and have also been discovered in a cave on the Isle of Skye in Scotland, a particularly poetic place, dating back 2,300 years from the present. The lyre is thought by some musicologists to be perhaps the oldest surviving string instrument. In *A Poet's Glossary*, a compendium of poetic terms and techniques, the poet Edward Hirsch writes that it was "the Greeks [who] defined the lyric as a poem to be chanted or sung to the accompaniment

of a lyre, . . . the instrument of Apollo and Orpheus, and thus [the lyre became] a symbol of poetic and musical inspiration." And so the lyre is everywhere in ancient civilization and since, depicted on sarcophagi, vases and urns, friezes and frescoes, celebrated in poetry from early Homeric hymns through Shakespeare and onward to *The Adventures of Rocky and Bullwinkle*.

In its earliest depiction, a fresco on the Mycenaean sarcophagus of Hagia Triada, fourteenth century BC, there is a lyre with seven strings, held by a man with a long robe in a procession of figures like a succession of summer thunderheads flowing across a Midwestern horizon. In a clearer, more detailed image on a Greek vase from the fourth century BC, a curly-haired and barefoot muse plays a lyre while she sits on the outline of a rock with the inscription *Hēlikon,* a mountain sacred to the Muses (and also the name borrowed for a $2,300 phono cartridge made by Lyra, a Japanese company). In the image on the vase, the lyre is shown, again, with seven strings attached at the top to two arms of a crossbar and then at the bottom to a curved sound box, apparently of painted or veneered wood. Strings were of gut, perhaps from the small intestines of sheep.

String resonance is not all that loud and needs amplification in order to be heard effectively. The resonating chamber at the bottom of the lyre is hollow like the cicada's abdomen and radiates sound transferred from the plucked strings above it. It was perhaps initially made of turtle or tortoise shell and receives the vibration of the strings via bridges at the bottom of each arm of the yoke. The taut string, anchored at each end, gets plucked or strummed and then vibrates, producing a fundamental frequency and accompanying overtones in a mathematical relationship to the fundamental. The string then creates pressure waves (like that Slinky) that emanate longitudinally across space, constituting a sound we define and hear as a musical note. The more rapid the string's vibration, the higher the frequency (or note) transmitted. The slower the vibration, the lower the note. A thick string would vibrate more slowly than a thinner one, creating pressure waves that are farther apart and of lower frequency. Loudness, in this case, would correspond to the amplitude of the pressure wave created, or, in other words, its abundance of energy as opposed to its frequency. Pitch is then a factor of

a string's thickness, tension, and length and is similarly created by all lyres, harps, zithers, and guitars.

A bridge, usually made of wood, placed near or on the sound box itself then transfers the vibrational energy of the string to the body of the lyre—the hollow sound box—where it resonates and gains amplitude, crossing the air and carrying its lyric message as a musical note. The sound wave, relayed from the tiny vibrations of the string by the bridge, transfers to the surface of the resonant sound box, exciting its surface so that it too vibrates and amplifies the string's initial energy. The minute and particular note struck on the string thus gains in volume and sweet affect, transformed from a buzzing gnat of sound into the fullness of a muse's airy note, which we honor with appropriate reverence.

The Sound of the Lyre

I wish to tune my quivering lyre,
To deeds of fame, and notes of fire.
                              —Lord Byron

I think I first heard a lyre when I was eighteen and drove up from Gardena to the Paramount Ranch in the Agoura Hills to the northwest of Los Angeles one weekend. I went to visit the annual Renaissance Pleasure Faire, the kind of thing blowing up in popularity back then. It was 1969 and I shambled around with a GI canvas gas-mask bag slung over my shoulder. I used it to pack rolls of black-and-white Tri-X film, extra filters, and two spare lenses (a telephoto and wide angle) for my Nikon, which I had at the ready to shoot candids of the emphatic faux Elizabethan and merry tie-dye colorfulness of the crowd in those High Hippie days. I vaguely recall that keeping my lenses free of the fine, California oak wood-land dust kicked up by patrons was an issue, and I was constantly brushing them and puffing on them with a rubber bulb.

Turning the corner of a booth for painting harlequin faces, I heard the serial plonking of an appealing, timbral sound I could not quite place. Then I spotted its agent if not yet its source—a

short, bare-chested athletic guy with a beard, wearing jodhpurs and red suede boots, and looking, therefore, a bit like a satyr—which must have been his intent, now that I think of it. He sat halfway up a roughly fourteen-foot-high ziggurat of bundled hay stacked against a fence and was plucking, with the fingers of his right hand, a U-shaped instrument he held against the wiry blond hairs on his chest with his left. The instrument was a lyre, and it was decorated with splashes of earth-red and gold paint over maple-colored wood. The air he played was unrecognizable to me—just stray notes and a few arpeggiated flourishes across the strings—but I recall its sound as something like that of a poorly tuned, plastic bass 'ukulele. It was not impressive. But it *plonked,* and its stuttering notes resonated over the hay-speckled air and fell upon my ears like the golden spatter from a gingko tree.

The second and only other time I heard a lyre in the live air was just a few years later, during my college days. In this instance there was no faire or revelry, but only a warm and slightly smoggy afternoon in Southern California. There was yet another bearded guy, a class ahead of me, curiously dressed—in carpenter's jeans and a puffy Renaissance shirt. This time, the collegiate scamp was up on the horizontally hanging limb of a huge sycamore with its mottled barkskins, striking a pose reminiscent of flutist Ian Anderson of Jethro Tull, lifting one leg up while the other dangled, and he twisted a bit on his ass, with his arms, torso, and chin set in a contrapposto, lyric pose. He held out the lyre stiffly in one hand as if it were a bow, while the other reached out, plucking single strings with doubled fingers as if he were arrowing them. There were girls around, of course, and a couple of other guys—one who was trying to yank him down from there, while our satyr du jour kicked him decorously away with a single, ungoatly foot. And the sound was nowhere near as memorable—in any case, it's now lost in a gouache of effortful pretension that does the instrument itself no justice.

I went to Rome in the early fall of 2017 and made a pilgrimage to the Cimitero Acattolico near the Pyramid in Testaccio, where I sat a reverent hour in the shade of an umbrella pine by the grave of English Romantic poet John Keats. Overcome by despair despite having penned some of the most beautiful and glorious poetry in

the English language, Keats thought himself a failure, and in a narrow room near the Spanish Steps, he died terribly of tuberculosis at the tender age of twenty-five. It was 1821. Almost two centuries later, in the breezeless quiet of the cemetery that hazy afternoon, I took a photograph of the four-stringed lyre in bas-relief at the crown of his headstone.

Keats's stone is simply fashioned, seemingly out of rough sandstone, his lyre a thin yoke and crossbar like a child's outline drawing of a French gentleman's head and his short-brimmed hat; the oversized rectangle of the gentleman's bow tie represents the soundboard, and, finally, there's a thin and narrow pedestal on which it sits. Below, these words:

This Grave
contains all that was Mortal,
of a
YOUNG ENGLISH POET,
Who,
on his Death Bed,
in the Bitterness of his Heart,
at the Malicious Power of his Enemies,
Desired
These Words to be engraven on his Tomb Stone.

Here lies One
Whose Name was writ in Water.
Feb 24th 1821

It was Keats who wrote in "Ode on a Grecian Urn," one of his final and now most treasured poems, the immortal lines "Heard melodies are sweet, but those unheard / Are sweeter . . ."

Recently, I spent an evening and a day listening to contemporary recordings of lyre music via iTunes on my computer and my head-fi rig—a combination of matched Chinese digital and amplification electronics, German headphones, and serious American cabling (balanced interconnects and USB). Previously, searching online, I'd found a wonderful English gentleman from Liverpool named Michael Levy, educated in philosophy at the University of Hull, who seems to have dedicated his life to the revival of the lyre and the collection of replicas of ancient ones he can play for numerous videos and recordings. Since 2008, Levy has produced thirty separate releases of lyre music, several available on iTunes, including *An Ancient Lyre* (2009), *Apollo's Lyre* (2010), *Echoes of Ancient Rome* (2011), *Ancient Landscapes* (2011), *Musical Adventures in Time Travel* (2013), and *The Ancient Greek Kithara of Classical Antiquity* (2016). The titles of the tunes are a combination of esoterica and whimsy: "Hurrian Hymn Text H6," "Awe of Aten," "Orpheus's Lyre," "Vapours of Delphi," "Procession of the Vestal Virgins," "The First Delphic Hymn to Apollo," "Song of Seikilos," "Shadow of the Ziggurat," "Lament of Simonides," "Ancient Greek Musical Fragment (Kolon Exasimon, Anonymi Bellermann 97)," and "Ancient Vibrations" (take *that*, Beach Boys). Levy plays these tunes on replica lyres crafted in Greece, done in the main, I gather from his website, by a workshop calling themselves Luthieros, being specialist lyre makers in Thessaloniki. Levy's lyres are tuned to various ancient modes: Greek Dorian, Hypodorian, Phrygian, Hypophrygian, Lydian, and Hypolydian; mystical Middle Eastern Hijaz; Hebrew Ahava Raba and Misheberakh; and an Egyptian minor pentatonic scale.

The tunes are a glorious assortment of the soothing (recalling young David's playing for King Saul), the stirring (appropriate for heroic hymns), and the pensive (provoking mild melancholia and inspiring mental travel through lyric time). By far the most compelling to me was "Hurrian Hymn No. 6 (c. 1400 BCE)," developed from an ancient Mesopotamian musical fragment. Not being a

musician, I cannot tell you the specifics of its melody or remark on its harmonic progression, but I can say it reminded me of boatmen's work songs I know from both the Russian and Japanese traditions. Think "The Volga Boatmen" or "The Boatmen on Toneh River," both barcaroles in steady, ponderous measures designed to accompany the labor of men walking a small cargo boat's deck as they resolutely pole upstream against a river's steady current. Like the Russian and Japanese songs, the lyre's tune is mainly dark, deliberate and halting at times, the bass strings thunking and buzzing as Levy plucks them. Yet there is an uplift of sorts as the melody moves through sweeter midrange notes and then back down the scale again, this time doubling the lower notes. It concludes with dancelike arpeggiated flourishes and strumming in its higher range, suggesting that hard work ends with some kind of celebration in joyousness rather than exhaustion.

The ancient tune was the only one found in nearly complete form out of thirty-six such inscribed in cuneiform on crumbling clay tablets excavated from the Royal Palace at Ugarit (Syria) in 1950, in a layer of earth dating from the fourteenth century BC. One of the tablets contained the Hurrian hymn to Nikkal, the orchard goddess of Canaan, daughter of summer's king, wife of the moon god, who adorned her with necklaces of lapis lazuli. It could be that her inscribed song was associated with the harvest of fruits in late summer, its rhythm mimicking the slow cadence of workers lifting heavy baskets, shouldering them, and swaying through the fields. An anonymous work first transcribed into Western notation by the scholar Emmanuel Laroche in 1955, the Hurrian Hymn is said to be the oldest surviving piece of notated music in the history of human culture.

There is definitely a sublime stateliness to the song as Michael Levy performs it on his album *An Ancient Lyre*. Its notes thunk resoundingly like a thick, magical cord beneath the ground, speaking as though from a dank hollow under an oak tree. Then, emphatic doubling of some midrange notes in a figure repeated from earlier in the tune creates a ritornello of poignancy to me, a sense of an ending as though time were closing. No, not *closing,* but *concluding.* It ends with a presto that is a scatter of rapidly plucked higher har-

monics and strummed arpeggios taking over the prevailing gloom in a kind of counterpoint of enervated ecstasy after deep sorrow. The song is most definitely ancestral, chthonic, of the earth and of the body *un-electric,* reverberant in the lyre's sonic bellyful of joy interfused with a mournful regret. It rattles and hymns.

The Aeolian Harp

I first became aware of the Aeolian harp, a stringed, rectangular box played by the wind, in a class on British Romantic poetry that met in the early morning (8:20 a.m., as I recall) during my sophomore year in college. We read "The Eolian Harp," a poem by Samuel Taylor Coleridge, about how the instrument inspired him to elevated moods and grand thoughts through the unseen mystery of its stirring into sound. I was a sleepy student, as I'd stay up, often past two a.m., trying to read everything assigned; there were reams of pages, it seemed, consisting of letters, essays on Romantic philosophy, and the poetry itself, which was written in a style of English then slightly archaic to me—convoluted, emphatic, yet blessedly *unrhymed.*

Most of my classmates were eager, bright, and alert, as they had no need to keep the late hours I did, having already read most of the stuff in prep school, even discussed it in their classes at Deerfield Academy, Phillips Exeter, Brooks, and Horace Mann. My public high school in Gardena was *not* like that. It locked books by Franz Kafka, James Joyce, J. D. Salinger, and Harper Lee's *To Kill a Mockingbird* inside a spinning glass case in the school library next to a wall with the American flag pinned to it. You had to have a note from a parent to get access and, to my recollection, no one ever did. But my blessèd liberal arts college had us kids from the EastLos, South Bay and South-Central, Zuni Pueblo, a Sauk and Fox reservation, and the agricultural fields of the Central Valley in California mixed in with children of the elite. I did not feel so much out of place as at a disadvantage from not having slept much. My classmates were exceedingly kind, and one who sat next to me, a curly-haired beauty from Tucson now herself a widely known poet, would gently nudge me awake if she spied our professor glaring at me and getting set to put a question. I would gather from my drowse, glance

at the stoical faces around the circle of desk chairs we sat in, and assume it was I who was to speak for all the pitying eyes riveted on me in the room.

"Mr. Hongo," Professor Frederick Mulhauser would say, pipe in hand, tamping the tobacco down with a pudgy thumb, "like our man Coleridge, could it be Music is still slumbering on your instrument too?"

Lovingly, the class would not so much laugh as titter, and I'd bestir myself and apologize. And the class would go on, Miss Bliss or Mr. Menand or Miss Hillman saving me with a comment of their own for our Yale-trained textual scholar. At our first meeting, he'd declared he was "a Victorianist" and not a Romantics specialist; consequently he would not lecture but teach via the Socratic method, putting questions to us, trying to inspire dialogue *dialectically*. He was cheery and soft-spoken, but intimidating as hell, with a slightly jowly face that went from cheerful to dour in an instant, depending on our answers. He had owllike, though unblinking eyes, a short nose, also like an owl's, and complete command of an immense erudition he'd display in baroque asides that cited Greek and German philosophy, British Hartleyism and French Cartesianism, Renaissance Jesuit scholarship, anecdotes about Sir Isaac Newton and Galileo, and conceptions of the universe from classical times through the Enlightenment and into the Romantic era.

How were we to dare uttering a syllable with Mulhauser as our judge? It was like we were all on a literary version of *America's Got Talent*, with our professor a bow-tied Simon Cowell remarking in clipped tones on our frail, tremulous thoughts as though they were notes sung slightly off-pitch and quaverously, but then, even more mercilessly, likening them to passages he'd recall to us from the disquisitions of Aquinas, the essays of Friedrich Schiller, and anecdotes from Augustine—all of which we *surely have read*—but only bespectacled Mr. Menand, a Brooks alum, really had. And, never a show-off, even Menand would not offer to speak unless called upon, sparing the rest of us the shame. But when he did, all of us would scribble down his remarks, underscoring them, noting their relevance and prowess. Now a leading critic and an English professor at Harvard, he holds at least one Pulitzer Prize.

In the poem, Coleridge is in the garden at Clevedon, his house at Somersetshire, where he is on retreat from London with his fiancée, Sara Fricker, whom he describes as pensive and possessed of a "soft cheek." She reclines on his arm beside a cot overgrown with jasmine and broad-leaved myrtle, while they take in the clouds "that late were rich with light" and smell the gay scent from a bean field nearby with the world "*so* hushed" around them and the evening star just winking on. It's an intensely pastoral scene when the title character makes its sonic entrance:

> *The stilly murmur of the distant Sea*
> *Tells us of silence.*
> > *And that simplest Lute,*
> *Placed length-ways in the clasping casement, hark!*
> *How by the desultory breeze caressed,*
> *Like some coy maid half yielding to her lover,*
> *It pours such sweet upbraiding, as must needs*
> *Tempt to repeat the wrong! And now, its strings*
> *Boldlier swept, the long sequacious notes*
> *Over delicious surges sink and rise . . .*

(from "The Eolian Harp," published in 1796)

Coleridge is describing the harp's uncanny sound, full of long, pliant, and continual notes that surge, as he says later in the poem, in a "soft floating witchery of sound / As twilight Elfins make . . ." The notes rise and fall in harmonics as the air blows over the strings of the instrument, a thing played by nature and not the hand of man. The poet declares that its sounds voyage on the wind "on gentle gales from Fairy-Land," its provenance from a magical realm and not quite of this earth, its melodies clothed in "honey-dripping flowers" like birds that never alight but hover wildly in air.

The kind of harp he might have been hearing would have been a box zither, popular in his day, made of thin, resonant pine cut to the length of a window sash with ten or twelve strings strung lengthwise over bridges on opposite ends and an aperture or two carved decorously into its top. It could have been placed upon a windowsill or out in a garden so the wind might pass eas-

ily across its strings, stirring them into its curious song. Tuned in unison, strings of varying thickness and elasticity would vibrate when excited, playing overtones—the octave, perfect fifth, major third, seventh, etc.—rather than the actual note of the string when plucked. In other words, *not* the pitch they were tuned to, but notes altogether arising out of a mysterious property of interaction with wind and a mathematical relationship with the tuned, fundamental note. The interaction produces a Pythagorean series—solely harmonic frequencies without the normal percussive aspect of plucked notes, spectral airs in crescendos and decrescendos that sound like sighs, small cries, and Bergman-like, *Seventh Seal* whispers that shift and morph in rhythm with each bend of the wind across the harpstrings. An article from an 1858 issue of *Scientific American* describes the sound this way: "When the air blows upon these strings with different degrees of force it will excite different tones of sound. Sometimes the blast brings out all the tones in full concert, and sometimes it sinks them to the softest murmurs."

You know that funny, plosive sound you get in your car when you roll the windows down driving at some speed? That *phumpp-phumpp* whack of a low-resonance pop with an accompanying pressure you feel in your ears and jawline? It's a phenomenon called the *von Kármán vortex trail effect* (named after twentieth-century Hungarian American engineer Theodore von Kármán), a repeating pattern of swirling and shearing vortices that arises when the turbulent flow of a fluid blasts around a blunt object or objects—in other words, the wind thudding around your car window. This is also what's responsible for the singing of telephone lines, as romanticized in the High Lonesome song lyric "Wichita Lineman," written by Jimmy Webb and made popular by singer Glen Campbell in a 1968 recording. Over a century before, in 1851, Henry David Thoreau heard it in the new telegraph line strung between Boston and Concord one afternoon when he made the walk down a new railroad track that linked the two towns. "It was as the sound of a far-off glorious life, a supernal life, which came down to us, and vibrated the lattice-work of this life of ours," he wrote. The math that explains the phenomenon is daunting; suffice to say, we're talking about wind eddies, resonance, and rows of vortices this kind of

turbulence leaves in its wake as it flows around a thing, causing it to vibrate. In a car, what vibrates is your curved window, which has essentially become a soundboard. In an Aeolian harp, what vibrates is the instrument's tuned strings, atremble in the circles of wind currents that shift from side to side as a mountain stream might its twin eddies as they swirl and break away behind a mossy rock. Thus, the ethereal music of the Aeolian harp became a heaven-sent metaphor for the Romantics, both British and American, and it makes appearances in works by Samuel Taylor Coleridge, Percy Bysshe Shelley, Ralph Waldo Emerson, and Henry David Thoreau (*why are nineteenth-century poets always three-named?*).

Legend has it that the harp shows up in Western lore when King David hung his kinnor, a kind of lyre, outside his tent at night to try to catch the wind, Donovan style, across its strings. But even before then, the Greek god Hermes is said to have discovered the lyre when he heard the wind sing across dried sinews stretched over a tortoise shell. Yet the first *crafted* such instrument comes much later, during the Enlightenment, when a quirky Jesuit scholar named Athanasius Kircher, whom historian Paula Findlen calls "the first scholar with a global reputation," depicted the instrument he had fashioned in his Latin texts *Musurgia universalis* (1650) and *Phonurgia nova* (1673). Kircher shares a line drawing of his invention, showing a rectangular booth with two diagonal doors or flies that fold out from its left side like wings of a standing screen. Then, opposite them, the box of the harp is affixed vertically to the right wall of the booth, sort of like a narrow portable A/C unit hanging lengthwise outside a window. On the far left of the drawing, floating in air over a desolate blank space, is a bewigged or curly-haired head (it's so small in my pdf page reproduction, I can't tell), presumably Aeolus, Greek god of winds, puffing black striations (zephyrs!) toward the open flies/doors of the booth. Under this all, another drawing, this of Kircher's harp itself, a large rectangular box (1:085 meters long × 0:434 meters wide × 0:217 meters high, saith *Scientific American* in 1883) with fifteen strings, two bridges, and a row of tuning pegs on its left edge. Three roseate clouds of notes emanate at spaced intervals above the strings.

According to a chronicle by his student Gaspar Schott, Kircher

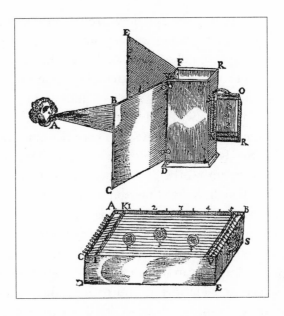

fashioned the harp after a description in *Natural Magick* (1540), an
even older Latin text by Giambattista della Porta. Where winds are
"very tempestuous," writes Della Porta, stringed or wind instruments would play by themselves. Like Della Porta before him,
Kircher considered the harp an instrument of "natural magic,"
belonging with the talking automatons and echoing mirrors in an
astonishing museum of curiosities he kept at his college in Rome.

As Findlen declares, Kircher was quite the Renaissance man—a
polymath, an obsessed though mistaken interpreter of hieroglyphics (and founder of Egyptology), and a clerical intellectual esteemed
throughout the Western world. He dabbled in geology (studying
Vesuvius, Etna, and Stromboli firsthand), biology, Sinology (once
asking to be assigned as a missionary to China), and made charts
and calculations that postulated the size, number, and distribution
of animals, and amount of feed contained in Noah's ark. "Nothing
is more beautiful than to know all," he once wrote. He could even
be considered the father of modern audio and acoustics, having
explored in his writings the transmission of music to remote places
(begetting radio, hi-rez streaming, wireless Beats) and the use of
horns and cones to amplify sound (movie theaters, hi-fi, home stereo systems!). He collected strange machines and gadgets that came

to him from throughout the world via his Jesuit associations. His Museum Kircherianum, founded in 1650 (and now partially preserved in the Pigorini National Museum of Prehistory and Ethnography in Rome), was what was then called a *Wunderkammer,* or a "cabinet of curiosities."

Kircher's museum, perhaps the most impressive such collection of its time, contained a magnetic clock, a sunflower clock powered by solar energy, a reconstruction of the flying dove of Archytas (the Tarentine philosopher), the first megaphone, various talking heads, a perpetual motion machine, numerous skeletons (human and animal), a preserved armadillo, paintings, and macabre art objects. Kircher used an early microscope too, observing that he saw "little worms" in flesh specimens gathered from deceased victims of the Black Plague, conjecturing that the disease was spread through tiny organisms (he was right, though what he saw were likely blood cells). With his eclectic knowledge, Kircher made friends with Italian sculptor Gian Lorenzo Bernini and helped him restore and then engineer the placement of the huge obelisk (with gibberish hieroglyphs of Kircher's design) adjacent to the sculptor's Fountain of the Four Rivers in the Piazza Navona in Rome.

When I think of Kircher's museum, all I have to compare it to is the Museum of Miniatures in Prague with banks of microscopes where visitors can observe tiny fleas under glass pulling a line of train cars, a caravan of camels parading through a needle's eye, a seed of grain etched with a reproduction of one of Van Gogh's paintings. It's not about the miniature for me (or an artist so obsessive he made his own instruments and learned to employ them between heartbeats so that the mere push of his blood through his vessels might not stray his hand from precision), but about the purely curious aspect of it all—a mind bothering to pursue the strange and provocative just out of the spirit of inquisitiveness. It doesn't matter if there might be *worth* to it in terms of knowledge. What matters is not the knowledge, it is the fascination itself, the way strange objects might bend the mind out of a stolidity inherited from any current zeitgeist. Kircher's collection is this Museum of Miniatures on a grand and more intellectual scale, his in fact perhaps the forerunner to modern museums of science and technology. Marveling at his penchant for

wondrous and bizarre esoterica, the Princeton historian Anthony Grafton claims "the staggeringly strange dark continent of Kircher's work [is] the setting for a Borges story that was never written."

Out of all this came the wind harp, whose harmonious sound, Kircher intimated, echoed the proportions of the universe. "In my Museum," he writes, ". . . it is listened to with great amazement . . . No one will ever suspect what kind of instrument it is, by what hand or pump or artifice it creates its melodious sound. This instrument will be so much the more recherché and worthy of wonder to the extent that it is more hidden or concealed." As William Blake wrote of the pacing tiger, *What immortal hand or eye, / Dare frame thy fearful symmetry?* The full story is that Kircher once played a practical joke on the minister of his abbey, hiding the harp, then sending the cleric looking for what he heard as an organ sequestered in one of Kircher's rooms. I imagine the poor man went from room to room, searching for that faintly crying baby of a sound, only to be stumped and frustrated, clutching his rosary and cross and murmuring Hail Marys.

Kircher named his elaborate harp the "Machinamentum X," a kind of *nom de cirque* that reveals what a prime huckster and showman its seventeenth-century inventor was. He never called it an Aeolian harp. That honor was left to a later scholar, Johann Jacob Hoffman, who gave it the name "Aeolium instrumentum" in another Latin text entitled *Lexicon universale* (1698). The harp makes just a few further literary appearances in notes, and most magnificently in the writings of nineteenth-century French romantic composer Hector Berlioz, who writes that the harp "possesses remarkable properties which act upon the nervous system and cause very different impressions, according to the temperament of those who listen to its accords."

During the eighteenth century, a group of Scottish poets and musicians, displaced to London, took up an interest in the instrument and wrote about it in poems, plays, and operettas. James Oswald, a member of that circle, started a small music shop and manufactured the instrument—now with seven or eight strings, rather than Kircher's fifteen—and their Society for the Temple of Apollo championed it as a delightful plaything for their readers. Oth-

ers in the group were James Thomson (author of "The Seasons," an extended landscape poem in, obviously, four sections, which I still teach to my graduate students), Christopher Smart, mad poet who composed the enumerative strophes praising his cat Jeoffry ("For he is the Servant of the Living God duly and daily serving him"), and the satirical novelist Tobias Smollett. One account, likely fiction, is that Oswald came by chance on a reference to wind-induced vibrations on a harp while reading a commentary on Homer by Eustathius and, already a maker of violincellos, Oswald fashioned his own instrument after the mention. Later, Robert Bloomfield, son of a poor shoemaker and a harp-builder himself, may have composed the first poem in English dedicated explicitly to the Aeolian harp, which begins:

> Oh breeze, when sleep'st thou? Come, oh, come,
> This languor of my frame dispel;
> Arise,—thy own loved harp is dumb;
> Arise, and bid thy chorus swell.

And closes:

> Thanks, charming zephyr,—Hark! That tone!
> Be true, sweet harp; hush all but thee;
> Perform thy task untouch'd, alone,
> And pour thy tide of harmony.
>
> ("Aeolus" by Robert Bloomfield, published
> posthumously in 1824)

I leave it to you to choose whether you prefer Bloomfield's charming, more innocent rhymed quatrains or Coleridge's unrhymed stanzas that make epochal claims regarding the relationship between nature and the elevated sentience of humankind. I like them both, Bloomfield's more like contemporary folk lyrics but for the *Harks* and *thees* and calls to *Arise*. (They're a little like Bob Dylan's *Come gather 'round people wherever you roam* from "The Times They Are A-Changin'.") And, come to think of it, had Professor Mulhauser quoted Bloomfield's ode to me that morning long ago and been perhaps blessed with an iPhone for sampling an electric guitar riff

from an Electric Flag song played by Mike Bloomfield (I assume no relation between the Chicago bluesman and the Scottish poet), I may have indeed arisen from my slumber sentient enough to comment on Coleridge's more mystical harp.

Again, Coleridge:

> *Methinks, it should have been impossible*
> *Not to love all things in a world so filled;*
> *Where the breeze warbles, and the mute still air*
> *Is Music slumbering on her instrument.*

> [. . .]

> *Full many a thought uncalled and undetained,*
> *And many idle flitting phantasies,*
> *Traverse my indolent and passive brain,*
> *As wild and various as the random gales*
> *That swell and flutter on this subject Lute!*

> *And what if all of animated nature*
> *Be but organic Harps diversely framed,*
> *That tremble into thought, as o'er them sweeps*
> *Plastic and vast, one intellectual breeze,*
> *At once the Soul of each, and God of all?*

Coleridge's harp is much the grander one, I think, capable of tricking down thoughts from "random gales" across its fluttering strings, inspiring the poet to speculate that all of the universe might be a collection of harps, variously framed, but all singing from one godlike intellectual breeze. Its music would be more abstruse and not as definable as a folk tune or guitar lick. There would be something of an evanescence about its song, a chromatic moan just barely audible, as though a choir might be singing from a distant riverbank upstream from where you stand, its voices carried by the air buffeted by the lurching back of flowing waters underneath, swirling vortices of faint cries and harmonies that might reach you as a perfect sound.

I came across such a thing one day while working for the Department of Water and Power of the City of Los Angeles during summer breaks from college. My job title was "seasonal meter reader." This meant that I spelled regular workers over the summer months so that they could take their vacations, about two weeks long. As the entire pool of meter readers numbered over two hundred, it also meant I worked throughout the summer.

My job was to walk a given route each day, taking me through just about every part of L.A., from the Van Norman Reservoir at the north end of the San Fernando Valley to the loading docks and canneries on Terminal Island in San Pedro at the city's south end, and log usage readings from hundreds of electric and water meters. As I walked, I hardly ever stopped, but kept moving so I could finish the route before one o'clock and get to the beach for a swim or a run before the winds kicked up at three. I walked the Chicano neighborhoods in the hills around Dodger Stadium and read water meters buried in the dirt. I strode briskly through Watts, where I saw children walking pet cockroaches on makeshift leashes of thread or string. I had a shotgun trained on me through a peephole in South Central, a policeman sweep his sidearm past me tracking a fleeing thief in San Pedro, and Dobermans and German shepherds and rottweilers foam at the mouth while pursuing me through the Hollywood Hills. I read meters throughout Laurel Canyon and saw a beautiful rock star stark naked, disporting with her pet Afghan hounds around her spacious backyard. On foggy Blue Jay Way, I read George Harrison's meters. I read James L. Jones's meters. I read the meters at the Hollywood Bowl. If you lived in L.A. at that time, I likely read your meters too.

While I was on the move, speed-walking from meter to meter—jumping fences, leaping over brick walls, cutting through a whole residential street's worth of backyards—I had a lot of time to think. And what I thought about entailed a kind of rhyming—squaring the experience of hard, blue-collar work against my liberal arts college courses in Shakespeare, British and American Romanticism, Chinese and Japanese literature, and the philosophy of Ludwig

Wittgenstein. What I didn't want to do was isolate one experience from the other. What I wanted was to *join* my life—one of work along the wide and narrow avenues of L.A.—to the great voices I was hearing in my head as I traipsed, in 95-degree heat, up a long hill full of apartment houses, dodging children and dog shit along the sidewalks. *They flee from me that sometime did me seek*—a famous line from a sixteenth-century poem—would echo in my mind as I glanced at the small dials on the face of an electric meter from behind a lavish bush of jasmine flowers, its redolence carried on an ocean breeze wafting over my route.

On a given day, I'd take my lunch in a park I'd spotted earlier, opening up my sack of sandwiches, chips, cut cucumbers and carrots. I'd have time for gazing deeply from under the mottled shade of a big-leaf maple tree out toward the end of whatever block to a confusion of billboards, street traffic, fast-food joints, and the sheen of yellow and brown along the belly of sky above them. I'd see past these to Othello standing under stars, raging in his folly; to Ophelia recumbent in a coffin of pond water; to lunatic Whitman yawping in ecstatic praise for all our peoples under democratic vistas.

I had that job over five or six summers. I liked it. It gave me a rhythm for my thoughts. It gave me the acquaintance of all of Los Angeles and its harbor. It gave me the start to all further ramblings and the ground notes to a barbaric song of knowings to come.

One day, in the early afternoon, I had been going slow and was out long past when I should have been finished with my route. It went through a pleasant neighborhood in Rancho Palos Verdes, a steep, hilly area at the southernmost tip of the city where it abuts the Pacific. It was full of three- to five-bedroom residential houses with large lawns and few savage dogs. But there were long, uphill treks I couldn't shorten by jumping fences or rock walls. I had to "square in" most every one, unlock a side gate, and read each meter up close rather than spy them through my pocket-sized pair of binoculars. This cost me an inordinate amount of time, and it was past two o'clock—an hour when I should've been done with work, by then already body surfing or browsing the stacks of my favorite bookstore in Hermosa Beach. I was approaching the side of one house, tramping up a lavish lawn thick with tufts of green grass that made

the going feel as if through Velcro. I opened a white-enameled gate, walked down the narrow side path crowded with a wheelbarrow, jade plants in little terra-cotta pots, and the green coil of a garden hose leaned up against a stuccoed exterior wall. I got to the backyard where my instructions said the meter box would be, and then I saw the brilliant ocean, tossing in cowlicks of waves where the ground fell away and the aquamarine waters extended to a ribbon of yellow haze clinging above the horizon. I felt little zephyrs swirl around my pant legs and shirt collar, cooling my sweating neck. I heard a faint sigh, a wraithlike metallic moan that seemed to warble and transmogrify from something insubstantial into being and then away again. I heard the odd groan and then the gentle harmonic murmur of an Aeolian harp. I turned. Its voice, frail and plaintive, arose from a V-shaped wind sculpture made of pieces of beach-worn, scavenged wood and painted metal mounted over a fan of red-earth-colored rocks where the aluminum whirligig of a rotary clothesline might've once been. Like a lyre, it had strings that ran up its yoke to a crossbar at its top. Like Hermes on a far shore, I heard the wind play on them, emanating its ghostly sighs. *I hear you singing in the wire. . . .*

The Guitar

Since I was a teenager and first heard one up close and live, plucked by the salesman at Wallichs, I never stopped loving the sound of the acoustic guitar, whether Spanish, steel-stringed, resonator, or Hawaiian lap steel. Something about the way its notes got amplified by its soundboard, pushed magically along by the sound hole, reverberating in air like a luscious smoke furling and bending and then gracefully decaying into disappearance. There is a romance to its sound for me, almost as though it were the voice of the gentlest singer, one who could be stirred to vibrant passion in an instant. Since those youthful days inspired by cheesy TV and movie theme songs, I've listened to guitars playing blues from the Mississippi Delta, Hawaiian *hapa-haole* songs, folk tunes from Appalachia, Tex-Mex *corridos*, Andalusian ballads, Scarlatti sonatas, and featured as the solo instrument in orchestral concertos.

In college, my dorm neighbor, whose major instrument was the violin (he is now conductor of a university orchestra), played classical guitar. He practiced Bach sonatas in the afternoon while I homeworked in German and Japanese, read Shakespeare and lit crit, and dreamed of writing my own poems. One evening after dinner, he knocked on my door, inviting me to listen to a new record he'd gotten. It was guitarist John Williams playing Joaquín Rodrigo's *Concierto de Aranjuez* with Eugene Ormandy conducting the Philadelphia Orchestra. He gestured for me to sit in a desk chair he'd set up in the middle of the room. He put the record on the turntable and adjusted the volume, turning it up so music filled his crib and we heard no extraneous dorm sounds. No shutting doors or voices in the hallway. No leakage of stereos from other rooms. He took his own seat in a wing chair in the corner. We listened the whole way through without speaking. We were especially still during the adagio second movement that opens with an oboe introduction, playing a mordant Iberian tune with the guitar slowly strumming somber chords in accompaniment. Then the guitar inherits the melody and takes it up in a shimmering, tremulous solo full of a tender, aching emotion that puts you on the verge of tears, while the strings drone murmurously and descant underneath. Evolving dramatically from stately to brisk, taking up flamenco-inspired rhythms and counter-rhythms, then giving way to a light, frailing arpeggio like a harp, the piece took my breath away, especially when the full orchestra weighed in with the theme just before the movement's end. Its moment is as heroic as the most grandiose sweep of uplifting soundtrack in any movie, war or Western. The "Rodrigo" remains one of the most magnificent and sublime pieces of music I've ever heard.

A resonator like the lyre or the abdominal chamber of a cicada, the traditional guitar has a number of tuned strings and a soundboard (the guitar's body). But it adds two more important features: a fretboard to change the length of the strings when plucked or strummed, increasing the sonic frequency of its vibrations in measured steps (defined by the spacing of the frets), thus producing a variety of notes from each string instead of just one; and the *sound hole*, a port on its body that adds another dimension of acous-

tic amplification beyond the transfer of string resonance to the soundboard.

When a guitar string is plucked, it vibrates and creates the rich spectrum of harmonic partials that is the source of its characteristic sound. But the string vibration is only the source—the instigation of the sound that then is transferred to the guitar body via air and the bridge mounted onto its body. As in the lyre, the bridge transfers the string vibration to the soundboard of the guitar. The soundboard is usually made of a springy, responsive wood like spruce, which naturally amplifies the string's harmonics through sympathetic vibrations, a pattern of sound waves that travel from the strings to the bridge and then to the soundboard or top of the guitar. Legbone connected to the hipbone, etc.

But there's an additional coupling and resonance effect performed by the guitar's design. As it has a hollow body, the air in the cavity of the instrument resonates with the vibrations of the string and soundboard. And, depending on the frequency of that resonance and the capaciousness of the body, the guitar's inner chamber acts like yet another resonator, increasing or decreasing the volume of the sound once again. This aspect of acoustic behavior is called a *Helmholtz resonance,* named after nineteenth-century German physicist Hermann von Helmholtz, who, in 1862, created a device that he used to discern specific pitches in complex music. An empty Coke bottle, when you blow across its top and hear a vaguely flutelike or pipe organ sound, is a Helmholtz resonator. A Japanese bamboo *shakuhachi* is one too (besides being useful as a good cudgel in a pinch). A conch shell held up to your ear is another. The air cavity of the guitar functions as a type of Helmholtz resonator, usually giving more oomph to the lower frequencies of the instrument's range. Luthiers, or makers of stringed instruments such as guitars, normally build their instruments to enhance the lower range rather than the upper frequencies in order to create a fuller sound.

Once a guitar is plucked or strummed, the volume of air inside the body gets excited, compresses inside, rebounds like a piston, and then shoots out through the sound hole, amplifying certain frequencies of its sound. But it also happens that, once the air expands, it's sucked back *inside* as the air in the cavity condenses

back (rarefies) from its momentarily expanded volume. This is the normal pressurizing and rarefaction that happens in a sound wave, but it's happening from inside the cavity of the guitar and through its sound hole, an area of exchange and passage for the alternating expanding and contracting piston of air. When the air expands, it enhances the launch of vibrations from the guitar as a whole, joined *in phase* with the vibrations of the soundboard, again, usually in the lower frequencies. When it contracts and the air compresses, it attenuates the sound and goes *out of phase,* against the launch of sound from the guitar as a whole, particularly its soundboard. The air inside the guitar oscillates in this way, driven by the springiness of the air excited by the vibrating body of wood, resonating sweetly or clanging in discord, depending on the skill of its player.

The Natch'l Blues and the Resonator Guitar

While surfing the net recently (I was looking at pictures of old bluesmen), I came across a 2017 video of contemporary blues artists Taj Mahal and Keb' Mo' playing an old song that I recognized from years past. They play guitars, passing licks and sharing choruses of the tune, alternating lead vocals, Keb' sometimes playing harmonica. It's a charming session, and what's astonishing to me is the guitar Taj is playing in the vid. While Keb' Mo' strums and plucks a traditional wooden acoustic with a big sound hole, Taj plays something else—a gleaming metal-bodied guitar seemingly without a sound hole but with what looks like screened vents checkered on both sides of the fretboard at the top of the instrument's body. There's a big T-bridge on the lower body behind the palm of his strum hand and a shiny tailpiece reaching out from the bottom. It looks to me like an original National tricone resonator guitar (or perhaps a replica of it, made of nickel-plated German silver), which was originally designed in Los Angeles in 1927 by John Dopyera, a Slovakian immigrant. And it made its debut with Solomon Hoʻopiʻi, master of the Hawaiian steel guitar, who was all the rage with the Hollywood film crowd back then: how about *that* for an all-time American multicultural mash-up? In the video, Taj wears a blue aloha shirt with swirl shapes that look like fish and lilies, while Keb'

has on a fancy dress shirt. They both wear caps and jeans, but Taj has on bowling sneakers and Keb' blue suede shoes. They glance at each other with affection throughout the session on the wonderful tune.

The song is a version of "Corinna," a traditional blues Taj reworked and recorded back in 1968 on his album *Natch'l Blues* (Columbia, 1968) and that I heard first in a live performance in 1969 during my freshman year. A crowd of us sat on the floor and filled the ballroom of the student union on a Thursday night when we might've been studying. We hooted, yelled, and applauded for Taj, then fairly new on the scene—a young Black bluesman who'd come with his band to play a concert. He's tall, maybe over six feet, and in those days was slim with big shoulders and biceps, looking a lot like a football player. He wore black jeans, a plaid flannel shirt, and a Levi's jean jacket. He'd wound a red bandana around his neck and topped himself off with a big, floppy-brimmed off-white hat with a red feather in its black hatband. He stood in front of his group of musicians, arranged in front of an idle grand piano pushed to one end of the ballroom.

"Play 'Corinna'! Play 'Corinna'! 'Corinna,' 'Corinna'!" someone yelled.

Taj turned toward the voice at the back of the crowd, held the shouter with his gaze a moment, then glanced around at his bandmates, who looked back, ready to play. He stepped forward, making a few kids in front, still seated on the floor, scoot back a little.

"You can't just call up the blues anytime you like," Taj lectured. "You got to live the blues. You got to *love* the blues. You got to feel it in your blood, people. Don't be yellin' at me for what you'd like me to play just 'cause you think I'm a jukebox. Be respectful. Listen to what we got to offer you. We come to share our music. We come to school you in the blues. We give you our blood boiled in the blues. Don't yell at us, please."

He silenced the crowd with that, all of us feeling chastened and instructed. We were college kids, most of us white, and what did we know? Music for us came off records we bought and collected, maybe shared, but who among us made it our lives? It was a grace, an indulgence, or a weekend concert that most paid for with money from their parents. If we wore patches on our clothes, they

were an affectation, something we picked up at Berkeley High to show we'd rejected the mainstream thing. If we even played folk music, we learned it from our precious collections of Folkways, Vanguard, and Smithsonian records—not in Appalachian hills and hollers. If we even played the blues, we got it mostly from records too, the albums by British players like Jeff Beck, Peter Green, and John Mayall, who'd learned it from records themselves—in their case, from records by African American bluesmen. Who could argue with Taj Mahal?

"'Corinna'!" the same guy yelled, adding a dry laugh when he felt the silence bury him.

The concert was sponsored by the Associated Students, led by an entrepreneurial classmate who cut a deal with the Ash Grove, an old dive out on Melrose in L.A. that booked blues and rock acts of a certain funky kind. Our college would get them for one night, usually on a Thursday when the Ash Grove was dark. That way, the musicians, from, say, Chicago or the Southeast, would get a full week's booking while they were out west. Before I got to college, I'd heard Ike and Tina Turner and the Ikettes there. Tina owned the room, dancing up a storm, her voice gritty and body gyrating, her big wig hat flailing the air as she spun, twined, and cool-jerked on the runway and stage of the small club. I felt the walls sweat when she pumped a lot of pain and glanced at me with a side-eye as she strutted to the music. Earlier that year, needing to rise by six a.m. for my job at DWP, I'd missed late-evening gigs by Howlin' Wolf, Gatemouth Brown, and Koko Taylor. Taj was younger than all of them, a kind of novelty in that he was urban-born, college-educated, and had gotten into the blues as a sort of revivalist, at least two generations removed from its originators in the Mississippi Delta.

"I'm-ah sing it for you," Taj said. "Not because you call. Not because you call or anything like that. I got honey in my heart is why I do."

He glanced then at Jesse Edwin Davis, the Native American guitarist, whose long black hair hung like a horsetail down his back, over his concho belt, and past his waist. Then Taj turned away from the audience and picked up a battered acoustic guitar that had been at rest on a stand. It seemed made of wood, with big

f-holes and a kind of dome-shaped, metallic top on the body just under the strings. Around the rim of the dome, there were squares or diamonds of perforations arrayed in a ring. They made it look like an aluminum vegetable steamer, ass-end up, had been screwed onto his old axe. Taj, instead of standing with his guitar, plunked himself down and sat with it cross-legged on the floor, just like the rest of us, placing the bottom of his instrument in the middle of the bucket his legs made. He began to pluck out that catchy three-note sequence, shaking his head back and forth so his hat shook too, the red feather swiping at the air in front of the black finish of the piano behind him. I saw that he plucked hard on the fifth string with his thumb, using a kind of apoyando stroke, resting it briefly on the fourth string, then plucked that fourth string and repeated the sequence many times before the rest of his band—Davis on electric guitar, an electric bass, and drums—kicked in, as Taj started singing the lyrics and melody with his gruff yet tender voice. The bass and drums locked together in a reggae beat, and Davis strummed syncopated chords just off it. With Taj playing flourishes on his curious guitar, the tune possessed a gut-thunk that slapped against the wooden ballroom floor and, simultaneously, a sweet chime that lit the air and chandelier lights above us. He sang in a Delta-style vernacular that he *got a bird what-a whistle,* then sang the phrase *got a bird* four more times in successive lines, each underscored by the same insistent beat. And that beat got accentuated in a syncopated rhythm played on the bass and rhythm guitar, which lifted the tune into a staccato shuffle. It stumbled, jerked, and then righted itself. Though that bird would sing, Taj declared, without Corinna, it *sure don't mean a natural thing.* The rhyme ended the verse and completed righteously on the beat too.

I'd never seen anything like his guitar before. And I had no inkling then that I'd already heard guitars just like it on my parents' Hawaiian records back home in Gardena. I went alone to the concert, and so had no friends to compare notes with that night. But the next morning, I described the strange guitar Taj had played on that one song, and a couple of friends who were guitar players made a few guesses. A classmate from Martha's Vineyard who owned a Martin 00-18 said it was "probably a Dobro," a wooden-bodied

resonator with f-holes and steel strings that was invented during the 1930s. Someone else said it was "a National," the first resonator guitar ever made, but that only a few were made of wood back then—most were "steel body." *How many of these kinds of guitars were there?* I wondered. *And what's a* resonator?

Though I can't be sure, on *Natch'l Blues,* his recording from 1968 with "Corinna" on it, I think Taj is playing a steel-body National Triolian, a single-cone (despite its name) resonator guitar made by the National String Instrument Company of Los Angeles sometime after 1930 and before 1936. And, in liner notes, he lists *Miss "National," Steel-bodied Guitar* as, along with harmonica, his instrument on the album. Yet there are photographs of him, taken around the same period, posing with a wood-body resonator guitar that could have been another National guitar or one built by Dobro, a company founded by Dopyera, the instrument's inventor, after having split with National. The photos are available online through Pinterest and on Taj Mahal's website as well. There were numerous iterations of this guitar design, as it turns out, Dopyera creating many versions to satisfy not only the desires of its varied players (who'd eventually span the seemingly separate genres of Hawaiian, blues, and country music), but the demands of his market- and profit-oriented business partner, George Beauchamp.

Beauchamp was a vaudevillian entertainer who came up with the initial idea of a resonator guitar during the late 1920s. He wanted a guitar loud enough to be heard above woodwinds and raucous brass in the orchestras that backed his act. His first thought was to take a Hawaiian steel guitar, a popular instrument of the time, place it on a pedestal that would serve as an external amplifying soundboard, and attach that to a Gramophone-like horn pointed toward the audience. Beauchamp hoped it would be an instrument of sublime sound and power. He took his idea to Dopyera, whose shop was a few blocks away from where Beauchamp lived. The craftsman and violin-maker Dopyera had his doubts, knowing a bit more about acoustics than Beauchamp, yet took the commission and made the contraption anyway, its horn bigger than a tuba's bell. As he suspected, though, the giant horn colored the tone of the instrument so badly, it sounded like a hornet with a cold.

Dopyera, an inventor of sorts who had several patents for other instruments (he made banjos and mandolins as well), then experimented with various alternatives, coming up with the idea for a resonator to be placed inside the body of the guitar. It would capture the vibrations of the strings and amplify them acoustically without the distortions inherent to a big horn. He made drawings, tried various materials, tried a single resonator, and finally settled on a design that used three small and convex six-inch aluminum cones that sat inside a circular metal basket mounted in a large cutout in the top of the guitar body. The cones were hand-spun in Dopyera's own shop, his brother Rudy working the lathe and spinning tool. The guitar's body was of nickel-plated German silver, fashioned by Adolph Rickenbacher, a Swiss immigrant who had a shop just down the street from Dopyera's. He was great at shaping metal to Dopyera's special requirements. For a top plate, Dopyera placed a round-cornered triangle of like metal with diamond- and pyramid-shaped screen cutouts over the cones, making for a gleaming, ornately constructed instrument unlike any guitar ever before. This was a tricone resonator with a stunning Art Deco look to its metal body, a honking big T-bridge, and a distinctive, metallically tinged sound bolder and louder than any other guitar of the pre-electric era. The three inner cones not only captured the sound of the strings, but resonated with each other and the bridge, producing a penetrating but sweet and rich sound full of upper harmonics. To debut Dopyera's instrument, Beauchamp commissioned the Hawaiian guitar virtuoso Sol Ho'opi'i, enormously popular at the time, to play it at a party for Hollywood insiders. Taj Mahal plays one almost exactly like it in the video with Keb' Mo'. It's a classic of its own kind of sound—full of gutsy punch from its lower strings and capable of gliding and quivering notes in its higher registers that can captivate a listener's ear.

I think I first heard it live when I was a child younger than five in Kahuku, Hawai'i—a town that neighbors Lā'ie (where my maternal grandfather had his store), today about an hour's drive from downtown Honolulu. Then, we were a more isolated plantation community made up of the descendants of Filipino, Chinese, Japanese, and Portuguese immigrants brought to work in the

sugarcane fields, which were owned by descendants of white missionaries to Hawai'i. *Kānaka 'ōiwi* avoided plantation work for the most part, but they were part of the community too, a few families with tons of relatives in neighboring Lā'ie and Hau'ula. Both towns were full of *kānaka 'ōiwi* whose living came from a combination of farming, gathering from the sea, and working jobs outside of King Sugar. On Saturday afternoons, we'd gather at one of the church lawns to hear music that the *kānakas* played, usually in trios and quartets—a contrabass, a *'ukelele,* and two guitars. Everyone in the group usually sang, creating marvelous harmonies that I think I can still feel in my body, especially on the high notes of a chorus. One particular day, though, there was a different composition to the usual grouping in the quartet—instead of two Spanish-style guitars, there was only one. In place of the other guitar, there was a boy, maybe middle school aged, seated on a stool with a guitar sitting sideways across his lap. When the group started up, I saw

his left hand gliding up and down the guitar's neck, holding something in it that he seemed to slide across the strings. The sound he made on his guitar was like that of a yodeling tenor singing a beautiful Hawaiian-language ballad. It was like a waterfall full of yellow flowers spilling across the mossy green face of the cliff in the Ko'olau Mountains.

Lele hunehune mai la i na pali
Lele hunehune mai la i na pali

*

The water sprays like lace down the cliffs . . .
The water sprays like lace down the cliffs . . .
(from "'Akaka Falls" by Helen Parker, translation mine)

The sound was so beautiful, I think I cried for the gift of having heard it. The Hawaiian steel guitar was like a human voice, quavering with motion in its highest notes, gorgeous throbs of music. At the age of five, I could have been devoted to that sound. But the moment passed quickly, the band launching into a *hapa-haole, holoholo* tune based on a *paniolo* cowboy rhythm, the lyrics half in English, half Hawaiian, about a road trip around the island. And the audience of villagers, parents and children, applauded in recognition and appreciation.

Kani ka pila! someone shouted. The throng joined in on the choruses, beating time with their hands that applauded and their bare feet that slapped and stomped the cool grass right on the beat.

I spent an afternoon with Keb' Mo' recently. His given name is Kevin Roosevelt Moore; he is a winner of five Grammy Awards over a recording career that spans more than four decades, and his oeuvre extends to over eighteen records from *Rainmaker,* his debut album on Chocolate City Records in 1980, to more recent works like *TajMo,* a wondrous collaboration with blues icon Taj Mahal

released in 2017 on Concord Records, and *Oklahoma,* released in 2019. We met one early November afternoon at his home near Nashville, where I'd traveled for a literary conference. Exactly my age (we were both born in 1951), Moore grew up in Compton, California, right next door to Gardena, where I did. If you don't know L.A., the racial shorthand was that Compton was Black, Gardena yellow. Bill Taylor, a classmate of Moore's from high school, was the friend who'd loaned me jazz LPs throughout our freshman year. Via text, Bill introduced us, and Moore and I spoke over FaceTime for a while to get acquainted (I was in France, while Moore was in L.A.). It ended up with him inviting me for a visit once I landed in Nashville.

From my hotel, I took a Lyft via pleasantly curving, tree-lined country roads over to Franklin, Tennessee, a swank suburb of mansions and big yards. It was the day after Halloween and there were big blow-ups of white ghosts, green dinosaurs, and black witches under the sycamores of the expansive front lawns. When I got to Moore's place, a newish three-story manse painted green and white, I entered through a side door to his basement, where the bluesman greeted me and ushered me into a big utility room, complete with a small kitchen and an area for a love seat, an easy chair, and a corner sofa. In the middle of the room was a stand for four guitars (one of them a resonator) and another lay on the floor in an open case. The resonator looked to me like a National Style O, a later single-cone model that used a biscuit bridge. On the wall over the sofa was a large postmodernist poster-like painting (à la Rauschenberg) with multiple rectangular images of the singer Amy Winehouse.

Moore is a slim, lanky man over six feet tall who was dressed in blue jeans and a long-sleeved black tee that day, a light green ball cap on his head. He wore high-top brown leather boots. Around his neck was a pair of breakaway reading glasses with black plastic frames. His voice is a pleasant baritone-to-tenor and he laughs frequently, his face breaking into a broad smile over a small salt-and-pepper goatee. His skin is the color of dark rum. He wears his hair short and cropped and he moves easily, like a dancer, gently stretching his long frame as he bends to pick up a guitar or set it down.

He tends to speak quickly, his voice so soft I had to lean forward to catch everything when I turned the topic toward his own upbringing and beginnings in music.

"I grew up in Compton. But my mom's from Hooks, Texas, near Texarkana, and my father's from Heflin, Louisiana. Roots in the South. Growing up in Compton, I hung out with a couple guys who introduced me to the blues. I heard Lowell Fulson and B. B. King records at my cousin's place where we'd go after church, but it was old people's music to me back then and I didn't pay attention. So, I came to the blues late. It wasn't until I was about fourteen, playing steel drum in a steel band, that I even saw a *resonator* guitar. I was waiting to get into the gig at the Troubadour out on Santa Monica Boulevard, standing outside next door to the club, when I saw all these shiny guitars in the window of McCabe's Guitar Shop. I said, *What kind of guitars are these with hubcaps on them?*

"A year or two later, my Vocational Drawing teacher Mr. McGee got me into a concert Taj Mahal was playing at my high school. Two periods, because the whole school couldn't fit into our auditorium all at once. You were supposed to go only to one assembly, but Mr. McGee got me into both. He was one of those guys who knew you before *you* knew you. I remember hearing and seeing something very different. Taj played a National resonator guitar—all metal and bright and brisk. He played 'She Caught the Katy and Left Me a Mule to Ride,' 'Paint My Mailbox Blue,' and of course 'Corinna.' I think he was playing songs that would later be on *The Natch'l Blues*. And it was like the universe was trying to point me there—to the blues. But it took a while.

"I was about thirty-three years old—much later, during the eighties—playing in the Whodunit Band with Charles Dennis, who later became B. B. King's guitar player, when I got curious about the power of the blues, the fact that it had *depth*. I was curious about the *truth* in the blues. I'd been writing songs, was a working musician playing around L.A., doing a few recording sessions. But I wasn't addressing truth and authenticity.

"Around 1991, I was over at my friend Nate Larson's house and he played two artists on the stereo. *Check this out,* he said. He played Big Bill Broonzy and I thought, *What the fuck?* Then he

played Robert Johnson. Then I bought the Robert Johnson box set. I borrowed an acoustic guitar. It was hard to play, man! I tried figuring out acoustic tuning. I called McCabe's Guitar Shop and found Fran Banish, a teacher who gave me lessons. I still didn't know that much. But I finally found myself in a search for a *tone* in my own voice that gave me a feeling like I had when I was a kid at my cousin's house—that *longing*. I started practicing slide—finger-picking and bottle-necking. I'd heard Big Bill Broonzy and Robert Johnson, you know, and I just couldn't go on being that shallow, L.A. pop music boy I had been up until then."

After a while, Moore busted out his guitar collection. He'd disappear, catlike, down the hallway and come back with guitar after guitar in their black cases, setting them on the floor, opening them up, and lifting an instrument to his side, tuning it, passing a couple along to me, inviting me to play. But I'd not touched a guitar in over forty-five years. The last time was the summer of '72, when I gave my brother my Gibson J-50 acoustic that I couldn't play then either.

"I got a lot of resonators," Moore declared. "You can play anything on a resonator. It's loud to compete with the horns. It's boisterous, with an *attitude*. That's why the Delta blues musicians took to it. Why I like it is the *steel*. The spider cone has a country sound. The chrome single-cone has the blues sound. I like the way it responds to the slide. It feels really down-home. The resonator gives that twangy feel like you're down South. But I'm more of a music fan than a guy who focuses on a certain guitar. The resonator gives me a tool I need as a songwriter. I love that space between the blues and being a singer-songwriter. The resonator helps create atmosphere. It's like I'm a carpenter and the resonator is the right *saw* I need to do the job. It has to do with the tones you use to get the point across. I've a National M-1 or M-2, a '33 Dobro, and a National Reso Rocket—that's my main one. I recorded Robert Johnson's 'Come in My Kitchen' with it."

We ended the gab session with Moore playing some blues. He reached first for the Regal guitar Taj Mahal had gifted to Moore after their Taj-Mo double bill tour together. A close replica of Dopyera's original tricone made of German silver, the kind Sol Ho'opi'i

first played, the Regal is metal-bodied too and of the same form factor (physical design) as the Slovakian's guitar, but is copper-tinged rather than silver-colored.

"Taj's got all kinda sweat on it. *It's good!* Funky like that," Moore said, standing up and handing me the coppery Regal tricone. Then he reached into another case and picked up a National M-2 mahogany resonator, its dark wooden body contrasting with the shiny metal cover plate. "I always wanted the mahogany National. I finally bought one," he said, settling down to play it.

He commenced a sweet, finger-picking blues, the resonator cone vibrant in the air, biting a bit, thunking and emphatic as it was plucked. Moore had started a soft, lyric line with gentle, rhythmic tapping of his left foot. Then, he began sustaining the bent and flatted notes of the tune as he fell into a rhythm. It was an old spiritual, "This Is the Way I Do," sung by the Black street singer Tee-Tot, from the musical *Lost Highway* about Hank Williams and his tragic life. As Moore sang the lyrics, down in his baritone range, his voice grew increasingly full of character and tenderly authoritative. I could feel how he opened his throat to the long, moaning way that he sang, stretching over the vowels and bending their notes.

> *This is the way I do . . .*
> *This is the way I do in my home.*

His playing was gorgeous, rhythmic, full of grace and precision, slide strumming and picking both. The single-cone resonance was a gentle thrumming against the guitar's wooden body, escaping like a light, audible mist into the air around us. Then, Moore picked a stalking, halting lead that emphasized the downbeat and filled out the measure with deft phrases. In his hands and in his voice, the spiritual felt like a Hawaiian, slack-key "porch song" to me, a tune for casual gathering, friends getting together *kani ka pila*—impromptu and spontaneously, sharing spirit.

As he ended the tune, Moore executed a set of abbreviated lyric repeats and then a sweet fadeaway with his voice and guitar. The bill of his cap flipped up and down as he glanced to his instrument and then up toward me, bringing the song to a ritardando close. He

laughed, and his hands, arms, and spirit all lifted lightly away from his guitar, liberated by something he'd just shared. Moore spun the instrument in his hands so that it lay flat on his lap.

"This is so cool," he said, pointing to the body, then the cover plate. "Good 'cause it's got that wood in it and a *straight* cone."

I sensed the session was over and started to take my leave. I thanked him, got up, and began saying goodbye. But Moore wanted to walk me upstairs. In the expansive parlor, we came upon his wife, his ten-year-old son, and a female friend, who all greeted

me amiably. It seemed they were like bandmates arranged for a record album photograph, asymmetrically dispersed around a green velvet, vaguely Victorian-looking sofa. His wife stood in front of it, the friend seated near her upon it, and his son sat sidesaddle on the far arm next to a large, floor-standing lamp with an elaborate silk shade that looked like the skirt of a large, translucent jellyfish. We exchanged light words of cheer, but it was time to go. Moore and I walked out to the porch to meet my Lyft ride. Before it pulled up, he said something remarkable.

"The blues originated in America," Moore said as we stood overlooking the gentle S-snake of his asphalt drive. "In South Africa, when I went there, everything was joyful. You really get the depth of what was robbed from you when you feel the joy in that African music. It's deep culture with all these animals—a tribal culture. When we were brought over, we lost all that. We got a different feeling then. It's in the blues. And the way I connect is not necessarily from the music per se, but from the feeling I get when I *hear* the blues—that longing. *In the body.* You go back to the field hollers—that lonesome feeling of being lost. *That's* the blues."

Hog-Calling the Blues with Etheridge Knight

Sometime in the early 1990s, I was at MacDowell Colony with Etheridge Knight, walking a dirt trail through tall pines spaced widely apart, cool New Hampshire summer air lightly billowing our clothes. Etheridge, about twenty years my senior, was wearing a dark indigo vintage Hawaiian shirt, big and loose on him, its pocket billowing where he'd tucked in a half-smoked pack of Pall Malls. He wore jeans and big shoes, stood about six foot three, and was built like a power forward with slightly smallish shoulders, a wide ass, and powerful legs. His bare black arms were muscular and worthy of a sculptor's study. His face was battered—a diagonal scar across one cheek, small flecks of scars near his eyes, and a puzzle piece of light, cream-colored skin that dropped down from his lower lip over his chin. He wore wire-rimmed glasses, a mustache, and a raffish stubble of beard over his jawline that stopped when it reached the puzzle piece. I wore jeans, sandals, and a light blue

aloha shirt myself. And I'd a scar of my own—a curved, barbed line like a centipede crawling across my chin.

We rounded a bend and exited the small forest into a field of tall grasses, their yellow-green stems doing a kind of wavy *hula* in the wind. Etheridge stopped at the far end of the field, near a patch where the tall grass had given way to gravel and stones along the edge of a dirt road that wound toward us from down a hill. He faced the crest of it, put his big hands together, and made a trumpet that he at first compressed and then widened around his mouth.

"WHOOOOO-WHEEEE!WHOOOOOO-WHEE-WHEE-WHEE-WHEE-HAW-HAW-HAW-HAW!" he shouted, his voice ranging from long howl to yelping, ending with a falsetto yodel and a series of baritone yawps that reminded me of John Lee Hooker singing "Boom Boom."

"That's hog-calling in Mississippi," he added. "You got a farm, you get your pigs in that way."

I'd asked him what the difference was between hog-calling and field hollers. We'd met some fifteen years before in Michigan, when he'd come to Ann Arbor where I was in grad school and had put

together a summer reading for him and a few other star poets. Etheridge had given a terrific performance, reading poems from his life in Indiana State Prison, where he'd been incarcerated for larceny and possession. He'd been an addict, he said, from taking morphine after getting wounded as a soldier. On the back of one of his books, he'd written, "I died in Korea from a shrapnel wound and narcotics resurrected me. I died in 1960 from a prison sentence and poetry brought

me back to life." And actually, "reading poems" hardly captures what he did. He intoned, he chanted, he wailed, and he droned. His body swayed to music in his words, it jumped and his shoulders shimmied. He preached at the podium, made jokes as though doing stand-up, brought laughter and smiles to everyone there—an audience of young Blacks, Asians, Hispanics, and Native Americans assembled to celebrate our triumph in the aftermath of a student strike protesting the university's cuts in financial aid to "minority" students. And he brought us testimonies of pain—the loneliness of his life as an inmate, his ferocity and defiance as a Black poet, his shame for having been an addict and becoming a thief to feed his habit, his love for kinfolk at a reunion and his sorrow for having only photographs of them taped to the walls of his cell. He brought us the mess of his life and the pure order of his singing that redeemed it.

> *Taped to the wall of my cell are 47 pictures: 47 black*
> *faces: my father, mother, grandmothers (1 dead), grand-*
> *fathers (both dead), brothers, sisters, uncles, aunts,*
> *cousins (1st & 2nd), nieces, and nephews. They stare*
> *across the space at me sprawling on my bunk. I know*
> *their dark eyes, they know mine. I know their style,*
> *they know mine. I am all of them, they are all of me;*
> *they are farmers, I am a thief, I am me, they are thee.*
>
> (from "The Idea of Ancestry" by Etheridge Knight)

"Field hollers are different," he said. "They come from slave days pickin' cotton and whatnot, workers in the fields calling to each other."

"Were they work songs?" I asked.

"Yeah. People sang them to pass the time, they say to the rhythm of work, but not always. I heard songs that were just songs. And I heard songs from prisoners too—chain gangs bustin' up rocks and building roads."

"How do those go?" I asked.

Etheridge paused a moment, threw a quick glance at me, maybe gauging the look on my face, cocked his head, and turned his eyes

like he was hearing something a long ways off. I counted a quiet breath or two. Then he started moaning.

> *Why don'tchu go down, ol'*
> *Hannah . . .*
> *Don'tchu ride . . .*
> *No mo' . . .*

He sang both call and response lines, the call taking a tenor's key and a lilt at the end, while the response was a baritone drone, trailing away mournfully and slow. The melodic profile was whipsaw— a steep rise in register, then a slow descent, another rise and, finally, it sloped away again. It had a heaving rhythm too and a kind of emphatic grunt to the baritone part that kicked in on the second syllable, shoving the note from the gut to the feet, then down through to the earth, digging in. It took strength to sing and I felt all of Etheridge's six-foot-plus body lurch into the song, as though shoveling, as though heaving up a mound of dirt and soul at the same damn time.

I was amazed he was teaching me this, allowing me to hear casually what it must have taken him knocks and hard labor to learn. Or maybe I was romanticizing and he'd just picked it up from kinfolk, hanging with them on a porch, sipping corn whiskey. I didn't ask. I just listened. Learning arcana was sometimes like this—it came almost with no expense except my earnestness, and was a gift from elders who wanted to give, for whatever reason, to ephebes like me who had only an innocent greed to learn what took their betters blood to know. I learned to chant the Heart Sutra this way, imitating the genuflection of the priest and then his guttural intonement from the belly, sun a gold glimmer through the slice of an opened *fusuma* in the temple. And now, among tall American weeds, I was hearing a work song sung by a Black poet who blessed me with his booming voice.

> *Well, I shook my-ah head*
> *And began to-oo moan.*

I thought back to the sugar plantations in Hawai'i, to my own family of cane workers, the songs and chants they made up while they stripped drying canestalks of their leaves, *hore-hore,* their hands covered in thick work gloves drenched with sap and oil. They flailed with hoes and machetes, wore straw hats like baskets to cover their heads, bandanas over their faces to protect them from breathing in cane dust. Etheridge moaned another verse, and I conjured a locomotive's black plume of smoke and the commotion of a mule team dragging a gang plow behind it, kicking up chaff from all the sugarcane lying over opened ground. Smoke and then dust, a whistle and then whipcracks and the braying of animals.

> *Hawai, Hawai to yō—*
> *Yume mite kita ga*
> *Nagasu namida wa*
> *Kibi no naka . . .*

> To Hawai'i, Hawai'i I came
> Chasing dreams.
> Now I weep long tears
> Amidst the cane . . .
>
> (translation mine)

Etheridge sang and his voice opened up the earth inside of me. It spoke a long holler as though screaming down the darkness of a well to the pool of generations of both our peoples.

A few days later, work done for the day, I strolled over to Etheridge's studio, about a ten-minute walk from mine. I went along a dirt-and-gravel path through woods that skirted an old frog pond and bent down a wider lane before dividing again to another path that led me by a field of tall grasses. Etheridge's studio was on the other side of it, and I could hear snatches of him singing. It carried over the wheat-like heads of the weeds that bent lazily in the wind.

It was slow, a mournful tune I didn't recognize, and not a pop song or a blues, quite, as it sounded smoother, despite an intricate shift in tempo that felt a bit like faltering before it righted its steady gait.

I'd realized, when I first saw him again, at dinner in the dining hall at MacDowell just after I'd arrived, that Etheridge was different from when we first met. The joyousness that had once run through his body and animated his voice seemed to have fled. He'd gained big weight around his belly, slowing him down, flattening out the bounce in his step. I didn't know why it was exactly, though I had guesses. Once boisterous in his speech, he was quieter, subdued, even a bit haunted by things. When he spoke, it was with more reflectiveness, a somber tone to his big, baritone voice, the high-life boom and brassiness of it evolved into the longer, elegiac notes of a tenor saxophone playing lyric ballads. His conversation came in quiet, finer tones, slower and intimate, even confessional. The dashing sparkle in his eyes had morphed into a shimmer and a kind of plea—when they fell on me, they'd stay an instant longer, searching rather than lit by what was within that had once fired him. Still I pursued him, needing his wisdom, drawn by the voltage I knew was still in his blood, the great power in the poet that I recognized he still remained.

When I rounded the field and approached the steps to Eth's porch, I saw that his front door was open, and I could see all the way through to a back door that was open too—like a Louisiana shotgun cabin. Through the back door, I caught glimpses of Eth swaying back and forth, rocking from side to side on his feet, his dark blue shirt dodging in and out of the door's frame, his head bowed, big voice booming out the tune.

It was "Willow Weep for Me," a jazz standard from the thirties new to me then, a slow, mournful blues written by Ann Ronell. Billie Holiday once sang it, among others. The way Eth was doing it, it almost sounded like the work song he'd intoned for me a few days before, low and moaning, dirgelike and valedictory.

I walked through toward the back and noticed Eth was singing into the opened door to his water closet, a small shack adjacent to the studio, joined to it by a wooden walkway where he stood, still swaying and singing, his voice amplified by the john and its tidy

chamber. His baritone boomed into the space and came soaring back out sounding even bigger—a rich torrent of blended and recursive echoing. I stood and listened while he worked his way through the melody and its muscular rhythms, its melancholy themes, the image of a willow's branches surrounding a solitary singer like a green shroud unfurling under a black sky bannered with a stream of stars. It was as though Etheridge had called into a cave cut into the earth, his voice traveling through an Orphic underworld before returning, made doleful and yet burnished by the passage. It wept, its sound saturated with a pain unknown even to my own ancestors.

Part Eight

Talking Heads
and Singing Platters

A PEEP INTO FRIAR BACON'S STUDY.

The Talking Heads

In the 2011 Oscar-winning film *Hugo,* directed by Martin Scorsese and based upon a superb YA novel by Brian Selznick, much of the plot revolves around an automaton—a mechanical head, torso, and arms—whose operation has ceased. Hugo, the young boy who is its keeper, believes the automaton holds a secret, perhaps a message from his deceased father, who'd made the machine, if only he can make it run again. Charged by a cruel uncle to keep the huge and magnificent clocks at the Montparnasse train station running, Hugo meets adventure, love, and sponsorship in the bustling environs of the station, while averting abandonment. And, miraculously, the pendant around the neck of the sweet girl who befriends him turns out to be an actual key that unlocks the lifelike operation of the mechanical man. It's a beautifully humane film with gorgeous cinematography and well deserves its recognition. And, though the mechanical man doesn't voice but rather writes its message to Hugo, I think it's about our wish that the world of technology speak back to us with an ulterior loveliness we cannot have predicted merely out of the rough components we've engineered to create it. We've long wanted its voice, speaking from the dark obscurity of cultural bricolage and applied science, to be reassuring, life-affirming, to buoy us up via the mysticism of mechanics from which it arises.

When the Greek historian Strabo wrote, in 20 BC, that the Memnon of Thebes, one of twin colossi built for the tomb of Amenhotep III (c. 1350 BC), sang to crowds gathered around its base, he was reporting as much from a wish of human culture that our built things had voice as from any verifiable worldly phenomenon. Others of the ancient world—Pausanius, Pliny, Tacitus, Philostratus, and Juvenal—all wrote of the structure's miraculous sound, like

"the string of a lyre" breaking, the striking of brass, or a soft, almost imperceptibly whistling sigh. And during the Middle Ages, Roger Bacon, an English philosopher and friar in the Franciscan order dubbed Doctor Mirabilis, built the first talking head, considered a product sprung more from necromancy and dark conjure of the dead than any empirical science. Between 1840 and 1860, Joseph Faber of Vienna constructed various talking devices, the most notable with the features of a woman, made to imitate human speech through a series of reeds, whistles, and resonators operated by a bellows. And by 1890, the American Thomas Alva Edison created the macabre invention of a talking doll, its insides fitted with a miniature version of his wax cylinder phonograph, operated by a crank sticking out of the doll's back. That it spoke nursery rhymes with a screechingly eerie female voice evoked both familiarity and distress—a kind of ur-Chucky. When Alexander Graham Bell premiered his telephone on March 10, 1876, he delivered the magically disembodied message, "Mr. Watson—come here—I want to see you" to his assistant on the other end of the line. When the tinfoil phonograph of Edison first spoke, it was a recording of its inventor reciting "Mary Had a Little Lamb" in odd, declamatory style, though his voice itself sounded wan and frail. Finally, the Bell Telephone Company at the 1939 New York World's Fair revived the talking man phenomenon and delighted crowds with a machine operated from an electric keyboard. Recognizable in various degrees, each was a kind of mysterious transmission at a remove from its source, and the creation of sound laced with meaning, both pure and impure, the embodiment of *vates* as though from a voice beyond the norms of human reckoning and yet discernible as meant for us.

Much of the early history behind these voices arising from the varieties of inanimate inventions of humankind is covered in the marvelous book *From Tin Foil to Stereo* by Oliver Read and Walter L. Welch (1976). In it, we read that the remarkable invention of the phonograph came via an accumulation of human effort around ideas of the acoustic transcription and reproduction of sound and then the transmission of them through electricity.

The Pre-Phonograph

Although it is general knowledge that the technology of the phonograph was invented by Edison in 1887, there was nevertheless a long, literary foreground to its idea. Fanciful and poetic, the notion of hearing sounds from the past without their source being present comes up at a lot of points in Western literature.

First, there is the classical myth of Orpheus, the poet-singer who journeyed to the underworld to retrieve his lost love, the nymph Eurydice. She'd been killed, while wandering amidst tall grasses, by a bite from a viper, and her death brought a lingering desolation to the poet's spirit. After consultation with a seer, Orpheus went down and got her by charming Hades, god of the underworld, who released her from death, moved by Orpheus's singing. But although Hades warned him not to, the poet couldn't resist a glance back at Eurydice as he led her in their ascent to the real world. He took the forbidden look and he lost her. Instantly, Eurydice got sucked back to the realm of the dead. But Orpheus never stopped singing, plucking at his lyre in lament well into old age. His incessant songs (and abandonment of womankind for the love of adolescent boys) made the Maenads so mad (they were female followers of Dionysus), they mugged him, separating him from his lyre; then, in a violent bacchanal, they tore him limb from limb and cast his disembodied head and his lyre too into the Hebrus river. The head floated on the waters there, still singing, long locks of its shoulder-length hair (albeit *without* shoulders) undulant on the surface currents, the strings of his instrument echoing strains of his elegiac and resolute voice. They drifted all the way to the sea together, washing up on the shores of Lesbos, an island where its native tribe was inspired into a crazy love of poetry.

In *The Histories of Gargantua and Pantagruel,* a great romp of a picaresque, the sixteenth-century French novelist Rabelais writes of a sea of frozen words that, jeweled in the icy cold after a great battle, came alive as they melted in the sun. Rabelais describes them as mysterious moans and indecipherable savage phrases heard by the giant Pantagruel as he voyaged on the high seas. His captain

explains they are shouts of men and the cries of women, "the slashing of battle-axes, the clashing of the armor and the harnesses, the neighing of horses and all the other frightful noises of battle" that had become frozen in the air, only to melt with the coming of better weather. The captain then tosses handfuls of these frozen words onto the ship's deck. Looking "like crystallized sweets of different colors," the words warmed their hands and "melted like snow" so they could hear them, though without their understanding a thing.

In *From Tin Foil to Stereo,* co-authors Read and Welch recall that seventeenth-century French author Savinien Cyrano de Bergerac imagined a story of men making a visit to the moon via rocket ship. This is the tale that inspired a 1902 French silent film adventure short (directed by Georges Méliès) that was partially reprised in the charming Scorsese movie *Hugo.* Published in 1649 and entitled *Le Voyage dans la Lune,* de Bergerac's early science fiction work describes books "made wholly for the Ears and not for the Eyes" that hung on the earlobes of the narrator "like a pair of Pendants" while he went walking about the moonscape. By our own time mankind not only had strolled on the moon but walked with Bluetooth Beats down our urban streets.

By 1867, most of Europe was falling in love with the novelty of the string telephone, the children's toy that can transmit a voice across a taut, thin cord running between two metal cans. Sound waves from one voice would rattle the bottom of a punctured can where a knotted string picked up its vibrations and sent them across the line to a can on its other end, making it vibrate in turn and reproduce, if more faintly, the sound of the voice that spoke. When we were kids, my cousins and I made one from empty cans of PET milk and kite string. We spoke the opposite of endearments into it as we stood under a plumeria tree or romped on the carpet of temple moss out in the Japanese cemetery in Kahuku.

Maxine Hong Kingston recounts, in *The Woman Warrior,* an old Chinese folk tale about a woman who confesses her erotic secrets into a hole she's dug in a dark patch of earth adjacent to her home village. Years later, a small stand of bamboo springs up from there, and her words of sexual longing are reborn and heard by her neigh-

bors, carried by the wind whispering through the bending stalks and softly rustling leaves of bamboo.

In 1855, though, Édouard-Léon Scott de Martinville, a printer and bookseller by trade, made a stunning contribution toward the realization of this wondrous and imagined thing when he came up with a device that could translate sound waves into movement. He'd captured stylus tracings on a piece of paper treated with lamp-black, the finely powdered soot gathered from the burning of oil furnaces or kerosene lanterns. First, he collected sound vibrations in a resonating chamber that converted air pressure and rarefactions into the mechanical movements of a diaphragm. The diaphragm then activated a pig bristle tipped with a stylus that traced the vibration patterns as an undulating line across a lampblacked piece of paper wrapped around a revolving cylinder. The cylinder was moved by a screw.

Scott de Martinville called the result of his invention a "phon-autograph"—in essence, a recording of sound waves. Though a star-tling visual depiction of sound, the phonautograph, alas, was mainly an experiment in acoustics and, at the time, could not be made to create a sound of its own. It wasn't until 2008, when scientists at the Lawrence Berkeley National Laboratory in Berkeley, California, scanned some of Scott de Martinville's phonautographs with a laser, that Scott de Martinville's squiggles were translated into digital audio files that could then be read and played back as sound. You can hear two of them on Wikipedia. One scan is of Scott de Martinville singing "Au Clair de la Lune," a French folk song, which initially got played back at a speed making him sound like a child or a young woman. Another is of him reciting lines from Italian poet Torquato Tasso's *Aminta,* a sixteenth-century pastoral drama of courtly love, set in an idyllic land of shepherds and maids. If you listen, the phonautographs sound very scratchy with lots of distortion, the disembodied voices barely recognizable as human—not unlike how Rabelais described the thawing jewels of ice that Pantagruel heard as releasing unintelligible, monstrous sounds into the air.

Possibly inspired by Scott de Martinville, the French poet Charles Cros came up with a design for what he termed a "Paleo-phone," or "voice from the past," in April 1877, some months before

Edison's patent application was filed in England. Slim and curly-haired, the mustachioed and dandyish Cros was a regular among the literary and arts crowd that frequented a Parisian cabaret called Le Chat Noir, perhaps the first of its kind. It was presided over by a charismatic, stout, and red-haired impresario named Rodolphe Salis. Initially, it was run out of Salis's parlor, but the club quickly outgrew the space and was moved to 84 Boulevard de Rochechouart

in Montmartre, soon to be famed for its concentration of artist residents. There and at subsequent locations it flourished, attracting, through its brief lifetime, the likes of poets Paul Verlaine and Jules Laforgue, composers Claude Debussy, Erik Satie, and Darius Milhaud (who served a stint as a waiter there), playwright August Strindberg, and painter Henri de Toulouse-Lautrec. Bourgeois

patrons were attracted by the ribaldry and high spirits of the place, the novelty of a boisterous music hall attached to a cheap watering hole that was frequented by artists and presided over by *conférencier* Salis who, dyeing his red hair green, acted as an ironic master of ceremonies, barking out false praise for the aristocrats as they came slumming, insulting regulars and mocking the performers as the show moved from

act to act. It's not Joel Grey of *Cabaret* I think of here, nor even Don Rickles, but more a comedian like the outlandish Rip Taylor screeching at guests and entertainers behind the salacious upswirl of his mustache. To this milieu, Cros contributed both a decorous but romantic poetry and satiric monologues as well, sometimes publishing with Stéphane Mallarmé and Verlaine in the literary weekly *Renaissance littéraire et artistique.* Reading it today, I find a certain sweetness to his writing in the sense of the *dolce stil novo* practiced by Guido Cavalcanti during the early Italian Renaissance. It is full of earnest gallantry for womankind and touches of tenderness for minute observations of the distaff and everyday. Here are six lines from something Cros dedicated "To Mademoiselle Nelsy de S":

> *. . . priestess who came with the Norman chiefs,*
> *It was you who made docile and dormant,*
> *Through your songs, the rebellious waves of the Seine.*
> *Escaped from old paintings, old novels,*
> *So your beauty amazes me on the stage*
> *Of the world today, poor in enchantments.*

(translation mine)

And here some sybaritic lines from a Cros poem entitled "Au Café," which I imagine might actually have been penned in a notebook at Le Chat Noir itself:

> *The dream is not to dine, but to drink, to chat, to joke*
> *When night falls, exhausting aperitifs.*
> *We laugh at the cypresses and yews shading the grave.*

(translation mine)

In April 1877, Cros wrote a description of his phonograph in his own handwriting and sealed it in an envelope with five big blotches of red wax, spaced along an X that formed two equilateral triangles. This he deposited with the French Académie des Sciences. He hadn't the scratch to hire a machinist who could build any kind of prototype or even to put funds together to file for a patent, but he wanted a record of his concepts and his own discussion of the technology

needed to execute it. In the note, translated by Roland Gelatt in *The Fabulous Phonograph: 1877–1977,* Cros writes that the process of sound reproduction performed by his machine "consists in obtaining traces of movements of a vibrating membrane and in using this tracing to reproduce the same vibrations, with their intrinsic relations of duration and intensity, either by means of the same membrane or some other one equally adapted to produce the sounds which result from this series of movements." Cros specified that the tracings were to be made on a disc of glass he proposed be treated with lampblack (like the head of Orpheus, it persists) and that photoengravings of its reliefs and indentations be made of these tracings in order to produce sounds that could be played back. The latter idea, of course, looked forward to the Gramophone disc of Emile Berliner that was to come ten years later. But, in writing his thoughts down and filing the document with the Académie, Charles Cros, a dreamy and impoverished minor poet, louche denizen of Parisian nightlife, was first to conceive of what we now call the *phonograph.*

Edison's Tinfoil Phonograph

In July 1877, inventor extraordinaire Thomas Alva Edison instructed his laboratory staff in Menlo Park, New Jersey, to rig up an indent-

ing stylus attached to a diaphragm that would itself be attached to
a telephone speaker. He'd previously used a short needle connected
to the diaphragm of a telephone receiver and thought to place his
finger over the needle. The tiny pricks of movement matched the
amplitude of the vocal signals he'd sent through the receiver, so he
reasoned that, if the needle could prick a piece of paper as it did his
finger, it would indent the paper with a record of his voice. Once
the new rig was ready, Edison shouted into the speaker, which set
the diaphragm in reciprocal motion, agitating the stylus up and
down. He ran a strip of paraffin-coated paper under it, and the
result was a pattern of irregular dot-and-dash indentations made
on the coated paper. When the strip got pulled back under the
same stylus, Edison heard the faint reproduction of his shouts, "a
light musical, rhythmic sound, resembling human talk heard indis-
tinctly" emanating from the speaker.

Edison took this result and, in December of that same year,
designed a machine that used a metal cylinder with a spiral groove

cut into its surface together with two diaphragm-and-stylus units (one for recording and one for playback), and mounted the cylinder on a turn screw operated by a hand crank. To the recording diaphragm, he attached a funnel-like mouthpiece. To the playback assembly, he attached a speaker. A sketch of his design survives and can be seen at the Smithsonian Institute, along with a later model of this phonograph. Edison then wrapped a sheet of tinfoil around the cylinder and set everything in motion, speaking into the diaphragm, which sent the stylus diving like a tiny jackhammer up and down on the tinfoil, scoring the sheet with a pattern of indentations that corresponded to the movement of sound vibrations in the air. Operating the second stylus-and-diaphragm assembly converted the marked tinfoil into audible sound—an analog to the live, acoustic event. Supposedly, Edison had shouted the verses to "Mary Had a Little Lamb" into the machine, and the wonderful machine played them back in something like his own voice. He wrote, "I was never so shocked in my life." Alas, the recording doesn't survive, but Edison made another some fifty years later on the anniversary of the event, and it circulates on the Internet, erroneously, as the recording from 1877.

Though Edison created the first working phonograph, pretty much each piece of its technology was already known before he put them together in his lab. As Read and Welch point out in *From Tin Foil to Stereo,* the trumpet, diaphragm, stylus, vibrating surface, feed screw, and wheel all predated the phonograph. During the Renaissance, Leonardo da Vinci had already sketched out the idea of a speaking trumpet in preparing a tube communication system for the duke of Milan. Since antiquity, ear trumpets had been used as hearing aids, functioning to capture and focus sound much like a cupped hand over the ear. Since the time of Hippocrates, with the dissection of human and animal ears, the tympanum (or diaphragm) became known, and it appears as a sound resonator in drums, banjos, and other instruments before the written record. Scott de Martinville himself designed his phonautograph around a vibrating tympanum. The stylus was around for the ancients of Assyria, Egypt, and Central America to use as an engraving or carving tool for mak-

ing pictographs and hieroglyphics. Edison put these all together, activating the stylus with a diaphragm, indenting an up-and-down pattern into a piece of tinfoil wrapped around a cylinder, scored on its surface by a spiral groove. To move the cylinder, Edison took the example of the lathe, operated by a hand crank on one end, essentially deploying the feed screw invented by Archimedes. And the origin of the wheel occurs deep in the prehistory of humankind.

Edison's contraption looked like a metalworker's lathe, with a platen mounted on a feed screw that attached to a crank with a revolving wooden handle. A sheet of tinfoil would have been wrapped around the platen, and over it, the operator flipped down a mouthpiece that looked like a big open piece of pipe. Attached to it was the diaphragm and stylus assembly as an undercarriage to the pipe, kind of like a stethoscope and kung-fu ring with a pointed spike on it rather than a jewel. There was a bit of madness in its look, but everything worked together to make sound recording and playback finally possible.

Edison almost immediately took his new phonograph to the offices of the *Scientific American* and got a big write-up in its magazine, the kickoff to a national sensation.

> *Mr. Thomas A. Edison recently came into the office,*
> *placed a little machine on our desk, turned a crank, and the*
> *machine inquired as to our health, asked how we liked the*
> *phonograph, informed us that it was very well, and bid us*
> *a cordial good night. These remarks were not only perfectly*
> *audible to ourselves, but to a dozen or more persons gathered*
> *around . . .*
>
> (from *Scientific American,* December 22, 1877)

In 1878, the inventor formed the Edison Speaking Phonograph Company, and in an article in *North American Review* (June 1878), he imagined his machine as capable of many things: an aid for letter writing, for listening to phonograph books, for teaching elocution, as a family record (e.g., capturing the last, dying words of a loved one), for educational uses, and even in connection with a telephone

(an answering machine!). He also envisioned the phonograph would have a role in the "reproduction of music" and as a music box, predicting the eventual eminence of his creation in the wider culture of the twentieth century.

But the tinfoil phonograph could only be used about two or three times before the foil degraded. Just operating it was a challenge on the order of the carnival act that tried to keep two plates spinning on top of different vertical rods. As the commercially released contraption consisted of twinned recording and playback devices mounted together on the same lathe, the user had to crank at a deliberate and steady speed for incising the foil wrapped around the recording cylinder, then detach the fragile foil carefully and remount it on the playback cylinder, moving over to the other drum and cranking again steadily to achieve sound. It wasn't fun. I see it as little better than detaching a fixie bicycle wheel from the front fork, remounting it on the bike's back end, then cranking the pedals by hand. It's a chore, with a piece of party-trick ridiculousness thrown in. All told, only about five hundred units were made, and after the initial sensation died down (it made most of its appearances in the parlors of the wealthy), it was seen more as a novelty than anything practical or giving significant leisure pleasure. Over the next few years, Edison abandoned his interest in his own invention and moved on to work on systems for illuminating entire cities—a project that was very lucrative and took up all of his time.

In terms of how these tinfoil recordings may have sounded, it's hard to say. Because of their extreme fragility and the penchant of their operators to tear pieces of them off to hand out at demonstrations, only a few survive. But at least one recording, made in St. Louis, Missouri, in 1878, can be heard. Preserved at the Smithsonian Institute, it's like a wide and long piece of unshiny Reynolds Wrap, a little torn at the edges and reticulated with ridges that were made when it was stored in an envelope for over a century. Think of a Dead Sea scroll, only made of tinfoil. In 2012, Lawrence Berkeley National Lab in California, through a process developed in the field of preservation science, digitized the shape of the tinfoil's preserved surface with an advanced microscopic process of optical scanning. The computer-controlled Berkeley scanner moved edge

to edge on the tinfoil sheet like a lawn mower, scooching back and forth, capturing images at two thousand per second, measuring the surface of the scored tinfoil in fine detail. In this way, scientists got an optical profile of its inscribed surface. Another computer read the scan and translated it into a digital audio file that could produce sound. Then, in an event at the Schenectady Museum of Innovation and Science, Carl Haber, senior scientist at Lawrence Berkeley National Laboratory (his specialty is particle collider science), gave a genial presentation that's still available on YouTube. After a long sound collage of old-time musical recordings (mostly gleaned from wax cylinders and Gramophone discs, I'm guessing), and a lot of talk about early recording history, Haber finally plays the tinfoil recording. There's a clacking *whump-whump-whump* I take as the sound of the folds and ridges as the scanner passes over the foil, then there are some unintelligible words of announcement (all these early recordings begin with announcements, kind of like sound gaffer calls for scene designations in unedited movie film). There's then what sounds like a really bad trumpet for twenty-four seconds. After that, someone recites two verses of "Mary Had a Little Lamb" (but it's not Edison) in a deliberate stretch of strained elocution. There are nine seconds of mad laughing, exaggerated and chesty, and for no reason that I can fathom. Lots of extraneous noise too, like radio static in a World War II movie about bombers, pilots, and ball turret gunners. It's thundering. Then there are eleven seconds of an extremely high-pitched voice screeching out a piece of "Old Mother Hubbard." At the end, it turns slower, lugubrious, and more like the voice of a man—the operator must have varied the pace of his cranking. Laughter again, then someone speaking more unintelligible words. All together it's seventy-eight seconds of almost macabre curiosity, like the desiccated finger of a corpse in catacombs turning to dust as you pick it up.

Volta Laboratory

Inspired by Edison's phonograph, Chichester Bell (Alexander Graham's cousin) and Charles Sumner Tainter set about to improve it with their own experiments at Volta Laboratory near Washington,

D.C. Working from 1881 to 1886, they came up with inventions and methods, some of which are still in use today. Their first innovation was to abandon tinfoil as a medium for recording and substitute wax instead. Rather than covering the recording cylinder with foil, they poured wax into the spiral groove of the cylinder to make recordings. The cylinder they came up with was of cardboard, light in weight and eventually makeable by a machine they patented. They also used a sharp stylus rather than the duller, Edison one so that the cut would be more precise and produce a clearer playback. Then, trying out various methods and materials (including an air-jet stylus), they settled on a mixture of wax and paraffin, to be used on a glass disc that rotated on a vertical turntable. Their sharp stylus dug into the wax, and instead of using Edison's "hill-and-dale" method for recording—the up-and-down indenting motion on tinfoil—they created a lateral cut, a zigzag motion of the stylus, so that the groove was incised side-to-side rather than with depth or shallowness. Their next innovation was electroplating the recorded discs with copper, then using the negative electrotypes to make stamped copies of playable duplicates, inventing the basic method still used to make LPs today. Finally, they came up with the rigid, pivoted playing arm, fashioning one that was capable of a 90-degree vertical motion so that they could remove a record or return the arm to its start.

The series of six recordings Volta Lab made that survive sound almost as bizarre as the first ones made on tinfoil. In 2011, using the same technology employed to digitize Edison's tinfoil sheet, the Lawrence Berkeley Lab scanned the discs and cylinders Tainter and Bell had deposited with the Smithsonian's National Museum of American History. Converted to audio files, these sound like messages escaped from a bottle washed up at Coney Island after years floating aimlessly on the sea. The first, recorded in 1881, is of a man executing trills—a favored method of the Volta Lab for testing sound—then speaking the numeric sequence 1, 2, 3, 4, 5, and 6. The second, recorded in 1884, is of a man, presumably lab assistant H. G. Roger, intoning the word *barometer* in a halting, syllable-by-syllable pronunciation. By the third, recorded in 1885, we finally get something a little more stimulating, though it starts

out bland as can be—a man speaking the word *phonograph* syllable by syllable. He then tries to recite "Mary Had a Little Lamb," but speaks it in broken verses. Either he forgets the words or else the machine malfunctions, and the man blurts out, "Oh, fuck!" It's the most interesting thing in all six of these recordings. The last three consist of someone reciting lines from Shakespeare's *Hamlet* and an almost unintelligible reading of an unknown short story. But, on one of these, someone utters one very clear sentence: "I am a Graphophone and my mother was the phonograph."

In 1878, Tainter and Bell took their new machine to Edison, thinking they might combine efforts toward perfecting it further, but the older inventor felt insulted and declined cooperation. Done with developing systems for incandescent lighting, Edison then set about improving his own machine, but not before adopting the better-sounding Volta Lab wax-coated cardboard cylinder for his own phonograph. The big difference, though, was in the style of cut each used, Edison sticking to his "hill-and-dale" vertical cut, and Bell and Tainter using their lateral, zigzag cut. Thus, their two systems remained incompatible—a situation that was to arise repeatedly in the history of recording. There came a sort of skirmish of commercialization, Volta Labs establishing the Graphophone Company of Alexandria, Virginia, which soon became the American Graphophone and then, eventually, after many consolidations, Columbia Records, and the phonograph's inventor creating the Edison Speaking Phonograph Company. For about six years, from 1885 to 1891, both companies issued numerous recordings in wax cylinders that could only be played on the dedicated machines of each company.

Both Edison and Tainter-Bell thought of their inventions mainly as instruments to aid business and provide tools for dictation, so what mostly remains from the period immediately after is a sound cloud (I'm tempted to say *babble*) of recorded voices, in varying levels of clarity and, alas, pertinence. We can hear a pitch by circus impresario P. T. Barnum, lines from Shakespeare's *Othello* (an entire soliloquy was too long to capture) recorded by actor Edwin Booth, snippets from Harry Houdini and Florence Nightingale, English Victorian poets Alfred Lord Tennyson and Robert Browning recit-

ing lines from their poetry (Browning forgetting them midway), and Wild West entrepreneur Buffalo Bill making an announcement about "the Rough Riders of the World" that predates Theodore Roosevelt. Of these, the most affecting moment is to hear the voice of Browning, who, at the age of seventy-seven, would die in Venice, just four months after this recitation of his verse in 1889. His voice is frail, the tone bewildered as he attempts twice to recall his own golden words. "I'm terribly sorry, but I can't remember me own verses," he declares. "But one thing I shall well remember me whole life is this astonishing moment by your invention." After a beat of respectful silence (heard by us as a ubiquitous whooshing of the cylinder whirling under the stylus), someone from the party shouts his name and then strikes up a cheer, "Hip-hip, hooray!" to cover the old man's embarrassment with homage.

As for musical recordings from this early period, the pickings are slim, though available on the Internet. There is a cylinder from 1890 of John Philip Sousa's United States Marine Band playing "The Thunderer March." A voice, likely Sousa's, announces the band and the name of the march, then we hear a welter of brass, with a piccolo piping sharply above it, playing in a jaunty rhythm this rousing, martial tune. There is a snare drum and, I think, a cornet solo, though the distortion is so bad, I can't be sure. From 1893, there is Irish American George Gaskin's rendition of the music hall tune "After the Ball," its playback marred by the circular *whoosh* of the cylinder under the stylus, and Gaskin's tenor voice straining through it with an affected gaiety and British diction. You can hear its nasal quality, the recording emphasizing his upper registers, likely the aftereffect of having its intensity concentrated and frequencies funneled by the cone of the recording horn he sang into. Performers would stand back from the horn and sing or shout into it, forcing the recording stylus to cut into the wax as firmly as possible. Yet, what resulted was that familiar, "old-time" distortion of the human voice, making it sound as though coming through a megaphone.

There are medleys of popular airs played by the Columbia Orchestra, some sentimental Irish tunes sung by a baritone, numerous songs written by Stephen Foster, and the attempt, in 1888, to

snatch a moment of "Moses and the Children of Israel" by Handel, a choral performance at the Crystal Palace in London with over four thousand voices and five hundred musicians. On the latter, there is the familiar character of a chorus, voices in harmonious unison blooming in air, high-pitched clouds of sound, an exultant throng translated into an airy smear. Yet there is the inescapable *whoosh-shh-whoosh-shh* of the rotating cylinder foremost of all.

For me, what saves what would otherwise be a dutiful survey of recordings from this period are rare captures of some other high-brow music. There is the great German composer Johannes Brahms in an 1889 parlor performance that seems impromptu. He speaks a few words, then plays on piano a few measures from his "Hungarian Dance No. 1," the sound faint, as though coming from a radio heard through the locked door to a suite of hotel rooms. Then there are baritone Bernard Bégué and tenor Ferruccio Giannini performing abbreviated arias (cylinders could run only for two minutes) in voices that sound like the faint wisps of feathery clouds. But the most splendid thing is Giannini's duet "O Mimì tu più non torni" with baritone Alberto de Bassini from a third-act scene in Puccini's *La Bohème* on a Columbia cylinder. This is Rodolfo the poet and Marcello the painter as each bemoans the loss of their love— the seamstress Mimì for Rodolfo, the saucy coquette Musetta for Marcello. They sing regretfully, recalling youth and passion, the virtuosic doubling and sweet harmonizing of their voices in the spell of Puccini's utterly romantic theme weaving through the chill of the story's relentless unfolding and the murk of these early years of audio.

For all of these recordings that survive, it's what we cannot hear that sometimes freezes my soul. For all the racist "coon" songs still extant from this period—whistling songs, laughing songs, and the various pieces of minstrelsy that express predominantly white attitudes about African Americans—we do not have the precious wax cylinders that Buddy Bolden and his band are said to have recorded one day in New Orleans during the late 1890s. Known as "King Bolden" among admirers, he was a fine cornetist who was the early and acknowledged nexus of the style of music that creolized Black

gospel, ragtime, the blues, Franco-American marches, and German waltzes into the completely new, catchily rhythmic thing called *jass*. From contemporary accounts, Bolden's playing was distinctively loud, improvisatory, and often set to a new, syncopated bass drum pattern. The drum figure, called the Big Four, is thought to arise from Africa and likely came from French marching band music transformed by the innovations of young African American musicians before or during Bolden's time. Bored with stuffy, unsyncopated marches, they slipped in a backbeat and turned lockstep Euro rhythms into Mardi Gras music, inspiring the colorful folk and community parades of second-line dancing. Bolden and others may have taken some of what they heard in Congo Square, listening to musicians improvise on gourds and other African instruments, and put that freedom into their new band music. As for Bolden's reputed volume, Louis Armstrong (who heard him when he was eight) said he played so loud, people could hear him blocks away. In a tribute to the brief period of his heyday (before he succumbed to alcoholism and schizophrenia, before his being confined to Louisiana State Insane Asylum), the charismatic Bolden is sometimes called "the founder" of jazz, his name invoked by the likes of King Oliver, Duke Ellington, and Count Basie. Willie Cornish, trombonist in Bolden's band, said they recorded an Appalachian tune called "Turkey in the Straw," and younger musicians recalled other tunes being recorded in that same session. But nothing has ever been retrieved. The cylinders he made are the holy grails of jazz, described by Donald M. Marquis in *In Search of Buddy Bolden: First Man of Jazz* as probably destroyed in Baton Rouge in 1967 in a shed belonging to a descendant of the sound engineer who made the recordings. Even tunes attributed to Bolden—"Tiger Rag" and "Funky Butt" (otherwise called "Buddy Bolden's Blues")—are known to us because others who came after him recorded them. Nicholas Christopher, author of *Tiger Rag,* a recent work of novelistic imagination that's a tribute to Bolden's brilliant and tragic life, declared to me that he is like Theseus or Romulus in Plutarch—both an actual historical figure and a mythical one. His precious cylinders, if they ever existed, have disappeared into the mythic space of jazz's *jes grew* beginnings,

and we may never know what Buddy Bolden played, what magic he made of human breath and the kiss of his lips against the mouthpiece of his cornet.

Emile Berliner's Gramophone

In 1887, Emile Berliner, a German-born immigrant and onetime dry goods store clerk, took ideas about recording that he'd gleaned from his French predecessors Edouard-Léon Scott de Martinville and Charles Cros and almost completely reinvented the new field. Berliner had served a long apprenticeship, having once been a bottle-washer for the man who came up with the formula for saccharine, spending his nights reading books on chemistry and physics in the Cooper Union Library, and eventually working with Bell Labs as a paid consultant. Inspired by Scott de Martinville and Cros and their descriptions of their recording processes, Berliner dispensed with the wax cylinders of Edison and Tainter-Bell, and instead took a heavy, plate-glass disc and coated it with a layer of lampblack, exactly as the Frenchmen described doing in their earlier devices. He then revolved the disc on a turntable (powered by a hand crank) so that it contacted a stylus mounted on a feed screw.

THE GRAMOPHONE, OR SPEAKING MACHINE.

The stylus defined a spiral pattern on the rotating disc. But, unlike the vertical motion of Edison's indenting stylus, Berliner rigged his, once attached to a diaphragm activated by sound waves, so it would move laterally (like Tainter and Bell's Graphophone) in a zigzag trace over the lampblacked disc. When the recording was done, he sealed it with varnish and had it photoengraved in metal, just as the poet Cros had described in the process he'd anticipated a decade before. The result was a recording Berliner could play with a stylus and diaphragm reproducer. The sound was louder than a cylinder's, yet harsh and grating, making voices almost indecipherable—but it was good enough to get him a patent. Distinguishing it from both the Edison and Tainter-Bell machines, Berliner called his a "Gramophone."

Once he'd secured his patent, Berliner set about improving his process, trying to create discs that sounded better than "the braying of an ass," as described by one contemporary complaint. He used zinc this time as his disc material and coated it with a thin, fatty film (some say beeswax) that would capture the movement of his stylus. Once the vibrations were inscribed in the film, he took the recording and placed it in a bath of chromic acid. The acid etched into the metal, leaving the delicate tracings—an analog to live sound—in laterally cut grooves upon the disc. And as Berliner had anticipated, he could have the disc electroplated, using it to create a metal "negative," and employ that as a durable, reversed master to manufacture innumerable duplicates that he'd stamp out of hard rubber at first, then a material called "Duranoid" later. Finally, in 1895, he settled on making discs out of a composite substance, combining cardboard and pulverized slate bonded with an organic material (made from the secretions of South Asian insects) called *shellac*. Though somewhat brittle, shellac records took well to the stamper and became the standard recording material for decades. Berliner had come up with the idea and the basic process still used today for making records out of vinyl.

Excited by the possibilities of his new record-making regime, Berliner scorned Edison's goal of a machine to aid business and office work; he thought of his Gramophone as an instrument for home entertainment and the key element for a new industry fueled

by easily distributed, prerecorded, standardized discs. He predicted that recordings would soon earn an income for performers based on a percentage of their sales—the first concept of music royalties. In 1893, he set up shop to manufacture his Gramophone in Washington, D.C.

Between 1892 and 1894, the new Gramophone discs were five-inch, then seven-inch, single-sided, hard Duranoid that sold for fifty cents each or five dollars for a dozen. By 1895, records were made of shellac, and you played them on a Gramophone turntable operated manually. At optimum speed (hard to reach, harder to maintain), it revolved seventy times per minute under a mounted or a swinging arm that used metal needles inserted in a sound box attached to a small metal horn. While the Gramophone's recording process afforded more acoustic power out of its horn than a cylinder recording, its metal stylus grinding on the hard surface of the disc combined with its acid-etched grooves to create an almost constant surface hiss and dulled clarity as well.

Undaunted, Berliner and his company brought to recording a small raft of opera singers of the time, both known and not much known. Among these, I've liked the tenor Ferruccio Giannini, whose recordings on Gramophone disc contrast quite a bit from his earlier, fainter ones on Columbia cylinder. His "La donna è mobile" from Verdi's *Rigoletto* comes through strong and clear (on a recording available on YouTube), with listenable top notes, his voice sounding robust and vigorous most of the time, though there's a slight breakup at the dynamic peak of the aria when he blasts the final "pensier" into the recording horn. Through the constant scratchiness of these recordings, you can tell Ferruccio was good enough for regional shows if not for La Scala or the Met. He sounds rhythmic and energetic on "Funiculì funiculà" (also on YouTube) from Rossini's *Il Barbiere di Siviglia* (an aria later made famous in a Bugs Bunny cartoon and fifties TV ads for Campbell's minestrone soup), accompanied by the Royal Marine Band (you can hear the tuba's earnest *oom-pah*).

Another important innovator was a New Jersey machinist whom Berliner contracted to build hand-wound spring motors for his Gramophone. Eldridge Johnson created several breakthroughs

in early recording that came soon after he started contributing to the manufacture of Berliner's machine. Johnson immediately improved the tonearm by coupling it to the metal horn without its having to support the horn's weight. In 1896, he dispensed with Berliner's approach (bathing a zinc disc in chromic acid) and found that a solid wax disc was a much better medium for recording and electroplating, after being coated with a fine layer of metallic dust—a crucial material in creating metal masters. There would be no acid-bath etching of metal in his process, as he put Berliner's masters under a microscope and saw the jagged edges the acid had left in the grooves, deciding it was this that contributed to the harsh sound of the earlier Gramophone recordings. And rather than come up with a new recording material himself, he simply melted down batches of Edison's better-sounding wax cylinders, pressed them into discs, and had a stylus inscribe lateral grooves directly onto the wax. This made a smoother set of impressions inside the groove, and once transformed into metal masters, the duplicates made from them had a sweetness of sound far superior to, though just as loud as, prior Gramophone discs, and with much less surface noise. Johnson increased the record size too, going from the seven-inch Berliner discs to ten-inch discs, expanding playing time from two minutes to almost three minutes (and close to the duration of a cylinder). Finally, he created the double-sided disc, with recordings on both sides (*who'd a thunk?*), instantly doubling the playing time of every Gramophone record and besting the cylinder too.

By 1900, Johnson had taken over from Berliner and renamed the business the Consolidated Talking Machine Company (this became the Victor Talking Machine Company in 1901), adopting a painting of small white dog sitting next to one of his players as a trademark. (The Victor dog was born!) Entitled "His Master's Voice," the emblem was stamped on the paper labels (yet another Johnson innovation) of every new record issued by the company after 1900. In 1902, Johnson established his record factory in Camden, New Jersey, home of the great American poet Walt Whitman. Finally, in 1906, Johnson introduced the Victrola, a new and fairly inexpensive style of player that did away with the big but gorgeous, japanned morning-glory funnel of the earlier Gramophone. Instead,

the Victrola featured an enclosed horn and, in fact, concealed all of its working parts inside a tidy box underneath the platter. It was cheaper than a Gramophone and made a sensation, selling not only to the rich, but to the aspirational. It soon overtook all other devices in popularity, the company's sales going from about 32,000 Gramophone units sold in 1900 to over 124,000 Victrolas sold in 1911.

Yet for me, the most momentous occasion in early recording came in March 1902 in the Hotel Milano near the opera house of La Scala. Enrico Caruso, a twenty-nine-year-old tenor with a powerful voice and dramatic flair, a new kind of operatic artist who sang with soaring emotions in his voice (introduced to the stage of La Scala mere months before), made a series of ten recordings on wax blanks, later duplicated on Victor shellac records. To maximize Caruso's voice, the engineer used a bell-shaped horn made of tin suspended five feet over the floor, and he hired an able pianist to provide accompaniment. The arias Caruso sang fit the capabilities of the new recording medium and method almost perfectly. The tenor's voice was so clear, it not only drowned out the surface noise, but sounded rich and vibrant, nearly all his range miraculously falling just within the tight frequency window of the technology. And in playback, the music came alive in a way that simply had not been accomplished before, either with wax cylinder or zinc disc.

When I was a kid in the late fifties, watching episodes of *Abbott and Costello* and *The George Burns and Gracie Allen Show* on a black-and-white TV set after school, what I remember most, other than the exotic New York and Jersey accents, was how mustachioed Bacciagalupe, the Italian baker in Costello's neighborhood, would listen to Caruso records as he put out his cakes and pies, the sound of a robust but altogether angelic voice floating over his cookies and cream puffs from a phonograph offstage within his shop. I saw how Gracie Allen would deadpan her line about a howling hound dog as "pretty good but certainly no Caruso." I saw how Sidney Fields, Costello's landlord, would chase the rotund comedian down the street trying to collect the rent, flinging shellac record after record at him, Bacciagalupe trailing and imploring, "Mamma mia! Non le *Caruso!*" and, getting on his aproned knees, trying to piece the fallen and shattered discs together on the sidewalk. Two gen-

erations before, Caruso was so famous, he'd become the common standard for vocal beauty that had saturated the culture, to the point where his very name had become a bit and a punchline for working comedians.

Seven of the ten arias that Caruso recorded in 1902 survive. I've a microgroove LP mono transfer from the original shellacs issued by Everest, and I play it from time to time, just to cast myself back to Caruso's earliest performances. As you might expect, they vary in quality, some of them faint as though coming through a bad telephone connection, but each is more than a document of his powerful voice. "Questa o quella" from *Rigoletto,* sung in a style more modern than Giannini's earlier recording, sounds close in dramatic rendition, rhythmic pace, and musical approach to any of a half-dozen tenors who are recording stars today, and the few archaic flourishes that stick out serve simply to "periodize" the style, mark it as of another time. Caruso's singing overloads the acoustic capacity of the recording chain at times, as he blasts his fortissimos with verve and a superabundance of volume that produce some distortion at peak. Cavaradossi's lament, the dramatic, even pathetic "E lucevan le stelle" from the final act of Puccini's *Tosca,* sounds completely affecting, even with the slight nasal distortion of the horn, Caruso shading the sweeping emotional progress of the aria slowly darker and darker, adding a melodramatic and throaty sob just before the final three descending notes. But the recording that best combines both sound and emotion is the aria "Una furtiva lagrima" from Donizetti's *L'Elisir d'Amore,* a piece I've heard countless times on contemporary recordings and at least twice live—once in San Francisco and again in Venice. You'd think I'd be inured, but Caruso gives it everything he's got, and I'm taken in, even through the disturbing and audible tick at the start, through the artificial warble in his voice because of the varying speed of the machine's operator, through the slight tunneled feeling of the sound distorted by the recording horn. What I focus on instead is what wrenches the heart, an aria that starts from a comedic situation—the sappy Nemorino, Caruso's character, spots a tear that drops from the eye of the woman he's adored, and concluding it results from the power of the love potion he's slipped her (the purported elixir of love),

he decides she finally has some *feels* for him. He believes, in the moment just as the teardrop falls, that their souls might at last merge and he could die as loved as he's wished since . . . forever. It sounds absurd and a lot like Jim Carrey's relentless, gap-toothed buffoon in *Dumb and Dumber* when Lauren Holly says his odds with her are one in a million (*So you're telling me there's a chance!*). But this aria from the *bel canto* ("beautiful voice") repertoire takes hold of any wish for ironic exit from Nemorino's sincerity in the insistent grip of its emotional lyrics, in the slow rise of the repeated musical figure (echoed deftly by the piano accompanist), in its heightening climb up the scale, before Caruso's voice swoops and descends back down on the words *death* and *love*.

"St. Louis Blues" and the Electrification of Sound

In early January 1925, Bessie Smith, already a star and called "Empress of the Blues," stepped up, sang into a Western Electric condenser microphone, and electrified the world of phonograph listening with the sound of her contralto voice moaning a tune called "St. Louis Blues." On the date with her were just two others— a twenty-four-year-old cornetist named Louis Armstrong, and Fred Longshaw, a keyboard player who played harmonium on the track. Written by composer W. C. Handy, the song was already popular with audiences in America and England and had even been recorded a handful of times by both white and African American artists. Yet nothing had the impact Smith's soulful recording did. It set a new standard for fidelity and for artistry, inaugurating a tradition in music, particularly jazz, that shaped the rest of the century and a bit beyond it.

Smith was no newcomer. She had recorded previous blues numbers like "Gulf Coast Blues," "Cemetery Blues," and "Downhearted Blues," had established herself as a headliner, and was the highest-paid Black entertainer anywhere around. She'd served her apprenticeship with a touring troupe of African American musicians called the Rabbit Foot Minstrels and had been mentored by Ma Rainey, a hugely important predecessor and influential blues

artist herself—think Cimabue to Smith's Giotto. Smith's own records were issued on Columbia's "A" list, not confined to racial or novelty categories (Irish, opera, bands, whistling, humor, and what were blatantly called "coon songs"). But she was later to be the first artist Columbia released under the new category of "blues."

Today, we can hear just about all of Bessie Smith's records on CD or computer files converted into sound via yet more technological wizardry of our own time. Her "St. Louis Blues" sounds indeed far away, filtered by the age—an ark of monophonic sound (albeit scrubbed of surface noise and minor clicks and pops) to our ears diminished in depth and nuance but with the unmistakable aura of human creation to it. There is a fabular grittiness, the feeling that these are natural figures conjured from a dance of the human soul across a cave of the mind. The recording opens with a brief

instrumental intro, the harmonium wheezing and Armstrong hitting a single, blaring note on the cornet that startles you awake. Then, almost lazily, Smith sings the classic line *I hate to see / that evening sun go down.* She sings it mournfully slow, at a steady pace but with a cutting power way above the volume level of her accompanists. Her voice is clear, bodied, solid as the trunk of an oak tree, but blaring sweetly as Armstrong's cornet at the top of her range. She often gives single words varied notes, bluesy mordents as she elongates vowels—(*suh-uhn, git-ah-ah-way, ri-ang, streeangs, sto' bought hay-yer, goin' no-way-yer*)—supplying them African American pronunciations, glottals and slides, and, just once, Black vernacular grammar too (*Ah fee-yulls*). And, when she hits the famous bridge (written by Handy to a habanera rhythm), it's hard to distinguish its compositional roots as Smith sings it like a big river flows, taking its time. The weary sadness in her voice just dominates.

What's also exciting is the way Armstrong trails her throughout the tune, often mirroring her phrasing but, increasingly, adding brassy flourishes in a clear exchange of call-and-response between singer and instrumentalist. Unfailingly, he hits clear, precisely separated notes with few slurs, and creates a thrilling accompaniment in his fills. The performance presages Stan Getz's tenor accompaniment to Astrud Gilberto on "The Girl from Ipanema." Armstrong punctuates Smith's artistry with jazzy, syncopated struts and beautifully bent notes that send the tune into another dimension, just as Getz does with Gilberto much later. The infamous harmonium on the track is plodding but not a distraction, sounding hurdy-gurdy, holding a shallow bottom suited to Smith's mournful pace. It's clumsy but appropriate, like Al Kooper's famous though amateurish organ fills on Bob Dylan's "Like a Rolling Stone" from the sixties. But Smith is the focus, her voice emerging out of growling depths and striking wails that pitch like muddy Mississippi waters roiled by a woman drowning in sorrow.

W. C. Handy says in his autobiography, *Father of the Blues* (1941), that he had three inspirations behind composing his song, the first blues to be published as sheet music and that reached the general public. While traveling on a train, Handy says he heard what he describes as a "lean loose-jointed Negro" plucking a guitar beside

him while he slept. It was played in the Hawaiian manner just becoming popular at the time, the man using a knife as a slide bar on the steel strings so that the notes he played quavered and flatted, "the weirdest music I had ever heard," Handy writes. Later, while Handy played a dance with his orchestra in Mississippi, the audience asked that three locals be allowed to play a few numbers. The locals brought the house down with a music that struck Handy as "primitive," raucous and monotonous, like the "stuff associated with [sugar]cane rows and levee camps." But he also found it haunting and was impressed by how it lit up the joint, members from the audience rising to their feet, gleefully dancing in time with the simple music he'd heard. He also noted that there were gaps or waits in the music where a singer could exclaim a word or two, accentuating the lyrics and mood of the piece with ad lib wails and shouts like "Oh, mama!" and "Lord, lord." Limited to three chords, this was a music more of feeling than of structure, Handy realized. He wanted his own compositions to inspire dancers to get up on their feet too. He writes, "When 'Saint Louis Blues' was written, the tango was in vogue. I tricked the dancers by arranging a tango introduction, breaking abruptly into a low-down blues . . . The dancers seemed electrified. Something within them . . . took them by the heels." Finally, he found the deep emotion for his tune after he witnessed a lonely woman singing on the street by the Mississippi. He'd been so broke, he'd slept on cobblestones by the riverbank and woke hearing a plaintive song about a woman having lost out on love. "Ma man's got a heart like a rock cast in de sea," the woman sang, her complaint about being bested by a pretty woman in "sto' bought hair" and diamond rings. Handy put the line and these details into his composition, calling on his own experience of being downhearted by the riverbank, and with a memory keen as Mozart's, easily transcribing snatches of melodies, lyrics, and rhythms he'd heard from the milieu of itinerant banjo and guitar pickers, honkytonk piano in bars, bordellos, and barrelhouses. It was wrapped in a structure that reflected native roots, a background in church spirituals, and syncopations from ragtime.

"St. Louis Blues" would eventually reach the status of a classic in American and jazz music. It would be re-recorded by an amazing

string of eminent artists such as Armstrong himself (with two different bands of his own), Fats Waller, Cab Calloway, Django Reinhardt, Art Tatum, Bing Crosby with Duke Ellington's Orchestra, Tommy Dorsey, Glenn Miller, Artie Shaw, Benny Goodman, Count Basie, Billie Holiday with Benny Carter's Orchestra, Earl Hines, Errol Garner, Dizzy Gillespie, Billy Eckstine, Ella Fitzgerald, Miles Davis, Dave Brubeck, Max Roach, and Herbie Hancock with Stevie Wonder.

But besides the considerable and inspirational artistry on Smith's 1925 record, the Columbia release also marked a watershed moment in the history of sound recording. Leading up to it were some important discoveries. In 1917, E. C. Wente had invented the condenser microphone, a device that could translate acoustic currents into electronic impulses. Two years later, scientists at Bell Labs came up with a recording cutter head that replaced the needle and diaphragm of acoustic recording with one that relied on electromagnetism to respond to signals the microphone had converted from sound waves. They also created an improved acoustical phonograph to play the new electrical recordings and fitted it with an exponential horn speaker, one nine feet long that was necessary to reproduce the wider range of sound made possible by the electrified recordings. Then, in 1924, Bell Labs invented the folded horn, working out the mathematics for fitting the horn into a compact cabinet.

But at Western Electric, a division of Bell Labs, scientists were still not satisfied. Sound reproduction (in the form of the horn) was still acoustic, and they wanted to take advantage of electronic amplification, de Forest's little Audion device that could amplify a recorded signal, a thing just beginning to be used for the implementation of radio broadcasting. They had studied the transmission of sound along telephone lines and worked out (based on Bell's earlier discoveries) what they thought were the basic properties of sound and its conversion into electrical currents. They used the condenser microphone to capture sounds and convert them into electric currents. They took an amplifier based on de Forest's Audion to increase the strength of these currents. And then they used another tube amplifier to drive the electromagnetic cutter head to make grooves in a record. The cutter head (with its own magnetized assembly)

was balanced to move precisely within the magnetic field generated by the amplifier so that the varying current controlled and directed the cutter's movement, transcribing sound waveforms onto the recording disc. Playback came from the movement of a needle in the groove interacting with yet another magnetic field within the needle's electrical pickup, which created the micro-currents then amplified into stronger currents by vacuum tubes. But this was not all. Finally, in 1925, C. W. Rice and E. W. Kellogg, two scientists at Western Electric, created the moving coil transducer. This was the dynamic loudspeaker that converted these electronic impulses into sound itself. The new loudspeaker was no longer just a horn attached to an acoustically activated diaphragm (that vibrated), but yet another electronic device that converted electrical currents into movements of its drivers, which were themselves suspended within electromagnetic fields. The scientists took a coil of wire and placed it within a magnetic field and found that it moved in relation to a current passing through it. They then attached the coil to a thin, rigid cone of material and connected that to a large diaphragm made of pulped paper. The entire diaphragm moved in relation to the amplitudes of the electrical currents being sent to it via the coil. The assembly re-created sound waves, compressions and rarefactions, originally captured by a condenser microphone. The drivers moved in direct relationship to the original sound waves recorded, providing a wider range of frequencies and more faithful reproduction of total sound than old acoustic methods could achieve. By 1925, just about the time Bessie Smith stepped up to the mike at Columbia studios, scientists at Western Electric had succeeded in creating an almost complete electronic equivalent to the mechanical system of amplification.

Saint-Saëns and the "St. Louis Blues"

Obsessive that I am, I've a set of CDs I got from England entitled *Leopold Stokowski: The Complete 1925 Electrical Recordings* (Biddulph, 2001) because they contain the first recordings ever made of a symphony orchestra once electronic technology was introduced. The set is a compilation of pieces by Saint-Saëns, Borodin, Tchai-

kovsky, Dvorak, and Ippolitov-Ivanov performed by the Philadelphia Orchestra under English conductor Stokowski's direction. The Saint-Saëns *Danse Macabre* was done before all the others, so it occupies a very special place in audio history. But it's not simply a relic. It still sounds *good,* especially on high-end equipment or over a pair of decent headphones (I've listened to it both ways).

Released by Victor on 78 rpm shellac and labeled "His Master's Voice," with the trademark white dog listening to a Gramophone, the recording lasts for just over seven minutes—perfect for the two sides of a ten-inch disc. It still has a vintage aura, of course, in no way as dimensional and refined as recordings done with today's technology, but the sound was startlingly clear and precise for its time. And it's important to remember that getting the full symphonic orchestra on recording just wasn't possible before. There was no way acoustic recordings in wax could capture the sound of fifty or more musicians playing either pianissimo or triple forte. The orchestral pieces that were done before Western Electric got in the game were made with reduced ensembles—a flute and a piccolo, a few violins (often with amplifying horns attached), a cello or two, one French horn, and maybe a tuba and a double bass. The full range and orchestral effect were reduced to an impressionistic approximation or just a few strokes of background noise even on the best acoustic recordings (notably Caruso's from 1914). But with electrification accomplished, recording a full orchestra suddenly became possible. And up stepped curly-haired Stokowski on the dais in front of the Philadelphia in July 1925.

The narrative conceit of *Danse Macabre* is that the character Death suddenly appears at midnight on Halloween, in a graveyard where ghosts and goblins abound. He plays a sinuous and chilling tune on a violin, calling forth other dead (as skeletons) to rise from their graves and dance until dawn. Saint-Saëns opens his tone poem with twelve pulsating chimes from a harp, followed by soft string sounds that give way to heavy pizzicati from a full complement of double basses. Then comes a discordant tritone from the concert master on violin, the *diabolus in musica* or "Devil in the Music" theme borrowed from medieval and baroque times. After that, a more sprightly violin passage takes over, and echoes of it come

from the orchestra in flowing, feathery accompaniment. The harp gets plucked again, and a haunting theme, reminiscent of Romani music, begins with the full orchestra joining in, lush and romantic, before the solo violin briefly reprises the tritone. The orchestra plays a crescendo and initiates a succession of emphatic passages with big, sweeping strings and woodwinds piping, all of which serves as the build-up for a xylophone obbligato (suggesting the tinkling of bones knocking together). A romantic violin solo follows, then rolling sweeps on the harp and sudden, tumultuous tutti from the full orchestra, with periodic brass fanfares and dynamic percussion. The whole composition becomes a magnificent tapestry of sounds. Violins are clear, accented at times by pleasing shrieks from a piccolo. And the tritone theme and clinkings from a xylophone keep the conceit of Death's dance going. The dramatic close of the piece thrills with alternately swirling and mournful violins, bass pizzicati, and tuba and oboe punctuations down low and up high. In this, the first of symphonic electrical recordings, Stokowski and the Philadelphia bring together the truest feeling of orchestral performance in audio up until that time.

Highbrow critics of the period weren't terribly complimentary, though, likely comparing this recording and Stokowski's subsequent ones of that year to the live sound of an orchestra in a concert hall. They called the 1925 recordings "abominable" with harsh exaggerations of sibilants, the sound of massed strings "atrocious," and the whole piece "a jungle of shattered nerves" and "a complicated cat fight." But, to my more forgiving ears, it's great to hear not only what an early electronic recording sounded like, but also to get a sense of the overall drama in symphonic performance from nearly a century ago. It's an acoustic wormhole to the past, my ears traveling back, imagining the gravity and sprightliness of another time, its sensibility and fanciful metaphysics. It's about human imagination and the legacy of our common culture.

⁂

Electrification was a big deal in other kinds of recording as well. In the acoustic era, for example, it just wasn't possible for an acoustic

guitar to be the lead instrument in a jazz combo, as its sound was too faint compared to reed and brass instruments, let alone piano and drums. And voices too were difficult to capture over the sound of a band or orchestra unless the singer happened to be a belter like Caruso or Bessie Smith. Electrification made it possible for singers with more modest pipes like Bing Crosby to shine, bringing about the advent of "crooners" who sang into the mike more softly, with dulcet tones fat or slinky. Electronic amplification meant recordings could pick up on the tonal qualities and performance nuances of more delicate, lower-volume instruments and voices—like Django Reinhardt's acoustic, petite-bouche Selmer-Maccaferri guitar, like Billie Holiday's teasing vocals, like John Fahey's intricate picking of a steel-stringed Gibson Recording King guitar.

The Quintette du Hot Club de France, with Django Reinhardt playing lead guitar and Stéphane Grappelli on violin, recorded a version of W. C. Handy's "St. Louis Blues" in 1937. In their hands, the tune gets jazzed up, has jump, and swings hard, with great solos by both famed principals. Reinhardt opens it up with a walking tempo on his guitar, plucking it almost lazily, then hits a bending note, followed by a brilliant arpeggiated fill, swift and delicately articulated, emphatic and startling. Next, he goes nuts, throwing off bluesy notes, going off the beat, strumming rapidly, hitting a big, downswept chord, flying into one of his crazy-fingers improvisations, making big, bent notes on the upper strings, with rapidly plucked treble accents on the lower, creating a call-and-response effect all on his own. On the tango melody Handy wrote as the bridge, Reinhardt throws in dissonant, bluesy chords, playing parallel octaves at times, then playing the Selmer like it was a balalaika or mandolin, plucking 1/16th and 1/32nd notes before the band goes back to 12-bar blues. After yet another solo, Reinhardt creates the song's climax with block chords and one of his famous rolls, descanting to accompany Grappelli's violin solo with more block chords. Throughout, there are unmistakably flamboyant elements of flamenco and Romani music. The "deep song" of *cante jondo* is never far away, the spirit of *duende* that Spanish poet Federico García Lorca celebrated as the devil that rises up through the bootheels of flamenco, the stomping foot of one who plays guitar. In Rein-

hardt, you can hear how it rises from the floor up through his body and comes out through the flashing eight fingers of his playing (in a fire, he lost the use of two on his fret hand). He strums, he picks, he double-picks, he feathers the strings, he hits a chord fortissimo, he plucks the E-string up top then arpeggiates the trebles, throwing off rhythmic changes all the while, blues and a demon at the root of his song.

I heard this style of playing in Paris one summer night. A Django festival was going on throughout the city over a weekend in July when I was lucky enough to be there, and I sauntered throughout Saint-Germain-des-Prés as snatches of trios, quartets, and quintets playing Django's music drifted out of every few clubs that I passed. Languages commingled in the air—French, English, Hindi (I think), and German—as I passed people seated at tables set out on the sidewalks, looking for a spot that might let me order a beer while I listened. I found a place with a blue awning and heard, from inside, guitars and a violin playing with that familiar jump rhythm. It was a quartet up on a small stage alongside the back end of the long corridor that was the café, a tiny bar near the entrance, with tables and bentwood chairs going back alongside most of each wall. There were two guitars, a violin, and a stand-up bass playing classics from the Quintette du Hot Club de France. I found an empty table near the back, opened my notebook, plucked my pen from my shirt, and started jotting a few notes about my day of wandering. A waiter came by and took my order. And then the music took over. It swirled through the club like eddies from the big river of blues it came from, taking me up in the gaiety of its current.

I think I first heard Billie Holiday when I was about twenty-five, living in Seattle and running a community theater group. Of course I'd heard *of* her, but I'd never much paid attention to her music until I heard it playing on a portable record player that sat on the floor in the apartment rented by a woman I was seeing. We'd been friends awhile and slowly slid into a physical relationship one summer. After a picnic up on the roof of her building, she took me

downstairs and put on a Billie Holiday record, telling me it was "a woman's blues" and that it spoke to her, particularly about how men *were*. I don't know why it was the first record she played for me, though I suspected it was a warning. "Mean to Me" is the tune I remember from that evening, but then everything faded out as we stripped and made love, sinking to the floor ourselves, the phonograph running through its tracks and repeatedly hitting the label until she got up, moved the tonearm aside, and turned the thing off so we could get back to business.

Billie Holiday recorded "St. Louis Blues" too—in 1940 with a jazz orchestra under the direction of clarinetist and sax player Benny Carter. In time, I knew what everybody knows—that she was called "Lady Day," the nickname bestowed upon her by Lester Young, the tenor sax man from Count Basie's band who was her good friend, until they had a falling out late in life. But this Okeh/Columbia recording is from near her heyday, before a grit came into her voice and age and drugs robbed her of the top of her range. It was intended to be part of a series of recordings in tribute to Handy that never got completed. On the date are some terrific musicians: among others, Bill Coleman on trumpet, Benny Morton on trombone, George Auld on tenor sax, and Carter himself on clarinet.

It begins with Coleman's trumpet intro, then there's a fill from Auld's saxophone. When Holiday starts singing, though, the tune takes off, her phrasing full of bluesy descants, a sibilance like a muted trumpet, with lots of subtle dips and lilts, mordents and glissandi, elongated vowels with the push on the back end of its enunciation. *Suh-uhn,* she sings, *too-mahh-row, fee-eel,* and *see-eee.* Carter's clarinet supports her, tailgating with fat, mellow notes. When they hit the famous Handy habanera bridge, Holiday rises in register, adding volume and bounce, sounding saucy and a little tart. The band plays gorgeous horn choruses while Holiday sings like she's playing a reed instrument, with smooth phrases and clean enunciation of the lyrics. Once they get back to the blues, they swing even harder, throwing more momentum behind Holiday's singing, taking a faster tempo than she tends to later in her career. Her voice is nuanced, lilting, her stylistic delicacy lifted to prominence by amplification. It's tactile, deft, and sassy, Holiday communicating emotional shadings

that just weren't around anywhere before. And Carter answers her with a clarinet solo as complex as her singing, blowing soft arrays of notes, accentuating a single one in a staff of them, making precise little turnarounds, then sounding like a trumpet for just one blasting note before he sashays away with soft squeaks and pipings dexterous as a Django run. But the power Holiday herself demonstrates isn't in volume or big dynamic accents; it's in the sneaking subtlety of her shadings, the sass with sadness that she puts into the lyrics, the plaintiveness mixed with the coy.

When I think back to the summer Holiday's music was introduced to me, things are mixed with the remembrance of the purple label on the LP, the light floral scent of my lover's perfume as it emanated from the warmth of the skin on her neck, from her hair, and the soft scratch from the needle of the phonograph as Billie's songs made their teasing circles around our bodies. By summer's end, like the blues prophesied, I'd done her wrong. My own heart was like a rock cast into the sea.

I first heard John Fahey's *Blind Joe Death* in the dorm room of my neighbor across the hall from me my freshman year in college. Robb was a man who rode a BSA motorcycle, wore cowboy boots and his hair long to his shoulders, and listened mostly to rock music from the Bay Area bands in California where he was from. But he had the best damn stereo of anybody—a Scott tube receiver and JBL speakers—and the largest record collection of anyone in the dorm. I drifted over there afternoons sometimes, hearing Steve Miller, the first Santana album, or Electric Flag (leader Mike Bloomfield had immigrated to San Francisco by then). It wasn't so much the music I liked as that I was *really* impressed with the stereo and its warm, natural sound. Robb ran the Scott receiver without its casing so the tube could release heat more easily, and while the music played, I'd stare at the luminous bulbs, the flame-red incandescence of their filaments beguiling as the sound they made.

One afternoon, he pulled a record out from the thick stack of his collection, which leaned on the floor between a bookcase

and his bed. It had an aqua-blue cover and a line drawing of faux Renaissance revelers with hair that looked carved out of marble and a pair of lovers (still clothed) nestled in a corner amidst sprays of leaves from an umbrella tree. This was *Blind Joe Death, Vol. 1* (Takoma, 1963) by John Fahey, a virtuosic acoustic guitarist who'd become a cult sensation just on the strength of this album's release. What was remarkable was that it was self-produced, on Fahey's own label, supposedly from money he'd saved by pumping gas in his Maryland hometown. The music was all solo guitar, Fahey playing a few of his own compositions as well as intricate versions of traditional songs, authored hymns, and classic blues he'd arranged for himself. The second track on the A-side is Handy's venerable "St. Louis Blues."

Fahey picks out a lot of the tune fairly close to the guitar's bridge, the notes sounding incisive, snappy, but at a slow, moderato march rhythm. He plays an alternating bass on the beat, also called Piedmont Blues style, that was designed to imitate the left hand of the piano playing rags and blues and stride. Chords are steely, emphasizing sharp attacks and long resonances, with big dynamic contrasts, particularly in Handy's habanera bridge section, which is gorgeous and bluesy with lots of bent notes. Fahey emphasizes notes either with volume, strikes, or conspicuous bendings sustained to the tails of their acoustic decay. His chordings are both tasteful and surprising, and he plays double octaves on occasion, bending both notes in unison. He snaps strings on attack, revels in the harmonics thrown off as a note decays, including a variety of acoustic guitar effects. The whole thing has the feel of a solo concert—like the performance of a baroque lutenist—Fahey's playing clean and precise as Julian Bream's perhaps. His approach to the music is indeed meticulous in the sense of a pristine presentation and a preservationist's homage to tradition. The music, after all, is American "traditional," dedicated to and celebratory of our blues, spiritual, and folk music heritage. Yet there is force and funk in Fahey's playing, then gentle picking and brushstrokes, inventive with the triplets at the turnaround, and a kind of reverence for each note struck. It's all like a man singing—slow, patient, and irrevocable as the coming of winter's cold.

I saw John Fahey live not long after hearing him for the first time on my friend's stereo. Fahey came to play a date at my college either later that year or the next. He performed in the chapel, a long, rectangular space with wooden pews on its main floor, raised tiers of benches along its sides, and an organ behind the lectern at the far end. Fahey tried playing from the stage at first, a big space in front of the organ, but then abruptly grabbed up his guitar and case and moved to one of the side rows of benches. I think he must've liked the acoustics better from there, the sound reflecting from the wooden stands stacked behind him, launched into the voluminous space with its vaulted ceiling. He used no amplification. Then he smoked. He took out a pack of cigarettes, shook one out, and lit it one-handed, twisting out a single paper match, pressing down on it, scratching it expertly. He took a drag, then leaned leisurely back on the bench, blowing a cone of blue-gray smoke into the still air. He murmured, he railed a bit, he excoriated us students, he blew more smoke. Then he grabbed his guitar and played like a demon or an angel, the brilliant sound of his Gibson reaching me like miniature clouds bursting with sparks and little notes of lightning. He was a performer like I'd never seen before, wearing just a dingy T-shirt, dirt-colored jeans, and sneakers. He was Blind Joe Death as a beefy and prattling skatepunk, magical only at his guitar, a haze of blue smoke and steel-string struck notes lingering in the air around him like the gauze of a scarf worn by a stout skeleton.

1948: The Microgroove LP and Full-Frequency Sound

I have two long-playing, 33⅓ records in my collection that I especially treasure because they mark significant developments in recording history. They are also distinctively different in the characters of their sound, demonstrating exact points when there were two cosmic shifts in hi-fi within the span of a single year. Both were issued in 1948. They are the Mendelssohn Violin Concerto in E Minor, op. 64 performed by Nathan Milstein with the Philharmonic-Symphony Orchestra of New York conducted

by Bruno Walter and issued by Columbia Records, and the Fifth Beethoven Piano Concerto in E-flat Major played by Clifford Curzon with the Vienna Philharmonic Orchestra released on the Decca Records label in England. The Columbia recording launched the completely new format of the twelve-inch microgroove LP that eventually retired the old shellac 78. The British Decca LP was the first recording made with "full frequency recording response," or *ffrr,* a new technology initially developed for submarine detection during World War II.

Columbia debuted the 33⅓ rpm, long-playing (LP) microgroove record in June 1948, at a gala demonstration at the Waldorf-Astoria Hotel in New York. At the demo, Columbia engineer Peter Goldmark placed exactly the same recordings in 78 rpm shellac and 33⅓ rpm vinyl side by side in two different stacks. The pile of 78s created a small tower about eight feet high, but the 33⅓s fit between Goldmark's hands in a neat, foot-high stack. A newspaper headline of the time read "ONE RECORD HOLDS ENTIRE SYMPHONY: New long-playing 12" disk holds 45 minutes of music." The old shellac 78 held only a maximum of ten minutes—up to five on each side.

There were other major differences. Shellac records—then made of a combination of pulverized slate or limestone bonded together with about 15 percent shellac—were noisy. There was an incessant hiss that the music had to rise above in order to be heard. And because the record constantly wore down, emitting loud pops and tics from grit against the needle making its way, shellac discs only lasted a maximum of a hundred or so plays. LPs, on the other hand, were made of PVC, called Vinylite, a new, unbreakable plastic initially used for radio cabinets and telephone parts. The material combined vinyl chloride and vinyl acetate, harder and finer than the shellac composite for 78s, which meant more grooves could be pressed into a comparable space—224–260 per inch compared to shellac's 80–100 per inch—the vinyl groove almost three times smaller, thus the appellation *microgroove.* Finally, made of finer and more pliant material, vinyl records were much quieter, playing with more smoothness, and used jeweled synthetic sapphire or diamond

styli instead of abrasive steel. All of these combined to eliminate hiss, lower the noise floor, and jump dynamic range so the finest pianissimos of an orchestra might easily be heard without being drowned out; even its boldest triple fortes could be heard without distortion.

It has always been the case that, in order to conserve groove space during lacquer cutting, bass frequencies have had to be cut (their waveform amplitudes reduced) and high frequencies boosted (amplitudes increased). Then, the reverse has to happen during audio playback in order to represent the entire audioband on any record. However, before an industry-wide standard equalization curve was introduced in 1954 by the Record Industry Association of America (known as the *RIAA curve*), the frequencies at which these cuts and boosts occurred, along with the degrees of cut and boost, were different for each record label. Pre-1954 LPs from Columbia, Decca/ London, Deutsche Grammophon, Capitol, Angel, EMI, Mercury, and Philips all used different equalization curves, requiring that tonal balances be adjusted in order for them to sound their best. Ideally, using the proper tone balance (equalization curve) results in a flat frequency response on playback—an evenhanded representation of the frequencies, without imbalances that can result in screeching or sour-sounding violins, opaque or chalky high frequencies, booming or nonexistent bass. In the era before the standard equalization curve was adopted, preamplifiers and receivers all had adjustable tone controls for bass and treble so a listener could "dial in" the proper balances. Today, almost no pieces of audio electronics readily available have these tone dials. So, for me to properly play vintage LPs from before 1954 (and numerous post-'54 outliers), I use a special Japanese phono preamplifier (designed by Kazutoshi Yamada of Zanden Audio) that has installed within its playback capabilities five different equalization settings—RIAA (standard), Columbia, Decca, EMI, and Teldec—accounting for pretty much all I need to make old records sound good.

The Columbia release of Milstein's performance of the Mendelssohn Violin Concerto was the first twelve-inch microgroove LP. It had been stockpiled with a small raft of other recordings (111 classical, 18 popular, 4 juvenile) that Columbia made in antici-

pation of opening a new market. My copy (I bought it used on discogs.com) has a deep, glossy blue circular label printed with silver lettering across the bottom that declares it is LONG PLAYING MICROGROOVE. Its sleeve is mostly a mustard-yellow inset with the fairly plain design of a white, vaguely Attic-looking podium within which the title and performer information are lettered. A collector named Ed Scrugge has scrawled his signature in pencil in the upper right-hand corner. But for all its significance in the history of recording, and for all I've done to replicate the Columbia EQ curve, its sound is disappointing to my contemporary ear. It's just not quite "high fidelity" yet.

Although Milstein's violin is lively—sprightly and silvery sounding, his playing nimble and joyous, with sinuous phrasings, thrilling cadenzas, and exquisite pianissimo trills throughout the concerto— the orchestral sound is diffuse, muffled rather than dynamic or clear, even muddy and sluggish at times. Woodwinds are all right, bassoons and clarinets possessed of their characteristic warmth and a palpable woodiness, but the strings, which might have been soaring, are recessed and damped, particularly the violas, cellos, and double basses. They lurk and they murk, and I miss the thrum and the glister, that feeling of the orchestra as an equal unit, vying with Milstein's soloing, alternating in brilliance and power, both a partner and a rival with his violin. The vinyl record is indeed "quiet" as a medium, yet I feel its representation of the music is flawed.

It is important to note that the Milstein recording was originally made in 1945 and issued at that time in an album of several records in 78 shellac format. Columbia then stashed the lacquer master as part of its plan to bank a number of recordings in anticipation of the debut of the LP. For the 1948 LP release, the company created a simple transfer from the archived lacquer 78s to 33⅓ format. But the Columbia microgroove vinyl did not have the advantage of *ffrr*—an even more significant development in recording that came along later that same year, across the pond in England, where British engineers had adopted a method of recording developed using wartime sonar research—the hunt for U-boats during the Battle of the Atlantic.

In his study *America on Record: A History of Recorded Sound,* Andre

Millard writes that "Full Frequency Range Recording" was developed by a research team led by Arthur Haddy during the 1940s, under a contract with the British Ministry of Defense. Their charge was to make disc recordings of submarine noises in order to train Royal Navy and Coastal Command sonar operators to recognize the difference in propeller screw sound from German U-boats and Allied vessels (both submarines and commercial ships). In order to do this, engineers determined that recordings needed a frequency response range far greater than any system of recording that yet existed. Human ears have a response between 20 and 18,000 cycles. Acoustic phonograph recordings of the time could only cover a range between 200 and 3,000 cycles. Electrical recordings stretched this coverage from 130 to 4,200 cycles, then from 130 to 8,000 cycles using more advanced microphones and cutterheads that came along later. Haddy and his British team took up the 1930s research and inventions of EMI engineer Alan Blumlein, using his moving coil cutterhead and developing an elliptically shaped stylus (rather than the spherical one that had been standard) to stretch their frequency coverage from 80 to 15,000 cycles—almost the complete range of human hearing. They also changed the electroplating process for making masters so that it preserved much more high-frequency information in the transfer to duplicates, improving the basic process invented by Emile Berliner and later developed by Eldridge Johnson in early Gramophone production. In these ways, more complete waveforms of sound could be transferred to the finer microgroove record so that recordings were not only bolder, with more bass and treble, but more nuanced as well. In 1945, *ffrr* was introduced by Decca Records in England and London Records in the United States. At last, a satisfying sonic replica of the living presence of an orchestra was able to be heard.

I've the actual 1948 Decca release of Curzon's Beethoven Piano Concerto no. 5 with the Vienna Philharmonic Orchestra conducted by Hans Knappertsbusch. Emblazoned on its bright orange label are the letters *ffrr* with a silver line drawing of a human ear adjacent to them. Below the Decca emblem are the words *Long Playing* and, to the right of them, a circle featuring the same words surrounding

a silver infield with a cutout that says "33⅓ RPM." The jacket is a rich midnight-blue with gold lettering naming the performers, declaring it is Beethoven's *Emperor,* and sporting a coronet of golden leaves. I dig it to death, just based on its looks.

From the first orchestral tutti, the recording has indeed a full-range sound. The orchestra just leaps out, midrange instruments pulsing forward from my speakers into the room, followed by Curzon's rippling arpeggios in liquidinous runs, punctuated by beautiful single-note accents, showing contrasts of touch and force. The orchestra is absolutely thunderous, with wonderful-sounding violas and cellos, winds superb in their piping passages, and mellow, even burnished French horns. In my notes, I wrote *Bass is here!* and the pleasing thrum of weltering orchestral sound blooms with great power and sublimity. There are dynamic contrasts to scale, sweet pianissimos, galloping crescendi, and largo passages that combine into a thrill that is alternately assaultive and seductive in its presence. Curzon's piano rings, reverberates, languishes in long tails of crystalline sustain. Overall, there is a shocking *aliveness* in comparison to the Columbia Mendelssohn Violin Concerto. The orchestras sound vastly different, Decca's *ffrr* so convincing, I surmise that it must have delighted its first listeners even more than it delights me. Curzon's forte entrances have real jump, almost startling, and the sonic images of his heavy chordings bounce like small tigers off the piano's soundboard. There is such richness, sensitivity, and hyper-focus on exquisite details of sound that I feel as though I were seated in a front-row center seat at a concert performance. I get the sock and brio of Curzon's playing, his emphatic dance rhythms, then revel in the orchestral ripostes to his long phrases that come in thumping, heroic passages. And, during the quieter *Adagio un poco mosso* second movement, Curzon demonstrates a balletic pianism, with gorgeously intricate, slowly evolving, and frequently descending runs that the orchestra answers with subtle swellings, the entire movement paced like a music-box butterfly slowly winding down. The fidelity is tremendous, the *as-if* of technological artifice banished, the listening experience a thing unto itself. Here at last we have the advent of the age of hi-fi. For its time, a perfect sound.

Alan Blumlein, the Inventor of Stereo

Back before the LP, microgroove, and *ffrr* of 1948, Alan Blumlein (1903–1942), a British electrical engineer steeped in the technology of sound reproduction (he'd worked for Western Electric and the Columbia Gramophone Company), developed what he called "binaural sound" while working at Electrical and Musical Industry's (EMI) Abbey Road Studios in 1931. Yet today, in discussions with most audio aficionados, Blumlein's name hardly ever comes up, his contributions obscured in a kind of vault of the cumulative past. I never heard it until Dan Meinwald, a pal of mine, a lover of music and a distributor of British audio gear, casually mentioned him as "the inventor of stereo" one day. *How could I not have known this?* I thought.

We were listening to Dan's system, a magnificent setup, painstakingly assembled in the living room of his home in Long Beach, California. Dan was showing me what a new pair of Marten loudspeakers he'd just gotten from Sweden could do. These were floor-standing three-way speakers with ceramic drivers made in Germany, each covered in a mesh cage. Ceramic drivers are said to be very resolving and fast, providing wondrous musical detail and possessed of a fineness in timing such that the most exquisite and delicate things without smear or confusion (getting lost in a hash of acoustic miscellany) can be heard as though in real space. Dan spun LP after LP (we mainly listen to vinyl when I visit), certainly showing off the grand qualities of his gear, but concentrating on the music, telling me bits of the history behind each recording—a marvelous and eclectic tour through albums by Sinatra, country crooner Ray Price, sixties rock (Julie Driscoll, Brian Auger & the Trinity), and a Stravinsky ballet.

"It's amazing how all this came out of one guy's head," Dan was saying, during a short break from Zubin Mehta's *Le Sacre du Printemps* on the London label.

"What do you mean?" I asked.

"Well, you know, Alan Blumlein's. The man who created stereo back in 1931."

"What?" I exclaimed.

"Yeah. Don't you know about him? A genius. The guy invented stereo at EMI in London way before anyone knew what it meant. He called it 'binaural' as opposed to monoaural and invented it after he came home from the movies with his wife. He didn't like it that the sound coming from all the speakers was as though there was just one speaker and without any spatial information. He didn't like that the sound had no sense of depth or movement or varied locations. So he took up a pencil and made it up!"

"You are kidding me!" I said, unbelieving. "You mean stereo was around in a lab since 1931 and it didn't get put to use until 1958? That's like thirty goddamn years!"

"Right," Dan said. "I told you no one knew what to do with it. It was a novelty for the longest time and wasn't commercial until the sixties."

"What happened to him?" I asked.

"Well, he invented a lot of things—the cutterhead for making two-channel records, the stereo groove itself, the playback system, stereo for the movies, and so on. And he wasn't just about audio. He died in a plane crash in 1942 when he was setting up airborne radar systems for the Royal Air Force. It was a tragedy, but wasn't reported for years because they tried to keep it secret from Hitler. I dunno. Stuff like that . . . There's a biography about him. I'll show you."

Dan motioned for me to follow him out of his listening room to the bookshelves in his study, where he pulled from a lower shelf an average-sized, hardbound volume with a man's portrait on the cover. In his thirties, I'd guess, this was Blumlein—a narrow-faced man with a strong jawline, slightly protruding chin, and high forehead. He was dark-haired, with thick beetle-brows and round eyeglasses (the kind once known as "British welfare issue"). The book was by Robert Charles Alexander and was titled *The Inventor of Stereo: The Life and Works of Alan Dower Blumlein* (1999). I ordered my own copy soon thereafter and learned more about him.

As Dan had told me, Blumlein was at the movies with his wife and felt frustrated that the sound lacked realism—there was no directional information. In that moment, Blumlein said to her, "Do you realize the sound only comes from one person?"

"Oh, does it?" she said.

"Yes," answered Blumlein. "And I've got a way to make it *follow* the person." (Italics mine.)

Audio was still only a monophonic system of sound reproduction. Whether sound came out of one speaker or several speakers (as in a movie theater), exactly the same signal was amplified in each of them and there was no illusion of the spatial placement of sound nor of its movement from one area to another, as with a train whistle blowing, an actor speaking when crossing a stage, or the luxuriant sweep of a theme in a symphony passing from the violins to the violas and cellos, and then echoed by woodwinds a measure or two later. Blumlein set himself the technological goal to create, in audio reproduction, a sound field that was much more realistic.

"Imagine a blind man looking at the screen," Blumlein wrote about that time. "He couldn't tell where anybody on the screen was speaking, although they may be walking across the screen, the sound is just coming from the loudspeaker. But I have a way, and I'll be able to make that sound move with the person across the screen, so although the blind man can't see the film, he could imagine the scene."

While working at EMI, Blumlein came up with the entire technology for stereophonic sound—new microphones, the technique of recording two channels instead of just one within a single groove of a record, a two-channel cutterhead for cutting this new binaural groove rather than a monaural one, and the amplification system based on what he called a "shuffling circuit," which reproduced the directional effect when sound was taken from a spaced pair of microphones. These were monumental developments that Blumlein demonstrated in a series of 1931 experiments and then on test recordings made in 1934 at EMI's recording studio on Abbey Road in St. John's Wood. The first stereo recording is a single track of Ray Noble's Dance Band. Browsing on YouTube, you can hear several tunes that Noble's band recorded from 1931 to 1935. Most are sweet and luscious British swing dance numbers, one of them, "Goodnight, Sweetheart," with a crooning and nasal vocal by Al Bowlly along the lines of what, to my ear, anyway, sounds a lot like Rudy Vallee. After recording the Noble band, Blumlein turned

to piano music—still very challenging for contemporary recording equipment to capture. According to his notes, one was Brahms's *Hungarian Rhapsody* (a series of pieces actually written by Franz Liszt) and the other an orchestral piece transcribed for piano, Wagner's *Ride of the Valkyries*. Finally, Blumlein led his research team to finish that first round of tests by recording Sir Thomas Beecham conducting the London Philharmonic Orchestra in a rehearsal of Mozart's Symphony no. 41 in C Major, *The Jupiter,* in preparation for their making a commercial release of a mono recording. It presented a much more difficult recording problem than the other music, more challenging in terms of presenting depth, multiple instruments arranged in space, critical timing, and the varied timbres of orchestral instruments.

Blumlein adjudged his effort "not bad" or "marginal" at the time, but these were the very developments that changed the course of audio and made possible the later eras of fifties high-fidelity sound and contemporary high-end audio. (The clips I've found of it are available only to European subscribers to the online listening service of the British Library at sounds.bl.uk, but the 78 rpm mono recording is available on YouTube and for purchase as an "EMI Classic: Great Recordings of the Century" on CD.) The sound of it is sprightly and "tipped up" by contemporary standards, emphasizing the upper registers of the orchestra, particularly the violins, and hasn't the warmth and depth of current stereo sound. But it isn't murky or muddied at all. It's clear and arresting, Beecham leading the Philharmonic through tight cues and precise, forceful attacks in the *Allegro vivace* first movement.

There are also vintage video clips of Alan Blumlein available on the Internet, showing off how his binaural sound works. One is of the sound of moving trains, called "Trains at Hayes Station," dated July 1935. Shot from a window at the EMI offices in West London, the forty-four-second-long black-and-white film shows a steam engine puffing away as it pulls a line of cars from right to left diagonally across the moving picture. What you hear, though, is the sound of its heavy puffing moving simultaneously, left to center to right, along with the locomotive's image putting out a trailing white cloud. The sound comes on from the right of the frame

before the train enters the picture. Then, as the train moves toward the center, the puffing sound feels closer and fills both your ears. Finally, the puffing fades a bit and travels to the left as the train moves that way too. Another video-and-sound clip, almost two minutes long, is perhaps even more illustrative, as it features Blumlein among his team of scientists speaking to the camera, moving around on an auditorium stage, taking up positions to the left, center, and right, then moving again. Called "Walking and Talking," the clip dates from July 1935. In it, the bespectacled Blumlein (dressed in a double-breasted suit and tie) and his team (dressed in shirtsleeves and lab coats) speak simple numbers, then days of the week, and finally the alphabet as they move about in front of a curtain. At one point, three of them speak at once and you get the effect of a varicolored cloth of woven sound, a kind of glorious babble of voices that you can nonetheless distinguish in space and hear moving about.

Though Blumlein clearly demonstrated the advantages of stereo sound in these tests and demos, the technology languished at EMI for decades before anything substantial was done to introduce it commercially. He went on to work on developing stereo for movie sound, the basis for the "ultralinear" amplification circuit (used in my Air Tight amp), and most notably of all, airborne radar systems for military use. As Dan had first mentioned, the great man died in an air crash in 1942 near the village of Welsh Bicknor, Herefordshire, while making a test flight of his system in an RAF bomber. The crash site was just over a hundred yards from the north bank of the River Wye, so dear to English poet William Wordsworth. Prime Minister Churchill kept Blumlein's death a secret for three years, supposedly not to give Adolf Hitler surcease from his mounting anxieties over the rapid development of British war capabilities. Blumlein was only thirty-eight, a kind of Mozart of technology.

Empire 398

Maybe the pride of my father's system was his Empire 398G turntable—a gleaming satin gold-finished aluminum plinth, platter, and tonearm with a tapered walnut base just slightly smaller at its bot-

tom than its top, giving off an aura of sleekness and chic. Of all components in his prized system, the Empire was the one major retail purchase—$175 in the Allied catalogue of 1963, its cost would be near $1,500 today. It was belt-driven (a new concept then), its compliant black band slipping out from the machined triangle of a topside housing and wrapping around the heavy platter to spin it at 33⅓, 45, and 78 rpm, depending on what position you chose for it on the trilevel pole of its pulley. And the tonearm was balanced—a gold wand that floated on a pivot with a counterweight at its back end, so that when you dropped the finger-lift and its headshell, it settled like a bird flapping invisible wings onto the record, light as a feather, catching the groove with a whispering tick. It was an instrument with a stylus force you could calibrate (via knob control) to less than a gram, not a bludgeon of a thing like the morbidly heavy, cast-metal tonearms on single-unit phonographs I grew up with. You could *feel* the Empire's precise engineering under the glamour of its shiny gold finish, the care and wonder of its design, a race car compared to the clunker sedan Silvertones and Webcors of our family's musical past. It was *cool*. If you handled it, it made you feel cool too.

And cool was completely new to me—a concept and a feeling I'd not had much acquaintance with, given the short term of

my life up till then. I was twelve and had just left comic books for the splendors of Jules Verne, given up "Hiawatha" in its Classic Comics edition for paperback James Bond novels I hid under my bed. In my mind, the gold Empire 'table, then, attached itself to the image of 007's Aston Martin, racing through cliffside roads near Monte Carlo, pursued by a tribe of assassins mounted on black motorcycles. Heaving trucks laden with brown heaps of cane, the tides at the shore rattling with white coral bones, and the emerald megaliths of the Ko'olau Mountains all withdrew from memory and dissipated in misty veils of forgetting as I imitated my father's new identity as a "stereo man," flipping on the magical hi-fi, cueing records with a curled forefinger, and sitting back on the sofa or on the carpeted floor in front of it to listen. Despite the crude music I preferred then—faux folk, overly produced soft rock (*ugh*, the Association), and, blessedly, a couple of albums by Johnny Cash and Nat King Cole—it was in this way I learned to live more in imagination than experience, less in memory than in sensibility. And stereo became both a shrine and a portal through which I could travel whole worlds of space and time—my own capsule burrowing through to the center of the earth, a kind of rocket ship to the magical landscapes on the far side of the moon.

This was not unlike the dream of the early inventors of the phonograph. In 1877, Edison wanted an instrument that would make a record of the human voice a means for the transference of a sonic image from one time to another, from a moment rapidly entering the past to a like moment entering an evolving future. It seems to me reminiscent of Augustine's meditation on prayer in *Confessions,* where the saint describes a temporal utterance becoming the access to a transcendent and sacred universe.

But there were technical challenges to all this potential and exotic travel. I remember that footfalls, my brother's racing around the house and stepping near the living room where the Empire was, would cause the needle to jump the groove, making it skip across the pristine surface of the record, and create thunking and garbled sounds through the speakers. My father would briefly rage, swear a bit, saying "Gunfunnit! I tole you not to run inna house!" and throw his folded-up *Racing Form* across the room, where it would

hit the wall with a splat or connect with my brother's forehead, sending my mother into her own protective rage. There would be arguing, the noise of strife, all the music cut off for the rest of the evening. And so I was careful to tread lightly around my father's stereo, walking softly on the edges and balls of my feet, never letting a heel hit harshly on our carpeted floors, settling myself down easy rather than plonking down on the sofa or floor before my father's great audio shrine. I paid *respect,* and my footfalls made light genuflections whenever a record was playing.

The stylus jumped because the Empire was an *unsuspended* table, one without spring supports under its plinth. Of a minimalist design, the machined aluminum platter floated only on its bearing, polished brass in a steel cup lubricated with just a few drops of oil, transferring (from the aluminum plinth and its walnut base) whatever vibration it picked up instantly to the featherlight touch of the stereo stylus at the end of the balanced arm. The whole mechanism would react then, the needle jumping from the groove, the back-end counterweight flipping the tonearm up like a seesaw suddenly slipping a child from one end, the cartridge-end rearing like a bucket on the end of a construction crane, then plunging down like a backhoe on the pretty black vinyl underneath, the stylus jittering as it made the sound of an angry guiro out of control. Like the toughest kid on the block, the Empire demanded a physical respect, our deference to its touchy and oversensitive foible.

And it wasn't a *changer* either. Unlike the Silvertone phonograph from Sears we'd had in the past, there was no automatic spindle and swing-out overarm mechanism that flopped a stacked sequence of records down on the platter. You could only play a single record at a time, giving each its own ritual of placement, the cueing of the tonearm, and the attentive needle drop. There was only a nub of a spindle in the center of the platter, a short, pewter-colored snub of a thing that centered 33⅓ LPs. For 45s, the smaller, seven-inch records that were the staple of Top 40s and R&B singles, there was a spring-loaded, pop-up adapter of the same shimmery gold aluminum as the Empire's platter, plinth, and tonearm. The machining of it all had an aura of precision and serene thought to it that was nowhere else in our lives, not even in our automobiles—at that time, cheap

and Detroit made with clunky column-shift, three speeds, stiff clutches, and lurching gears. Handling the turntable was in itself an expression of grace, a performance that reminded me of the way my grandfather handled incense sticks when he chanted and prayed, lighting one, then two, holding them poised delicately between his thumb and first two fingers, warbling something sweetly from his throat that escaped like smoke from his lips, then placing them in a tiny brass bowl on a shelf of the compact lacquer shrine he kept in the living room under the cutaway window to the kitchen just a few feet from the stereo. They both knelt when they performed, my father and grandfather: my father perched on his knees like a roofer just to place a record on his turntable, my grandfather sitting *seiza*-style, his legs tucked under him, as he intoned a passage from the Lotus Sutra. In different ways, both of them practiced an exemplary attentiveness I did not fail to notice.

Besides all of this, I think my listening to my father's stereo was my introduction to *longing,* wishing for a different world or hoping that the world *within* might be different. However dated or cheesy the music I heard, within another year or two, it brought me to a border in my adolescent life and then pushed me across it as though I were passing through its audible window to something just barely defined by the splendor of sound. After I'd run through the Brothers Four and the Tijuana Brass, after many evenings listening to the gooberish plaints of fifties slow-dance 45s, after movie music and 101 Strings, Mitch Miller and Peter Gunn, I found Sam Cooke and his song "A Change Is Gonna Come." Cooke was a crossover artist from the R&B world, who up until that time had been pumping out hit after hit to the mostly white, Top 40s audience. But there was something about this new song that made it stand away from all the party and dance music that was the rest of Cooke's *Best of* album, with its budget, yellow-colored jacket, songs like "Shake" and "Cupid," tunes both cutesy and clever or sassy and full of strut. When Cooke sang that long note right at the song's beginning—*I was booooorn by the river*—the stretch and continuous portamento took me somewhere, its swoop of sound signaling this was something else, the rising top of his voice, with more than a hint of gospel to it, the music come out of a childhood of ecstatic worship

and faith in providence reaching me, a kid from a different ghetto, locked in my own nascent longings. I played that song over and over again, getting up from the sofa and re-cueing it just at the gap before the track, dropping the needle precisely so that I'd have time to cross back to the couch and lie down, captivated by the sound of Cooke's voice, which carried another kind of longing, and one that tutored me.

Did I know its background in incidents of discrimination suffered by Cooke and others? Did I realize its connection to the history of civil rights protest? Was I aware of the marchers who crossed the Edmund Pettus Bridge in 1965? I didn't then know about the beatings by police. I didn't know the legacy of Black gospel and the struggles for equality and desegregation. I only heard the song's pure hopefulness, the expressive musicality in its wishes for a better day. The soul of it lifted me, plucked me from where I was lying down and bathed me in its feeling.

Cooke sings about going to the movies and how somebody tells him he can't be there, the implication being he's "colored" and not allowed. In the bridge of the tune, soft taps from the timpani underscoring the somber mood, he sings about going to his brother asking for help, and his brother knocks him to his knees. He says he's been running ever since he was born, that it's been too hard to live, but he's afraid to die. There have been times he's thought he couldn't last for long, but now he knows that, no matter how long it takes, *a change is go'n' come*. It's a song barely three minutes in duration, but it carries centuries of pain in its deep, plangent notes (surrounded by orchestral strings) and recommends the resolute practice of faith (signaled by soft, heavenly fanfares) in the presence of obstruction. And it was perhaps the first time something in music asked me to look within for solace and for power. It asked me to be somebody that I'd not been, step in the call from that river of music and let it help me imagine myself made otherwise by the pain and wisdom of another.

Part Nine

Wandering Rocks, 2

June 2009

The cardboard box was taped up and seemed rolled into a ball when it arrived. Via TNT Express, it had been sent from Mumbai, India, to Eugene, Oregon—roughly eight thousand miles. But though it was still intact, the box had come through serious slaughter. Shipping had not been kind. The corners were bent round, the top was caved in like a sunken cake, and the whole thing seemed as limp as a run-over dog, held together only by the ribbons of duct tape swaddled around it. When I'd lifted it from the porch, the box rattled. From the inside. Things were loose in there, and I had the dark thought that heavy parts had battered metal against metal, that sharp pieces had scoured polymers with brass, that brittle bits had cracked and shorn apart. I dreaded opening it up, but did it anyway, slicing through the sticking cloth of the tape with the blade of the box cutter in my hand, prying the two wings of floppy cardboard apart as though they were a rib cage, then finally looking inside to carnage.

It was a disaster like no other I'd yet experienced. My seller, an exchange trader in Mumbai, had simply thrown the precious turntable into the cardboard box, completely unanchored, and not bothered to disassemble it or insulate its parts from one another. Shipping a turntable demands tremendous care—an insulated container, a clamshell interior, cutout foam nests for its various parts, at the very least Bubble Wrap, and Velcro strips or adhesive tape to anchor whatever might work loose during shipping. But my seller had none of these—just a cardboard box, no longer sturdy. In transit, the heavy black platter—two and a half inches thick, weighing thirty-five pounds, and specially made out of a resin impregnated with particles of copper—had slipped off its spindle, skittering off-kilter and scouring its own underside against metal while in transit. Meanwhile, the brass armboard had worked loose too, twisting off

its anchoring screw, clattering around freely and battering against the once-smooth sides of the plinth and the platter, gouging them, and the screwhead making little teeth marks or divots a tiny sand wedge might have made all over the shiny black surfaces. There was a big, zigzag scratch about five inches long on the rim of the platter that looked like the jittering index of a day's trade on the stock market. How apt, I thought at the time, thinking of the broker in Mumbai. But the cheap tonearm was undamaged, disassembled and tucked neatly into its rectangular box. Amidst all the wreckage, it must have floated amiably above the fray the whole while. The plinth's footers had come loose too and were scattered throughout, shiny pucks of silver metal with the weapon of a screw sticking from each of them, little round daggers of enmity having stabbed away freely during the entire trip, jabbing all exposed surfaces. One of the daggers was bent from impact. My beautiful German table, bought cheaply and shipped internationally, had come to me in pockmarked, gored, and, I thought, unsalvageable pieces.

I sat a long time next to the opened box, the damaged parts scattered around me. There was shock, grieving, and dismay, cold blood running through my body. After a while, I think I stood up from it all, stretched, and took in the scene, just to get a wide-angle shot and catalogue the losses: *Platter scratched, top and bottom, big zigzag gouge along the rim. Plinth pockmarked. Armboard battered. Spindle good. Bearing likely unharmed. Footer bent. Motor good. Motor controller good. Tonearm intact.*

I emailed the seller in Mumbai and said we needed to talk. He'd been very communicative all along and might even have written me, asking if the 'table had arrived. We spoke the next day, and I explained. He was shocked and made apologies, claiming he'd been distracted by all the excitement of the "doubling" of the Indian stock market the week before. He said he'd become very, very lucky. He said, without prompting, that this was his fault entirely and that he would refund me the full sale price. This was a shock to me too, as most sellers of used gear would never take responsibility so quickly, so easily, with so much contrition. The man was rare, I thought, a gentleman perhaps. And, I guessed, he'd made so much

in the float of all Indian boats that the refund was nothing. He felt generous and I was lucky.

I'd purchased a lot of used gear from audio market websites before. It was a way to get great stuff at about half the retail price sometimes. Granted, the gear might be years old and have some signs of wear—small scratches on casings and facings, tired audio tubes that needed replacing, screws loose or missing, dings and dents sometimes. But, generally, in "good enough" shape you could live with whatever small cosmetic or repairable mechanical shortcomings might be there. Yet there were small disasters too—a pair of prime speaker cables with ravaged terminations that needed replacing, an Italian CD player with a Lucite top plate that had cracked into pieces, a donut-shaped preamp in black powder-coat with a long diagonal scar across its face that had never been disclosed. Sellers could be dishonest, even perfidious, refusing to discount for damages, to take back their compromised gear, or to refund you a dime. And you'd have to get tough back, appeal to anonymous and uncaring website masters, engage the PayPal dispute procedure, engage the eBay one, or worst case, sic Visa or Mastercard or American Express on them. That usually ended in your favor, though you'd made an enemy for sure. But my man in Mumbai put me through none of this. My funds were credited back the next day.

"What should I do with the turntable?" I asked. He said for me to keep it. He had no use for it. That maybe I could get it fixed. His message included the email of the manufacturer—Thomas Woschnick in Munich, Germany.

I'd bought a TW-Acustic Raven Two—the "Two" meaning it was fitted to hold two tonearms, perhaps one for a stereo cartridge and another for a mono one. That had been my plan. I'd gotten curious about pre-1958 mono recordings, wanted to hear older recordings, expand my territories of sound.

And I wanted a better turntable. The Nottingham tonearm kept picking up stray radio frequency signals that crept into the

sound of the system. I'd even called the manufacturer in England about it and the guy on the line apologized, saying it was probably due to the way their rig was designed—an electrically ungrounded arm that, while preserving a kind of purity of signal, was prey to unwanted magnetic and radio frequency interference sometimes. It so happened I lived in an area where there was a lot of it. In audio, it's always *something*.

So, I researched and identified the TW-Acustic Raven as my target. It was a fairly new product when I found it, only a couple of years on the market, and had been getting terrific reviews for its richness of sound, smooth operation, and stunning looks—black as a raven's wing all the way. The designer, Woschnick in Germany, was a precision machinist who'd turned his skills and a section of his shop over from machine parts manufacturing to making the turntable. Besides his fine machining, his "trick" was the special quality of the platter he'd designed, using a moldable but heavy resin impregnated with fine, granular particles of pure copper that gave it dense mass and repelled resonance, making for a steadiness of spin, a relative purity of signal, a constant momentum that provided for both clarity and sensuousness in its sound. And Woschnick used a pulley-and-band method to spin his platter, an external motor placed next to the turntable's base, its plinth, that connected to its platter via a black band of synthetic rubber. It was belt-driven just like my father's Empire 398.

Retail, the Raven Two was out of my budget, over $7,500 or so at the time. But I was almost immediately lucky and found an ad for a used one at just over half that price. The catch—it was in India with all the attendant complications of a long-distance, international purchase. Yet all of that turned out to be easy. The seller was not only amiable, answering all my questions, willing to ship, but he also let me bargain the price down a few hundred dollars—as one would have hoped with any deal. We dickered only a little about shipping. FedEx and UPS were out of the question for him, as his brokerage had arrangements only with DHL and TNT. We settled on TNT, which offered an amazingly low price—just under $300—to ship to Oregon. My error, though, was that I didn't interrogate him about packaging. Being still a novice in analog, it just

hadn't occurred to me. But as most audiophiles know, packing a turntable is a most difficult and intricate process. It's best if you've original packaging always, the manufacturer having designed specific procedures for its disassembly and protections for its shipment.

I called Woschnick in Germany. He was brusque and straight to the point. He said, "Ship the broken parts to me and I will rebuild it for you." He asked how much I got refunded. I told him. He said, "Give me the same and you will have a thing like new. Two weeks."

I wired Woschnick the money, losing a bit in the exchange, but what else could I do? Why quibble over a few hundred when a complete restoration of my hopes was at stake?

In three weeks, the refurbished plinth, platter, and feet came back in a big, trilevel box with foamed cutouts for all the parts. The top layer was for the motor, motor control unit, footers, spindle, and belt. The middle was for the heavy platter and armboards. At the bottom was the big, winglike plinth, like a black-sand island with two podlike peninsulas where the armboards were to be attached. It looked like he simply replaced the feet. I lifted out the layers and marveled. The plinth and platter were both smooth as silk—no scarrings, no gouges, no little marks of teeth or divots anywhere. Woschnick was right—it looked brand-new.

Carefully, I set up the entire mechanism, and, using a little strength (the whole thing, assembled, weighed sixty-five pounds), placed it on the top of my rack. I mounted the cheap tonearm last, using a protractor that was inside its own box. I hooked it all to power and adjusted the two speeds (33⅓ rpm and 45 rpm) of the motor that spun the platter. I pushed a button, and the motor and belt engaged the heavy wheel. I put a record on. I lifted the tonearm from its catch and dropped the stylus into the lead-in groove . . .

> *That thou, light-winged Dryad of the trees . . .*
> *Singest of summer in full-throated ease.*

It sounded great.

System: TW-Acustic Raven Two turntable; Jelco SA-250 nine-inch tonearm; Zyx Airy 3 MC cartridge; Herron VTPH-2 phono stage;

deHavilland Mercury 3 preamplifier; deHavilland KE50A monoblock amplifiers; Von Schweikert Audio VR5 HSE loudspeakers; Cardas Golden Reference and Verbatim interconnects (RCA); Verbatim speaker cables with jumpers; Cardas Golden Reference power cords; Box Furniture five-shelf equipment stand.

August 2010

I was home in the cramped study that doubled as my listening room, playing Ravel's *Boléro* on my analog system. The recording was by the Los Angeles Philharmonic Orchestra, conducted by Zubin Mehta on London LP, a tour de force of a single long crescendo of fourteen minutes, from pianissimo to triple forte, famous and agonizingly slow, its insistent theme repeated again and again until the piece ended in a thunderous climax. What makes *Boléro* compelling throughout is how the theme is taken up by various instruments in succession, demonstrating a variety of achingly glorious timbres and exotic performance flourishes on a vaguely Iberian melody. As an undercurrent, the incessant, march-like bolero rhythm is played on a snare drum, with staccato accompaniment of trumpets and occasional rousing accents from the horns, bassoons, and timpani. The new preamplifier I was using, designed by Valve Amplification Company (VAC) in Florida, captured the piece's feeling of an ornate, equestrian processional, medieval and military, as though emerging from a near horizon, then gradually approaching to eventually fill the immediate foreground with musical pageantry as the entire parade, muscular and disciplined, tramps by mere inches away. By degrees throughout the whole piece, the snare increased in volume from pianissimo to a rattling immediacy, joined at the climax by a second snare, their duet accentuated by braying fanfares and explosive thumps from a bass drum. The music gradually built up the tension and excitement, with growing forcefulness and unmistakable authority. Almost every sound, whether loud or soft, was clear and precisely timed. The character of the playback, exceptionally clean, made starkly apparent the timbral differences among all the instruments: oboe from flute, trumpet from trombone, and timpani from bass drum. It was a grand march of perfect sounds.

System: TW-Acustic Raven Two turntable; Tri-Planar VII Ultimate II tonearm with Zyx Airy 3 cartridge; Herron VTPH-2 phono stage; Valve Amplification Company (VAC) Renaissance III preamp; VAC Phi-200 stereo amplifier; Von Schweikert Audio VR5 HSE speakers; Cardas Clear Beyond biwire speaker cables; Cardas Clear RCA and XLR interconnects; Cardas Golden Reference power cords; Balanced Power Technology Clean Power Center; Harmonic Resonance Technology SXR stand with S1 platforms, Nimbus footers, and damping plates.

September 2010

It may not be obvious that I've been much in love with Italian culture. I am knocked out by its rich history of art from medieval times up to the present, the fabulous textures and tastes of its food, the articulate landscape of its bucolic countryside, and most of all, the manner in which *all* of Italian culture seems subtly interfused with every event that takes place there. For example, the lavishly delicate taste of a spiny, spotted fish called the San Pietro (*Zeus faber,* aka St. Pierre, aka John Dory), usually caught in deep Mediterranean waters, recalls to me the sixteenth-century architecture of Andrea Palladio. And I've enjoyed Italian audio equipment too, suspecting that its quality and design features have been influenced by the country's long tradition of classical, baroque, and opera music. Through the years, I've not only admired Italian audio gear, I've owned two pairs of Sonus faber speakers and still have a VAIC-badged Mastersound amp and a Viva amp too, all made in Italy.

In fall 2010, I was in Italy touring Lombardy, when I broke off from my trip and took trains from Milan to Vicenza in Veneto, the northeast region that comprises roughly the area from Verona to Venice. After I arrived, I spent a morning crisscrossing the city by car, gazing at a magnificent villa in the hills above, then driving to admire the grand façade of an ancient civic building, and finally stopping to circumambulate a palazzo's imposing rotunda. All of these were Palladio designs. Later, just past noon, I strolled the boulevards and pedestrian walkways of the Centro Storico on my way to the restaurant where I ate the San Pietro. I savored the freshness of the Pinot Grigio, like a liquefied autumn morning ordered up

by my Italian host. The Italians have a certain way of *integrating* and enjoying life. "They takes they time," as a basketball player said to me once, an American forward in the Italian pro league, while we spoke on a trans-Atlantic flight. "And they likes they food slow."

Through my years listening to hi-fi, I've come to believe that the best reproduced sound is like this too—subtly interfused with the tactile splendors of the music, all to be slowly savored, as one might a fine meal. I sat over such a meal in a restaurant at Bassano del Grappa, with its wooden bridge (designed by Palladio) spanning the Brenta River. I was visiting with Amedeo Schembri, chief designer of Viva Audio.

Viva has become well known not only for its superb tube electronics, including zero-feedback single-ended triode (SET) amplifiers with point-to-point wiring, but also for great industrial design featuring custom automotive paints and bold, M-shaped chassis that never fail to impress, visually and sonically. I'd been curious about Viva ever since I first heard their Solista integrated amplifier in a showroom at Top Audio (once the major Italian trade show for audio), and since then have checked out their gear at every national and international trade show I've attended. I've never been disappointed.

"It's about the sound," Schembri said as he poured red wine into my glass, watching it swirl and spume up inside the hiplike curves of the glass's bell, then settle back before he finished pouring. We were having lunch outside, under a latticed arbor of curling vines, the warm noonday sun seeming to light everything from within. "You can research and discuss and make up any kind of *surrrkits,* but you must test this against the ears to be sure you are not *cray-sie,* and you get *somethingz* that is *music,* and not just some noise from Electric Street."

We'd driven a few miles from Schembri's home in Dueville, a new, residential suburb of Vicenza, to a lovely restaurant at the end of a church street on the outskirts of Bassano. We were joined by Joelle De Jaegher, an accomplished and busy architect and (at the time) Schembri's partner. It was Joelle who was responsible for the sleek, postmodern look and vibrant automotive colors of Viva's electronics. As we spoke, Schembri took photos with a vintage Contax

rangefinder—a film camera. For a youngish man—I'd guessed him to be no more than forty-five—he had a definitive way of speaking; an insistent, discriminating manner of expression characteristic of many accomplished audio engineers. He spoke authoritatively about cameras—SLRs, rangefinders, digital versus film too.

Our talk was also of the fine dishes we were eating—*linguine nere,* a pasta rendered black by the inclusion of squid ink; the homey regional dish *baccalà,* a dried cod reconstituted in milk; and a delicious *astiche,* or lobster—and of audio. De Jaegher encouraged us to try all the different dishes, pointing out their colors as well as their tastes.

Lunch concluded two hours later *(they takes they time)* and we split up, De Jaegher returning to her work designing a chain of boutique shops, and Schembri and I driving, in his Volvo, back to Dueville. Upon arrival, he ushered me downstairs into his basement—he calls it *la tana dello stregone,* the sorcerer's den—where we were to hear what he called his "experimental system." But then Schembri turned hesitant, humble, even worried.

The basement was no artfully decorated, palatial listening

room, but a crowded warren of speakers and electronics at one end, overlapping area rugs amidships, and, at the other end, racks of LPs, stacks of CDs, stools, folding chairs, and a fat, worn-leather easy chair. Light streamed through a row of papered-over windows that tilted open over a long bench strewn with electronics, and spider-webs of wires spread into every cranny and corner. This indeed was Schembri's workshop: eight prototype tube amps laid out on four unfinished chassis (two per channel), their transformers exposed; an assortment of speakers made by hand and scrounged from history; and a Frankenstein's machine of an analog rig, comprising a Lenco motor and drive mechanism, a custom turntable, a twelve-inch Japanese tonearm and phono cartridge. His CD player was a Viva prototype too. This was a two-channel, eight-amped, four-way system.

"We are making our own speakers, and these new amplifiers, as well," Schembri said, gesturing to a large stack in the front right corner. He declared that he was studying the influence of power supplies on the various amps. Then he explained that the speaker array consisted of a fifteen-inch woofer shared by both channels; two unfinished, handmade midbass horns; a pair of old Vitavox midrange horns (the black beehive famously used by Pink Floyd); and, finally, two magnesium tweeters from Fostex. Tucked behind each speaker stack were the amplifiers: each chassis housed two amps, 845 output tubes driven by 211 tubes. It seemed a system almost better suited to a concert hall than a listening room—each speaker stack was larger than a conventional home entertainment center. But I reserved judgment. *It's all about the sound,* I reminded myself.

Schembri put on Miles Davis's *Kind of Blue* LP—"The original," he proudly announced. What I heard were pleasingly saturated notes, rich timbres, and distinct separation of the soloists—John Coltrane's tenor sax on the left, Cannonball Adderley's alto on the right, Miles's trumpet emerging out of sweet silences. But, I thought, sweet and susurrous though it was, this was still small-combo jazz—no sweat. After all, I'd heard this recording sound good in small rooms with low-watt solid-state amps and single-driver Fostex horns.

Then, no longer shitting around, Schembri brought out a

recording I did *not* know—Mussorgsky's *Boris Godunov,* with Herbert von Karajan conducting the Vienna Philharmonic. I'd often read about this opera but had never heard it. I was in for it. Nicolai Ghiaurov's bass voice was so stunning, rich, and powerful that, I have to say, I'd never heard anything like it before—except in concert. Schembri's experimental system reproduced the entire range of Ghiaurov's bass without any "letterboxing"—that is, the bottom wasn't lost to murk or imagination; the sweet, sub-tenor top wasn't etched, and caused no ring or overhang; and all vocal harmonic overtones were rendered with breadth, equanimity, and on time. To say it sounded "realistic" seems silly, to say it was "musical" even sillier. It was emotional—*thrilling.*

"Like life?" Schembri asked.

"No," I said. "Like *art.*"

And so it was with recording after recording throughout the entire afternoon: a chromatic density to the piano and effortless orchestral crescendi on Sviatoslav Richter's recordings of Beethoven's piano concertos with Charles Munch and the Boston Symphony; smooth, rich, full-bodied choral voices from René Clemencic and his Clemencic Consort in selections from the original medieval manuscript of *Carmina Burana,* on which Carl Orff based his famous modern version; and delightful, three-dimensional bells on a recording of ancient Greek music.

I asked Schembri to put on Renée Fleming's eponymous disc of famous arias, with Sir Charles Mackerras leading the London Philharmonic. The Viva system gave her voice more of an old-timey feeling to me, sounding more midrangey and embodied—sweeter and purer somehow, with less emphasis on the top notes than in the pentode sound I was used to from my own reference system—more relaxed, yet still uncommon.

"And so, do you discover *life*—the Viva sound?" Schembri asked.

"*Certo,*" I said. Of course.

We finished the demo just as the sun had intensified, leaking through cracks along the papered-over windows of the basement shop. It was already early evening. We went upstairs, where Schem-

bri helped me figure out how to accomplish my next adventure: a trip to see the Giotto frescoes at the Scrovegni Chapel in Padua, about an hour away.

What I remember is a precise combination of different qualities of light—the illumination from the screen of Schembri's laptop falling across his bearded face, the warm infusions of late sunlight slanting in from outside to penetrate each moment as if it were sound from a more articulate traffic in the street. It seemed to me that in Italy, audio was so very much like cuisine—each element fussed over yet never out of concert with the overall measures of the meal for flavor or of the music for enlightenment. When we stepped outdoors, the autumnal light, as though through an amber, reticulated glass, streamed down on us. *Eccola!* It was the world declaring itself—in light, in sound, in an aura of thorough plenty.

System: Viva prototypes entirely.

March 2013

On an impulse, less than two weeks before its start, I decided to fly out to AXPONA '13, the first major audio show in Chicago since the *Stereophile*-sponsored one was there in 1999. I wanted to get out of Oregon for a while, hear new audio equipment making its debut, and take in Anna Netrebko and Joseph Calleja singing the iconic roles Mimì and Rodolfo in *La Bohème* at the Lyric Opera of Chicago. I'd never heard either of them sing live—neither the gorgeous Russian soprano nor the burly, "Caruso-like" Maltese tenor, and both had been among my favorite recording artists for years. When I went to the opera's website, there were only three single tickets left. I nabbed what I thought would be the best seat—one in the first balcony—and then booked a plane to O'Hare.

I'd never been to AXPONA, and likewise never to the famed Lyric Opera. Neither venue disappointed. I heard that there were over four thousand AXPONA attendees, and I spotted fellow audio scribes running around everywhere at the conference hotel—the DoubleTree by Hilton O'Hare in Rosemont, Illinois. The show was spread out over five floors with many of the ninety rooms

sponsored by local dealers. And though there was lots of great gear everywhere, I mainly focused on the knockout absolute best exhibit in show, put together by George Vatchnadze, a concert pianist who teaches performance at DePaul University and has the sideline of an audio dealership called Kyomi Audio. In the hotel room, he'd partnered the demo with old friend Dan Meinwald of EAR USA and Ken Stevens of Convergent Audio Technology. Besides *La Bohème,* the other reason I came to AXPONA was expressly to hear the CAT JL5—a new triode stereo amplifier that was supposed to rival, if not equal, the performance of the grand, all-class-A, JL2 Signature Mk 3.

The JL5, like the JL2 I'd heard in 2007 at my first audio show, has an output of 100 watts per channel, but according to Stevens, owner and chief engineer of CAT, it was to be a physically smaller version of the behemoth stereo amp I'd heard years past at CES. Once casework was completed, the JL5 would weigh roughly half the bigger amp's 170 pounds. I'd already heard it was being called "the CAT Mini."

The Kyomi demo was expertly set up on a diagonal by Meinwald. It featured a CAT preamp along with an Esoteric SACD player with master clock, wires by Magnan Cables, and a pair of Marten speakers. The CAT JL5 was still in prototype, fitted to the larger JL2 Signature chassis and using only half of the JL2's sixteen KT-120 tube sockets (four per channel rather than the JL2's eight). Meinwald seemed to be in charge of the demo, so I got him to play my tried-and-true demo track of "Kyrie" from Mozart's Mass in C Minor, performed by Le Concert d'Astrée and featuring Natalie Dessay (soprano). He called it "a real system fucker," as its dynamic range and demand for power is immense, with chorus and full orchestra and operatic soloists to boot. The JL5 sailed through it all, sounding full and authoritative, with great orchestral drive and timing and the male and female choral voices distinct. The period strings were very clear and never ragged, the violins as forceful and synchronized as the offensive line of the Pittsburgh Steelers in playoff mode.

"Just about the real thing," said my friend Stuart Dybek, the fiction writer and poet from South Chicago. He was in front of me,

nodding, with a big grin on his face, sitting in the prime seat in the front row. He turned to speak to his girlfriend, a neuroscientist from Miami, who sat next to me in the second row.

"With stuff like this," he said to her, "we'd have a concert every night!"

Stuart had come to the show looking for components to create a new system for her condo in Coral Gables, and I thought for a second he'd buy it whole-hog right then. In the end, along with him, I stopped listening critically and just lost myself in the magisterial beauty of the piece.

We split up. I finished up early on Saturday around five o'clock, walked a few long and lousy blocks through the gray weather to hop on the Blue Train downtown, then cabbed from the station to the Chicago Civic Opera building and got to my first-balcony seat with about fifteen minutes to spare until curtain. What can I say? The opera, about young artists scuffling on the Left Bank in Paris during the mid-1800s, was everything I'd hoped for and more. Netrebko made for an all-too-lovely Mimì, looking more robust and busty than frail and consumptive as the role calls for, but she sang so gorgeously, with a plushness to her high notes that belied the slight wiriness I hear in her recordings, that I just did not care. Though the distance from the stage to me was about that from the right-field wall to home plate in Wrigley Field, I thought I heard every precious note of "Me chiamano Mimì" as she sang her opening, gently declarative aria. And Calleja was fantastic as Rodolfo—with a gleaming top to his tenor voice that cut like a brilliant comet through the hall whenever he sang. His opening aria, "Che gelida manina," sung as he takes Mimì's hand in his in his freezing apartment, is an anthem to love, poetry, and survival through tenderness rather than ferocity. There is just too much to say about it.

When I came back after the show, I ate a wondrous ten-ounce medium-rare tenderloin at Gibsons Bar & Steakhouse next to the hotel for all the calories I'd spent crying through Puccini's opera.

System: CAT JL5 prototype; CAT Renaissance 2 preamp; Esoteric K-01 SACD player with G-0s master clock; wires by Magnan Cables; Marten Django XL speakers; Townshend Audio rack.

April 2013

On a whim—no, it was more out of a hunch and growing suspicion, really—I switched out my favorite monoblock amplifiers for a single, stereo amp one day. I'd been feeling a touch disappointed with the tenor of the sound I was getting, its character a shade thinner than I imagined the music should be, especially with orchestral recordings and on those troublesome dynamic peaks and sustained high notes of opera. I had very carefully assembled these electronics, choosing them deliberately over all others and especially for what I heard was a fullness and nimble way with amplitude—those passages of instantaneous orchestral tutti wherein the entire ensemble would play out at max volume for a thunderous and dramatic effect—a cresting mid-sea wave pitching up like a dread hydraulic mountain during a sudden storm. In a way, I lived for these moments in music and the subsequent surprising peace of sweet, restored calm among the woodwinds and maybe a plucked harp punctuating the elapsing measures as the symphony or concerto moved through its alternating emphatic and pianissimo contrasts. These amps of mine were champions at this, great at the forceful smackdowns and even better with the delicate sugar plums of music. But, lately, with the newest set of speakers I'd acquired—a pair of Von Schweikert Audio VR-44 Aktives—I'd been missing these widest capabilities of the amplifiers to cover the breadth of musical articulations that were there to be had in a Mozart piano concerto, say, or the triple fortes and pianissimos in Rimsky-Korsakov's orchestral tone poem *Scheherazade*.

So, increasingly dissatisfied and then suspicious that the output power in my beloved custom amps was below that demanded by the music and new speakers, I started eyeing a VAC Phi-200 stereo amp, also a tubed affair, but with twice the power of my monoblocks, capable of 100 watts per channel, a piece of equipment that, for only silly reasons, had been sitting idle and yet beautifully displayed (as in a showroom of static gear) on the top shelf of my console (next to the twenty-inch Hewlett-Packard monitor and Seagate HD of my computer-based audio system). It is a magnificent amplifier as well, and yet I felt its character (with its stock tubes) was too often a bit chilly for my taste, darting out the notes of an aria or jazz

ballad with precision and alacrity but without the ease and warmth that were the principal advantages of my monoblocks. And yet its superior power had to be investigated.

I'd had this notion for two weeks before I acted on it, though, as I'd a bad chest cold that kept me suffering and loopy for ten days and duties at the university that limited my free time. These sorts of changes need to be done thoughtfully, invested with an expansiveness of consideration so that one could take in the full measure of the change, interrogating scores of listening parameters besides just raw power. You'd hear that power right away, of course, but there might be trade-offs like that hint of coldness I mentioned, the lack of sinuous tails and leading edges to the notes that made a piece of music sound natural, an emphasis in the midband versus a rolling off in the higher notes (or vice versa), which might throw the balance of frequencies off kilter and made the music *sound weirrrd.* You had to listen for those elapsing subtleties and listen for them over a range of musical genres, and that took time. I think the process is even more intricate than a wine-tasting session when you line up several bottles of choice Bordeaux or Super Tuscans, say, and try not only to make judgments, but to discern, through the complexities of what each presents to your palate, what it is exactly you're left with on the tongue and in memory. Does it produce an *explosion* of flavor like a California Cabernet? Or does it present an aggregate of ephemeral things, wending around and overlaying each other in a finely attenuated mixture that demands more discernment in recognition of their evolution in time than a purely instantaneous *ga-ga* response? This is what music does too— cultivating the finer tones of sensibility and an interplay between memory and a momentary sensation, kicking you out of a Neanderthalian brutishness and into the brisk Venetian baroque of Vivaldi, the lap dance of Miles's modal jazz, or the impure whomp and stomp syncopations of the P-Funk All Stars, each of these calling for a sophistication on the order of apprehending Paleolithic cave paintings illumined by torchlight in combination with their natural phosphorescence.

So, I waited a full fortnight to pull the tubed monoblocks out of the system and replace them with the VAC stereo amp, a big-ass

black thing heavy at the back with potted transformers and sporting a frontal tray of eight vacuum tubes in three rows. What I heard instantly was what I expected—more authority with the music, speedier transient sounds on the leading edges of notes (particularly evident on the attack bowings of violins), and a tighter bottom end, which is to say the bass notes were more sharply defined and didn't linger beyond their welcome, imparting to the music a sense of cleanness and precision. I listened to it through about half an hour of warm-up (tubes need time to reach optimum performance) and then started to lean in, hearing more critically, listening for shortcomings.

And there it was, lurking in the shadows of sound, a kind of distancing simplicity that lacked what my other amps were so good at—the organic qualities of music. I missed a natural sinuousness and shapeliness to the notes, as though they'd lost their little flags and tails that bannered in the spaces between them, the undulations that connected them all together like a field of tall grasses bending in semaphoric sequence as a sashaying wind tousled over them. But I was not lost for what to do here.

The VAC amp used two pairs of an eight-pin octal tube designated as 6SN7 and initially used in radar equipment during World War II. Probably designed by RCA, it was later produced by a lot of U.S. and British companies, which encased the inner workings of their rugged metal electrodes in thick, snub-nosed glass bottles darkened with blackening on their insides to prevent excess electrons from building up. In the fifties, over six thousand of these were used

in the ENIAC computer. They showed up in early television sets and in movie theater amplifiers too. In fact, they were all over in early hi-fi equipment before the smaller, nine-pin micro-tubes became the rage. But I thought these stock

tubes in the stereo amp might be the source of the "coldness" problem that was vexing me. The amp-maker had used current production Chinese tubes in his amplifier, very well tested for reliability and proper function, but I knew from experience (and "tube lore") that one could fuss with and perhaps sweeten the sound of a tube amp simply by swapping in vintage tubes made by the old manufacturers in England and the United States.

So I rummaged in the cubbyholes of my listening room, the ones I keep filled with miscellaneous gear—coiled interconnects, damping feet (to place under components), cans of air under pressure, and a treasury of spare NOS (for "new old stock") tubes. I'd gathered most of them just after I'd acquired the PrimaLuna amp and begun my frenzied dive into audio tubes and their remarkable, articulate glory in shaping sound. The output tubes I then swapped in, German-made in the sixties, did indeed sound better in the amp than the stock Chinese tubes. I ran it that way for a few weeks, until I got curious about swapping out its miniature signal tubes as well, and bought a few pairs of vintage substitutes for them. And *they* made a nice difference too, immediately producing a sound easier on the ears and more liquid as well. The music was much more sensuous for just these little changes, tubes from the fifties and sixties sweetening up the amp's sound, their tiny fires glowing red underneath their little bottles of glass and behind chalked labels that spelled out their exotic origins of manufacture in England, Germany, and Holland. For I'd somehow eschewed those made in the USA—those Sylvanias, GEs, and RCAs so familiar to me from my father's coffee can collection—and found myself swooning at the call from Europe that started with those Dutch-made Bugle Boys I remembered were once cradled in the palm of my father's hand as he readied them for installation in his own homemade amplifier. I bought a pair of those (from Kevin Deal in Upland) with their bumptious cartoons of tubes blowing bugles on their shapely glass. These were labeled "Amperex—Made in Holland" in letters of white chalk that smeared off all too easily if you weren't careful in handling them. Then I bought Mullard tubes, stamped with a chivalric shield that said "Made in Great Britain." Finally, I bought a pair that had diamond shields that read "Telefunken—

Made in West Germany." And every pair, after I swapped them in, sometimes for each other, sent me into fresh, softly radiant circles of delicate hearing. The Amperexes were rich, warm, and punchy, making the horn choruses of big bands sound full and harmonically complex, I thought. The Mullards were even richer, but smooth rather than punchy, great for the sound of orchestral violins, but perhaps glossing over the inner details of string sound, homogenizing them. But, *oooh,* were they great for operatic voices, capturing their shine and darkness both, their ululations and gleaming top notes. And the Telefunkens? They were smooth too, though not as rich as either of the other two, perhaps more "neutral" and without glorifying any particular thing, playing voices as well as violins, saxophones, and a Hammond organ with equivalent realism and an unburnished attractiveness.

It wasn't long before I moved on from that PrimaLuna that was my primer in tube amplification. I got an even more powerful amp—the Cary SLI-80 that looked to me like an old hot-rod with its cans of capacitors sitting right on top of the chassis next to its tubes (like an engine exposed). It used front-end tubes called 6SN7s.

I bought even more tubes to roll into it, went hog-wild over some months and bought more than a half-dozen pairs of 6SN7s— smoked glass, black glass, mil-spec (made for the military), chrome top, round plates, and "tall bottle." Every one of these made the amp sound different. I tuned my ear to their capabilities, parsing the music with each one, noting their special character and way with sound, how one made things more hearty, another more detailed and finely nuanced, yet another added more bass or a superior refinement of treble. That I most coveted a tube that was best with operatic voices narrowed the field to the two most opposite—the round plates the most rich and evenhanded throughout the audible frequencies and the tall bottles having the sweetest and most ease at the top of a soprano's range. The specifics of their brands aren't as important, though, as knowing that I was creating a catalogue of sound attached to each tube's make and vintage, an act akin to a wine collector's arranging his cellar as much by taste as date and *terroir.* At the same time that I was tweaking and geeking, I was

acquiring a vocabulary of sound and its manifold variations according to the gospel of the vacuum tube.

So, I went back to my stash of 6SN7 tubes and its accomplished
vocabulary of sound and rolled in a few changes. I replaced the
middle row of stock, current production Chinese 6SN7 tubes with
a special pair with reinforced and taller electrode structures known
for sweet and extended sound. These were U.S.-made Sylvania
6SN7GT tubes nicknamed '52 Bad Boys for their prowess and date
of manufacture. They made the amp's sound immediately sequa-
cious, more organic and fluid in its tracking of notes in succession. I
stayed with that change a full evening and the following day until I
figured I still heard a lack of shapeliness in the "Casta Diva" aria of
Angela Gheorghiu, too much bite in the string attack on a recording of music by Vivaldi entitled *Concerto Veneziano*. I changed the
first line of Chinese 6SN7s for Sylvania "chrome tops," known for
their sprightliness in the treble range without etch or bite. I listened, again for a night and a day, before I decided this didn't quite
work either, the sound lacking warmth and an ease in the midrange, despite how nimbly it handled operatic top notes and string
attacks. I swapped in the ones called round plates, listened, and
then, after yet another day, changed them out for a pair of British
black glass military tubes that at last gave me everything I wanted.
A sound of earth, of clouds commingled with angels, of the harvest
of souls among the bending wheat tumbled under a heel of wind.

*System: Cary 303/300 CD player; TW-Acustic Raven Two turntable;
Ortofon RS-309D tonearm with Ortofon SPU Mono GM Mk II,
Ortofon Cadenza Mono, Ortofon Anniversary SPU cartridges; Tri-Planar
Mk VII Ultimate II tonearm with Zyx Airy 3 cartridge; deHavilland
Mercury 3 linestage; Herron VTPH-2 phono stage; VAC Signature IIa
preamp with phono stage; deHavilland KE50A monoblock amplifiers;
VAC PA-100/100 stereo amp; VAC Phi-200 stereo amp; Von
Schweikert Audio VR-44 Aktive speakers; Siltech 330L speaker cables
with 330L jumpers; Siltech 330i interconnects; Siltech Ruby Hill II power
cords; Audience Adept Response aR6-TSS line conditioner with Audience
Au24 powerChord; Box Furniture S5S five-shelf rack; edenSound FatBoy
dampers; HRS damping plates.*

September 2014: A Visit with Michael Fremer

Michael Fremer, All-American guru of analog, led me down a wide well of carpeted stairs to his basement. We turned a couple of tight corners by a laundry room to a makeshift corridor between a wall and several racks of record shelves packed with LPs, tchotchkes, and miscellaneous audio gear. There were cables and packages of tiny metallic talismans—spades, bananas, RCA barrels and plugs piled and scattered over every flat surface like rice at a wedding. The atmosphere was like that of a decrepit yet charming submarine— tight quarters crammed with thousands of records and memorabilia and Fremer my own Captain Nemo guiding me through its compact maze, stepping down to where the action was at the far end of the dimly lit basement. He pointed to a pair of chairs, then picked up binders, invoices, and audio junk stacked on one of them, inviting me to sit. He turned and turned again, not finding another place to set it all down. He sighed and dropped the floppy heap onto the chair that had been vacant.

"Well, *I* don't have to sit," he said.

Swirls of audio detritus circled up from the floor like miniaturized masses of the worshipful moiling around twin Mayan ziggurats the color of faded gold. These were the darTZeel monoblock amplifiers made in Switzerland—$170,000 per pair retail and highly prized by audio cognoscenti. They rose up from the littered floor like towers from the plain of Mordor. Soft jazz—a guitar and piano—played in the background.

Writer-editor of *Analog Planet,* a commercially sponsored website where he posts daily blogs, Fremer is arguably the most well-known reviewer in audio, equally famed for the brash humor of his articles for *Stereophile* and for championing vinyl over CD sound, analog over digital. Born in Manhattan in 1947, he grew up in Jamaica, Queens. Since high school, Fremer has been a record hound and into hi-fi. He built a Dynaco Stereo 70 from a kit while in college. He claims he can show you records he bought as a kid in 1957.

"I've been into analog only since *forever,*" he said.

About nine feet away and maybe ten or twelve feet apart, across

from me, were two speaker towers, each taller than Fremer, seemingly hunched over in the corners like phenolic sculptures of Capuchin monks. These were a pair of Wilson Audio Alexandria XLF speakers ($200,000 a pair) in black with gunmetal-gray trim. Along the wall to the left of the chairs was an elaborate, four-shelf, side-by-side audio rack made of aircraft-grade, black anodized aluminum. In the rack was a vivid splendor of electronics—a majestic preamplifier made by Boulder Amplifiers (in for review), an Ypsilon phono stage from Greece that had two huge outboard step-up transformers looking like twin, chrome-plated hydroelectric engines, a USB DAC and CD transport made by Simaudio in Canada, and three smaller, tidier gizmos that dealt with digital files, converting them into analog "and vice versa," Fremer said, echoing Yogi Berra. Finally, next to these was the famed Continuum Caliburn, a monster-sized turntable that floated on its own suspension rack, all glittery and silver. The visual would have been stunning except that, for all of its audio glory ("My stereo system is worth more than my house!"), this assemblage of precious gear seemed like a washer and dryer set crammed between the wall and some laundry shelving stacked with records and electronic miscellany.

It was late September, and I had driven four hours from New Hampshire (where I was in residence at an artists' colony) to visit. Near the end of my drive, I exited the turnpike and drove by countryish estates, then through humbler neighborhoods of mixed ranch-style homes and two-story mini-manses. Fremer lived in suburban Wyckoff, New Jersey, on a short, dead-end street. His home is a two-story white rectangle with large double doors painted black. After I rang the bell, a trim man of about five foot six opened the door, sweeping his arm and saying "Enter!" He was dressed in crisply ironed khaki shorts, running shoes, and a black tee with white lettering that said DOG IS MY CO-PILOT across its chest. He wore glasses and had wavy gray hair trimmed short on the sides and standing up on the top of his head like soft spumes of rolling combers. This was Fremer.

Taken by many as the godfather of vinyl, Fremer got his start in audio reviewing about thirty-five years ago with Harry Pearson, the legendary editor of *The Absolute Sound,* which was, at the

time, one of the two most influential audio magazines in the United States. *TAS* (as it's known), with Pearson as editor, championed *subjective* listening impressions of audio gear as opposed to touting technical specifications, insisting that the perceptions and sensibilities of the listener take precedence over any technical claims of the manufacturer about the gear. When *TAS* first started publishing in 1973, manufacturers had held sway over the markets, and reviewers were mainly shills for the engineering achievements of the audio companies. Pearson's savvy and influential writing (honed as an environmental reporter for *Newsday*) started to change all that, through a combination of his listening discernment, straight talk, and personal charisma.

"Pearson was looking for a pop music editor," Fremer explained. "This was around 1986, and I'd been writing about music since the late sixties. I sent in some writing samples and he loved them because I was just regurgitating stuff I read in *TAS*. So he hired me."

Fremer wanted to write equipment reviews too and was eventually assigned some speakers, but it took almost a year before what he wrote was published. Pearson had given his work copious edits, red-penciling things from opinion to phrasing, sending it back,

acting through the mails as an indomitable mentor, gruff and yet somehow endearing. The relationship grew, Pearson always ending each piece of correspondence with a verbal wink or lift of his martini glass, and after that gauntlet, Fremer was appointed to the regular review staff of the magazine.

After some successful years, Fremer sought a change and wrote a fax message to the competition—*Stereophile* editor John Atkinson—asking for a job. Atkinson faxed back immediately, offering music reviews, but Fremer said he wanted to write about equipment and be a staff columnist. Atkinson asked, what kind of column? Fremer proposed he write about *vinyl*. Atkinson responded skeptically, saying Fremer would be writing himself out of a job because *vinyl was going away*. If vinyl was going away, Fremer replied, then he would too because he didn't want to be in the business if it didn't include vinyl. Atkinson caved, and within a few short years, Fremer's "Analog Corner" and reviews using vinyl as his primary source became the most popular things in *Stereophile*. "Analog Corner" featured writing both witty and descriptively memorable about what Fremer heard as characteristic of a product's sound. He attracted a staunch following of readers and manufacturers alike.

"Look," he said, reflecting on the evolution of his column into a website. "It's like this—I started with a Corner and now I'm a Planet. I didn't plan anything. I just kicked the can down the road."

While he appreciates all kinds of music, his vast collection of LPs ranging over all genres and the world's regions (he especially likes Europop), Fremer seems most passionate about vintage rock from his own youth. He'd played me a few cuts from premium reissue classical LPs, a track of solo jazz piano by Thelonious Monk, and another of Tony Bennett singing "Lullaby of Broadway." But he was particularly voluble and enthusiastic when he played cuts from rock groups like England's Small Faces (Rod Stewart's band before his solo career and Ronnie Wood's before he joined the Rolling Stones), Motown's Supremes, and early Beatles in mono.

Fremer's analog rig is perhaps one of the most elaborate and costly in the world. Central to it is the Continuum Audio Labs Caliburn turntable manufactured in Australia and designed by a team

of scientists trained in aeronautics and software development rather than audio. It arguably represents the ultimate at the luxury end of the audio market. Taking turntable technology to an extreme of both design and expense, it incorporates magnesium alloys as a major building material, uses sophisticated computer modeling and magnetic levitation of its bearing (on which the platter spins) rather than viscous lubrication, employs a system of vacuum hold-down of the record onto the platter, and houses a scary maze of hydraulic lines, pipes, and hoses under its casing. Fremer likened these innards, in his own *Stereophile* review, to the workings inside a jet plane's wing. The Caliburn weighs 160 pounds and sits on a companion stand made of channeled, aircraft-grade aluminum X-braced for rigidity and tensioned with nautical thumb buckles. Called the Castellon, the stand weighs 176 pounds and isolates the Caliburn 'table from its environment.

Currently, the Caliburn turntable, its Cobra tonearm, and the Castellon stand carry a retail price of two hundred thousand dollars. To date, about sixteen have sold in the United States. Fremer told me his was his most prized material possession. He emphasized that he owned his own gear because, by contrast, he thought a lot of other audio reviewers got their equipment on long-term loans that were essentially donations from various audio manufacturers.

"I've got my own skin in the game," he said.

In his reviews, Fremer likes to be critical, he told me, as he believes no matter how good a piece of gear might be, every product has its flaws that, if observed, he feels obliged to point out. He once gave another costly turntable from a noted American manufacturer what he thought was a rave review insofar as how it sounded. But Fremer also criticized the piece for the way it looked. The maker was outraged and sent him an email saying that what Fremer wrote would put them out of business. Six years later, the manufacturer is still around "and doing better than *before* I wrote the review," Fremer said.

He thinks it's a real problem for audio magazines that reviewers are so positive, their articles read like shills for the industry.

"One month, there's this amplifier that's the world's greatest

ever, and according to the same reviewer, the next month there's one that's even better. There's no criticism there," he said. "It's all publicity and that, that's *bullshit*."

Fremer speaks rapidly, emphatically, halting on a word or phrase for a moment, italicizing it in speech, giving it space, holding the whole sentence back while formulating its close, and then letting it loose. It's difficult to capture exactly what he sounds like, as it's a kind of antic chatter so much of the time. It's like he's got three alternative sentences forming in his head before he speaks. He starts one sentence, then switches tracks and picks up the other, then jumps away from that to the one he wants to ride to a period or exclamation point. The track he chooses is usually full of momentum. It rushes out with a pent-up power that rattles by like a subway train.

Fremer slid a record out of its jacket and sleeve and placed it carefully on the platter of the magnificent turntable. It was a boutique re-pressing of a classic RCA Living Stereo recording of *The Royal Ballet: Gala Performances,* with Ernest Ansermet conducting.

I noticed there was a hum coming from the Caliburn. Fremer explained that it was the vacuum pump trying to suck the record flat.

"It'll stop in a moment," he assured me.

Fremer's patter filled the air with a little more light comedic schtick, and then I heard, amidst all that clutter that was his basement room, a gorgeous and delicate playback of Tchaikovsky's "Dance of the Sugar Plum Fairy." There were beautiful orchestral strings rendered with superb and pleasing resolution rather than the glassy or etched kind of sound that normally plagued inferior systems. Attack transients, that leading edge of sound that is the extraordinarily prized, almost super-real effect of high-end audio, came speedily without any hint of harshness, easing into midrange notes that were mellow and yet alive with timbral contrasts among the brass, strings, and woodwinds. There was a precious clarity throughout all the frequencies and fine, textural detail to everything. I wrote *delicacy* in my notes and then circled it. I thought a splash of cymbals in one passage was exactly that—a gentle shimmer with only the slightest tang of metal and more like a brief,

sugary burst of sound. The whole thing seemed as clear and lacy as a puff of cigar smoke curling in still air—tactile, sensuous, and achingly 3-D in an expansive soundstage before the massive Wilson speakers.

When I asked him what he tries to listen for, he countered my question. "I try *not* to listen for anything. Over time, you play familiar recordings and maybe see what's changed with a given piece of new equipment. Do I hear a difference? I trust myself at this point and I'm rarely wrong. People rely on you. They'll buy a record I said sounds a certain way and then, if they find that it does, they go, *You're right, that IS what it sounds like!* It's the same with equipment. They go to hear a piece in a store, and if it sounds like what I said, whether they like it or not is not the issue. I'm not here to tell people *what* to like. The stupidest reviewing tactic is trying to push people a certain way. If it sounds like hype, it's poison. What I do is say, *Here's what it sounds like.* And, usually, the great products cause a creativity boost that allows one to break free of the usual audio clichés. The more organically whole the sonic performance, the easier it is to describe. Here, just listen to the tambourine in this . . ."

The rattles clattered with soft, metallic impacts in space, started by a tap on the stretched skin of the drum and then resonating with it in the soft shakes of a miniature dance of exquisite sound.

"Never ceases to amaze me," he said.

We talked about different pressings sounding better or worse than others. Fremer seemed to prize clarity and ease of detail, the sweetness of transient responses (those leading edges of sound) and the way they interacted with the harmonic blooms of notes and the rhythmic envelope of a piece of music in its entirety, all of which, to him, made recorded music sound real.

Talking vinyl, Fremer knew the history of most any given recording he spun for me, who the recording engineers were, the chain of mastering and the process that was used, the different pressings, the re-releases and how they sounded compared to the original and each other, and how each one was mixed, down to the specific board that was used, its location, and its reputation and provenance if it was moved from one studio to another and reas-

sembled or modified. It was ridiculous. He knew the *lineage* of the sound engineers. He cited these facts not so much like a collector, which he is, but more like an attorney arguing legal precedent and citing historical rulings in front of a judge. There was a certain intensity of knowledge in what he said, a veritable epistemology of vinyl in his head. Put another way, he knew the vintage, *terroir,* and cellar storage of most any given LP in his possession.

Fremer estimated he owned over fifteen thousand LPs. These were organized into categories for blues, Motown, Stax-Volt, imports, with prized classical segregated by label (RCA, Mercury, London Blue Back, etc.) and other classical by vintage and reissue. He told me most of it was alphabetical within a category, but that the organization was like a recipe box—he knew where everything was by how it related in his mind to the next thing. When I asked him what he thought it was all worth, he said he'd give himself a call one day for an appraisal.

"What got me jump-started was when CDs came out, so it wasn't all bad. People were getting rid of LPs like they were carcinogens."

Fremer played a Beach Boys record next, a test-pressing with only the vocal tracks of "In My Room" on it. This was a particularly serene choral number recorded during the group's early years. Sweet, sighing clouds of three- and five-part harmony filled the air, angels of sound dancing amidst all the clutter and intelligent, curated chaos in the chapel of the basement room.

When the track ended, Fremer told me about how he once fixed Brian Wilson's turntable. Wilson is the falsetto voice and composer of the most famous Beach Boys songs (like "Surfer Girl" and "Good Vibrations"), often called a genius by the rock 'n' roll press. But undiagnosed mental illness and drug abuse led Wilson to fall deeply into a kind of manic obsessiveness with editing techniques in recording short, interchangeable fragments of music, isolating vocal lines and splicing tracks together, pursuing an elusive perfection. He suffered from paranoia, and while working on a tune called "Fire," worried that it had indeed sparked a fire near his studio one night. He later claimed he'd destroyed the tapes of that album (one he called *Smile*), fearing they were "bad magic,"

but in reality, the tracks were locked in a vault and all the while protected. The unfinished recordings languished for years, wasting mountains of reels and what some assumed was the masterpiece of Wilson's career. Chastened, Wilson went in a completely different direction—he and the Beach Boys recorded a much less musically ambitious collection of tunes called *Smiley Smile,* a record that even loyal fans found a disappointment. Wilson disappeared from public view, confining himself to his bed and under an unethical psycho-therapist's care for several years. He was forced to sit at a piano and compose the whole while, rewarded with a cheeseburger each time he completed a tune. He became a cautionary tale and an urban legend of rock.

At the time of Wilson's seclusion, Fremer was living in Los Angeles, and a physician friend invited him along on one of his house calls. The physician said there was "a real goof" he had to see for himself. The doctor explained it was his job to give certain prescription meds to Wilson's nurse. Fremer and the physician then went into the Santa Monica hills to a mostly empty house. Inside, there was an expansive living room, also empty, except for a turn-table and stereo system at one end. Dreamlike and ominously, there was also an adult-sized high chair sitting in front of the stereo with burn marks on its plastic tray-table and the wooden floor below it. Brian Wilson came in, wearing only a diaper, greeting them, ask-ing if Fremer wanted to hear some music. The doctor hurried off to the kitchen with the nurse, leaving Fremer alone with Wilson.

"I'd play some music but my record player's broken and nobody will fix it," Wilson said. "Why won't they fix it?"

It was a German-made Dual turntable that Fremer knew "back-wards and forwards." After a quick inspection, turning it over to expose its undercarriage and checking it out, he could tell that it was just locked. It had to be lubricated, so he found some Vaseline, put it in, and got the cam device inside to unfreeze so the 'table could operate. Then Fremer looked at the cantilever of the cartridge and saw it was bent like a crooked finger. It was made of metal and not boron, which could break, so Fremer got a pair of tweezers and carefully straightened it out. He reset the tracking force, asked for a record, and put it on. It worked. Wilson started crying. The Beach

Boys composer sat down in the high chair and rocked back and forth. He said, "Thank you, thank you. I can hear music . . ."

System: Continuum Audio Labs Caliburn turntable with Cobra tonearm and Castellon stand; Transfiguration Proteus MC cartridge; Ypsilon MC-10L and MC-16L step-up transformers; Ypsilon VPS-100 phono preamplifier; Simaudio Moon Evolution 780D streaming DAC; Boulder Amplifiers 2110 line preamplifier (in for review); darTZeel NHB-18NS preamplifier; darTZeel NHB 468 monoblock power amplifiers; Wilson Audio Specialties Alexandria XLF speakers; ASC tube traps; RPG BAD, Skyline, and Abffusor panels; HRS Signature SXR stand and HRS shelves; Finite Elemente Pagode amplifier stands; line conditioner unknown; cables various and unknown.

Part Ten

Among
the Bohemians

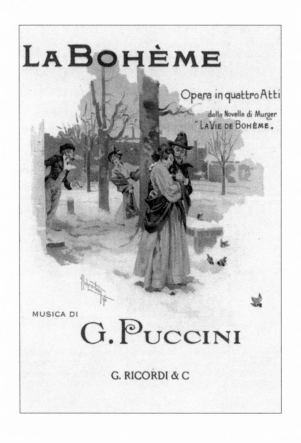

The Johnny Winter, Doc Watson, and Glenn Gould Trio

In 1969, back in Gardena when I was eighteen and my brother fifteen, we'd get up every morning and blast my father's stereo with a blues record I found in a local store after reading about it in *Rolling Stone*. It was the first Columbia album (Columbia CS 9826) by an albino player from Texas named Johnny Winter. Celebrated as a novelty, Winter played electric mostly, fast and hard-charging licks on guitars, with a raw-edged and shouty sound to his singing. But there is one cut on the first side that was completely different from the rest. He played an acoustic resonator guitar, and the track started with Winter speaking into the mike. "Lucky Thirteen," he said, his voice low and damped. I think he was referring to the thirteenth take on the tape. And it was a humdinger. He played solo and acoustic, plucking, strumming, and bottlenecking a guitar that sounded like no other guitar I'd ever heard. It had that characteristic, faintly metallic, yet pleasing sound full of upper harmonics I'd learn later to recognize as the product of a steel-bodied resonator.

> *Goin' back to Dallas, man, take my razor and my gun.*
> *Lots of people lookin' for trouble—hee-ee!—sure gonna give 'em*
> *some.*
>
> (from "Dallas" by Johnny Winter)

After some low opening notes, Winter immediately screamed a burst of slide-slurred ones, furious and rapid, the guitar sound full of as much drawl and bite as Winter's own caustic and raunchy style of singing. His solo wrung the neck of the guitar, as he plucked the fifth and sixth strings so hard that their sound boinged off the back of the cone and into the mike, then picked and strummed almost

at once, squeezing down on the steel strings, letting you feel the slide of the smooth glass of his bottleneck as if he were pressing it on your eyeballs. His high notes possessed a glinting, shardlike sharpness like musical sparks being thrown off as he shivered his slide way down on the neck. His instrument on the take was likely a later, single-cone resonator National called the Duolian or Triolian, that style of guitar initially invented as a three-cone version by John Dopyera. It's a zitherized type of Spanish guitar with a metal cone inside its body that makes it resonate so that it's louder and can sustain notes longer than any regular acoustic. A steel bridge raises its strings so far over its fretboard, the usual way to play is by using a metal bar, knife, piece of pipe, or bottleneck to slide up and down it, making for its characteristically wavery, glissando-like notes and chords. The single-cone was more strident, could sound more percussive, and had more bite than regular guitars. Its sound could sail over the hubbub of sweating revelers crowding the dance floor of a barrelhouse on Friday night. It's funkier and bluesier, an instrument perfect for the music played by African Americans in the Mississippi Delta that inspired Winter.

Imagine my delight and surprise when, years later in the late seventies, as a graduate student in the UC Irvine poetry program, I discovered Winter was the first cousin of my teacher, the poet Charles Wright. He never said, but I read about that in an interview Wright gave in a literary magazine and took notice. Wright's mother's brother was Johnny Winter's father, who'd gone off to Texas from Tennessee, where Wright and the main part of the family were from. I came across the Winter album again when I house-sat for Wright one summer, finding it among about a dozen other records nestled on the floor between the sofa and a wooden box full of liquor bottles. The LPs included the Carter Family, Doc Watson, Bach's Goldberg Variations by Glenn Gould, and the first two albums by the Band. Country, roots rock, and baroque piano music leaning up against bottles of Jack Daniel's, Smirnoff, and Wild Turkey.

Wright was an extremely private person, seeing us graduate students only in class and maybe at afternoon poetry readings and the brief wine and cheese receptions on campus. I'd occasionally spot

him loping across campus in that easy gait of his or see him in my rearview pulling up behind me in his blue VW bug as we both hurried home to try to beat the traffic flow away from school in the late afternoons. Slim, a bit under six feet, with wavy brown hair he wore fashionably longish, just over the ears, he looked a bit like Peter Fonda in photo-gray glasses and a brown sports coat over a dress shirt, jeans, and cowboy boots. Ever casual, he was also distant if not remote. It seemed to me that, most of the time, Charles was in a world all his own and that he liked it. In the poetry workshop, he was rarely the most talkative. He was, instead, probably the most taciturn, sitting with us in the discussion circle, presiding while others held forth. He'd gently make a small suggestion here and there, saying wise and pithy things like "The short line depends on the image, the long line on rhythmic integrity"; "This poem of

yours ends before it ends, you know?"; and "Many are chosen, few are called."

Aside from that, you couldn't get a sense of his own aesthetic principles except through his poems. Charles did not preach. But if you got down on your knees and put your nose to the ground, there was a forest snail's glimmer train of revelations that he gave in his stray remarks and in a few published interviews. In the one I read, he said he wanted his own poetry to be a combination of Emily Dickinson and Walt Whitman—the nineteenth-century spinster mother and the gay father of American poetry. He wanted his poems to be "sneakily spiritual" and unabashedly lyric as Dickinson's tiny poems but also to have the rhetorical "carry" of Whitman's long verse line. He said that in music it would be like putting together Doc Watson on guitar and Glenn Gould on piano. I got that he liked Watson's country style of guitar picking and Gould's piano technique, and I saw what he meant about the two poets right away, yet Watson and Gould mashed together left me baffled. Doc Watson I knew from friends in college who played his records and sang a couple of his songs, but Gould I'd only vaguely heard of. The Palestinian literary scholar Edward Said, a visiting professor, had once mentioned him admiringly during a lecture.

When Wright asked me to house-sit for him, I finally got an inside view into his elusive aesthetics. In the evenings, I started to play the Johnny Winter record a lot, of course, but since I was already familiar with it, I moved along to the other records, curious what they might reveal about my maestro's taste and affections. I pulled Bach's Goldberg Variations out of its sleeve sometimes, trying to figure out why Charles thought so highly of it, both as music and in terms of poetry. Gould played on piano what Bach originally wrote for harpsichord, a far less dynamic instrument. I noted the recording was in mono rather than stereo, the LP issued in 1955, a couple of years before stereo came into the picture. Putting it on the phonograph—kind of a battered tan-and-beige thing with a heavy tonearm and speakers on the side—the record sounded at first soothing, attractive, a bit simple on the first track, called "Aria." It had the guilelessness of a child's lullaby with just a touch of fussiness, even meticulous trills on the piano the way Gould played it.

But the melody was easy to follow and it felt like a primer to me, an entryway into the complexity of Bach's contrapuntal sequence. After that, Gould really takes off in Variation One. It's like he's given the tune rocket fuel with a dash of meth. The single melody is stripped so that just the bass line and chord progressions remain (as mainly so with all that follow), chopped up into two rapidly played contrapuntal lines, and the notes of the Variation charge by like thoroughbreds through the starting gate. Variation Two takes it a bit more slowly, elegantly, a three-part contrapuntal piece, and you feel just a bit lulled, beguiled by its sweetness. Variation Three, a canon, speeds things a touch more, the left-hand line lifting under the right, like a skate of surf under a boogie board. Four is a charming dance at the tideline, with little turns that change at each repeat. By Five, you're into deeper water with a fast, high chop, as both the pianist's hands race furiously with rapidly struck notes in two-part variations. Six lulls again, smooth sailing in canon form. Seven is a drawing-room dance, a French gigue, with lots of cute trills and buttery arpeggios, decorous and proper, easy to follow. Eight is up-tempo, cataracts of patterned sixteenth notes rushing along, slowing only at the end. Variation Nine is another dainty interlude, a canon moderate in its pace, with two clear, contrapuntal lines that weave in and out, easily distinguished. Ten amps the complexity once again, but pleasantly, a four-part, miniature fugue with attractive ornaments and deft trills; and you feel as if you're finally "getting it," recognizing the harmony, the feel of counterpoint, reminded of the beginning aria. But then Eleven comes, a tocatta's brisk blizzard of notes blown by varied winds. Twelve is like a culmination of all, two hands of counterpoint hammered heavily, ending with a light farewell touch. Thirteen is a slow sarabande, an interlude of peace and musical serenity with a contrapuntal complexity that pleases in its grace. And on the variations go, thirty of them, alternately sedate and thrilling, Gould demonstrating both an astonishing virtuosity and a sensitivity that can be sublime. He plays contrasts, Bach's compositional skill juxtaposing tempi and mood, fantastic intricacies succeeded by musically engaging charm, so you get a mini-education in baroque music just by listening through the variations. In this one sequence, you're introduced to the varieties of baroque

experience Bach has to offer. It's indeed a primer of alternating simple and complex approaches.

What did I take from this as far as poetry goes? It was an abstruse lesson that puzzled me at first, having heard no other performances of the Goldberg Variations before. I've since thought that it could be about interpretation—hearing the tradition-bound language of poetry anew, with your own speech rhythms, eccentric accents, involuted and vernacular phrasings even as you might clash against canonical English meters. Gould had heard Bach in his own contemporizing way, twisted what others had heard as musical givens and regularities into stunningly renewed versions, even violations. Eventually, I was able to transpose Bach's musical forms over to the problem of *meaning* in language as well, to develop the thrill of Bach's contrapuntal compositions into a kind of mystical anagoge, translating Gould's assaultive approach to the Variations into a series of provisional poetic dicta: that a master must hear multiples, in sentences and alternate phrasings, in just one line of thought; that there are myriad ornaments of language one can add to anything; that you must hear more than a singular music, but that it must be both the high and the low, the earthbound and paradisal in one breath's utterance. Nothing can be simple without the potential for a radiant complexity. Nothing complex cannot be captured in a straightforward, sincere sentence as though spoken by a child. This was a lesson in intellectual polyphony; multiple strains and variations all inhere, nascent in what might be mistaken for a simple theme. And, finally, I understood that the world of genius and that of innocence are corporeal twins tied together by insight achieved through human imagination, itself a mysterious blessing, the sand sublime as stars whirling in the dark, midnight sky. That the nacreous surface of a pearl freed from its oyster reflects a celestial river of lights. That, pilgrim, you must raise your hand in witness because, up in here, all things possess the turning of heaven as they wheel and as they hum.

Listening to Charles Wright's chosen music and inhabiting his space on earth, I was being guided through a corner of his visionary universe. His house was in a ridgeline neighborhood called Top

of the World, up a long incline that wound above Laguna Beach, a town that is part of what's known as the California Riviera. It stood at the end of a dead-end street along a short shelf bulldozed into the earth with a huge vacant hillside field next to it, a gigantic pepper tree drooping over the asphalt driveway. From his front windows, you could see the Pacific, like a blue-green sheet with embroideries of white combers stretching out below the long, lavish rumble of a canyon his house was perched on. The ocean and the orange glaze of sky were visible through a large picture window over a camelback sofa upholstered in pearlescent white brocade. Every morning, a shush of sprinklers doused all the vines of ivy, vinca blossoms, and pumpkin blooms surrounding the front deck, a raft in a sea of botanicals. After the tacky apartments or tiny beach cottages I'd been renting, Wright's place felt like another world. Each item around me—every leaf, every flower, every birdcall and shine of light in the trees, every cymbal splash and *gut-thunk* from the kick drum of the rock 'n' roll band practicing in a garage down the street—seemed curated, an icon in the Charles Wright Museum where I was caretaker that summer.

Those weeks I spent in Charles's house, the quiet, his few records, and the bucolic surroundings got hold of me. A willow tree in the yard let down its spiny green lattices before my eyes, and I quizzed myself on the metaphysical meaning of Bach. I realized all around were Charles's lines and his poems—his deck, the shrubs and flowers, the weather and hillside, and the Pacific below were all characters and figures in his own lyric dramas.

> *Sun like an orange mousse through the trees,*
> *A snowfall of trumpet bells on the oleander;*
>
> > *mantis paws*
> *Craning out of the new wisteria; fruit smears in the west . . .*
> *DeStael knifes a sail on the bay;*
> *[. . .]*
> *A wing brushes my left hand,*
>
> > *but it's not my wing.*
> (from "Dog Day Vespers" in *The Southern Cross*)

Laguna Beach was the stage for Charles's post-Romantic epic, albeit a skeptical one, of salvation and redemption. I'd stumbled into his private *Paradiso,* each scene surrounding me an illustrative panel of his somber and sometimes wacky devotion to pursuing spiritual questions in our time.

The rest of that summer I just went about my business like a janitor, flipping my keys, making my rounds, watering the plants and grounds, drinking up the boss man's liquor. But never mind the whiskey—I was almost constantly drunk with Charles's spirit. Sometimes I'd drive the lane away from his house, bomb down the narrow road and its moguls, race to get groceries or on some errand or other in the town with its bustle, its aprons of surf and tide pools, then hurry on back up the hill to rest in how cool it was up at his place, how glorified the universe seemed to be as it wrote *the silvery alphabet of the sea* whenever I glanced down from the deck and out to the gray sheet of the Pacific.

⟂

> *Let it rain, let it pour.*
> *Let it rain a whole lot more . . .*
> (from "Big River Blues" by the Delmore Brothers,
> adapted by Doc Watson)

When I played the Doc Watson record in Charles's collection, I'd lie down on the camelback sofa and try to take it easy as Watson seemed to, picking out a lazy, walking line on his guitar as he sang. The country tone in his baritone voice (sort of a better Burl Ives) was beguiling, relaxing, and it settled me down into an ease I didn't myself much possess in those days, high-strung on anxious ambition and worry about my place in the world. Classmates from college were finishing law school or med school, starting residencies or clerking and getting married, while I foundered on the strange aspiration of becoming a poet—a piece of adolescent indulgence, most of my relatives thought. "Why don't you take the civil service exam," my mother would say, "get a job with the City?" At family gatherings on holidays, cousins my age would twit me for "still

being in school," *the perpetual student.* Yet these were nothing compared to the pains of doubt I inflicted on myself. The philosophy of country music gave me a ready riposte to all that. And, in a lovely irony, it squared with Eastern teachings I'd received.

Wu-wei, I thought then, the Chinese Taoist principle of "letting be." And once I could do that, the doxological mysteries of Charles's poems at last unfurled for me, their images like small pellets of colored paper that, if placed in a bowl of water, bloomed into lavish flowers of artifice and imagination. The "Darvon dustfall off the Pacific" that Charles described was a stunning gateway into his memories—of childhood and adolescence in Tennessee, driving winding country roads across the county line to fetch a bottle of gin with his brother; sitting on bricks of a walkway in Venice, letting his legs dangle and watching the ochre reflections of a palazzo glint, wither, and resurrect on the crepuscular surface of a canal's lapping waters; laughing with army buddies in a Florentine bar; witnessing the "spiked marimbas of dawn rattling their amulets" on a Dantescan hillside in Hawai'i. And then each of these turned to a deeper, more esoteric and ephemeral meditation—the product of his observations and abundant quietudes, as Charles says, *the far side of the simile / the like that's like the like.* His poems are reflections not only of the earth and its properties, the mind and its acts of affectionate and somber memory, but conclusions and speculations regarding insubstantial things—the ghosts and frail gods interfused in an infinite, transubstantial music that was the actual subject of the man's work. Inside his poems, I traveled with him through the creation around me, to memory and its attendant regrets and joys, to minute and everlasting confrontations with the Absolute.

When that summer ended, I gave Charles an old hardbound copy of *The Sacred Harp* I found at a yard or library sale on Balboa Peninsula. It's that book of shape-note singing—hymns and spirituals for church services. The lyrics were printed above notations in a kind of Western cuneiform of images making up a mnemonic lexicon of musical notes corresponding to a simple, seven-note scale. Always performed a cappella, it was a tradition of choral singing that first sprang up in New England but then took root in southern Appalachia for country folk who couldn't read music and for

churches that couldn't afford a piano or organ. Doc Watson must've known quite a few songs from this tradition himself. I hear it in the simple melodic lines he engages in the soft baritone of his voice, lightly plosive, in the drawls he makes over some of his words, in the mournful lilt at the end of his verses, in the easy way you can figure out how to sing along.

> *And who shall wear the starry crown?*
> *Good lord, show me the way.*
>
> (from "Down in the Valley to Pray,"
> traditional, adapted by Doc Watson)

And when he accompanied himself on guitar, Watson had that easy gait to his playing, a smooth and relaxed touch to his finger-picking, a clear and deliberate manner to the way he flat-picked his six-string Gallagher guitar. His rhythms came from combining the notes he plucked, the phrases and chords he struck, pairings of hard and soft, light and heavy, balanced and clean, the punctuations he added with single notes, the combination of subtly different tones, a kind of polyrhythm of complexity presented as though from a moment's thought, a casual summa of sound. He never seemed to rush through a tune, but lingered over its notes while he sang, elongating vowels like a hound dog howling, a country holler crooner if ever there was one. When I thought about it, I could hear one of Charles's verses draped like a long, silk runner over a Doc Watson guitar line, the both of them eased back in rocking chairs sitting on a wide porch near a shady grove somewhere in the Smokies.

The House of Blues

The single-cone, biscuit-bridge National guitar, perhaps the loudest acoustic ever made—a later variant of the original Dopyera tricone, was fashioned out of steel or yellow brass and sounded three to five times louder than a standard acoustic. It was quickly adopted by Black American blues musicians in the South after its invention in 1929. Sometimes just called a "biscuit" by those who played it, its strident sound could be percussive as well, having more bite than

the earlier tricone resonator, less of a sweet sustain, and definitely more funk. It became the instrument of choice for legendary players like Tampa Red, Bukka White, Blind Lemon Jefferson, Blind Boy Fuller, Kansas Joe McCoy, Memphis Willie B., Memphis Minnie, and Sister Rosetta Tharpe. You could say that a style of instrument originally designed for Hawaiian steel guitar music found itself repurposed for the blues once it hit the hands of these musicians from the Delta. They likely found the guitar through the Sears catalogue, where the price was cut from $32.50 retail to $29—a relative bargain that kept National alive through the Depression. Over thirty-two thousand of them got sold by 1937.

But unlike a Hawaiian steel guitar, which was placed across a musician's lap and athwart both knees, the single-cone blues guitar was generally played in the traditional and more recognizable position. It was hung with a strap that ran across one shoulder and wound around the neck, or it rested with the lower bout in the well of a sitting player's lap and leg. The guitar was strummed and picked while pressed against hip and torso, the guitar and its underlying resonator facing forward, toward a listener, projecting its distinctively metallic sound horizontally in space. Thus, the guitar had great "cutting power," able to be heard above all the dancing, singing, and jubilation in the juke joints and crossroads shacks or at neighborly picnics where blues people performed. With this single-cone resonator, blues artists created not only their own kind of music, but an entire ethos out of its percussive, steely sound, producing myths and storied moments out of tough, impoverished, yet extraordinarily spiritual lives.

Of these, I think there was no one more spiritual than Eddie James House Jr., who performed and recorded under the name Son House, usually playing one of three National single-cone guitars he possessed during the course of his career—a Triolian, Duolian, or Style O resonator. Born in Riverton, Mississippi, in 1902, House grew up on a plantation between Clarksdale and Lyon in the Delta, moving frequently between Mississippi and Louisiana throughout his childhood. His father, Eddie James Sr., was a musician in a family brass band, perhaps blessing him early on with the seeds of love necessary for a life in music. The younger House grew up

preaching, though, delivering sermons in local Baptist churches by the time he was fifteen. During his twenties, the future bluesman served as pastor of his local church in Lyon and preached in the Colored Methodist Episcopal Church too.

At some point around 1922–23, House found work in Saint Louis, Missouri, and East Saint Louis, Illinois, discovering the thriving local musical scene there that would later be the early proving ground for Tina Turner and Miles Davis, among others. He taught himself the guitar along the way back to Lyon, hooking up with other bluesmen at house parties and dances, barrelhousing his way to a local fame. House played guitar and sang in Dr. McFadden's Medicine Show but then ran into some trouble, shooting a man dead after the man had shot House in the leg during a performance. House got tried and sentenced to Parchman State Farm, perhaps the first in a line of bluesmen incarcerated there, working on a chain gang, living the blues life, hearing prison songs and their words of hard luck and trouble chanted to the rhythms of hard labor. But after only a year (or two, depending on oral history, as no records exist), House was released and ran into Willie Brown, the legendary bluesman celebrated in Robert Johnson's "Cross Road Blues." House worked with Brown at juke joints, chicken shacks, and other venues, eventually running into the even more legendary Charley Patton and finding work with him too, running up and down the Highway 61 corridor between Clarksdale, Mississippi, and Memphis, Tennessee. He performed solo sometimes, working country dances, bars, and levee camps, soaking up the life of itinerancy and the blues, a thing perhaps parallel to sharecropping, laying down the soundtrack and accompaniment to its otherwise uncelebrated suffering. Son House scuffled around in towns like Robinsonville, Jeffrey's Plantation, Lula, Lake Cormorant, Tunica, Ruleville, and Clarksdale before he got invited to perform for the Paramount label up north in Grafton, Wisconsin. He was recorded in a small room where he likely faced himself to a corner with the mike positioned behind him, his guitar, and the chair he sat in as someone ran the recording rig, a machine laid on a wobbly table behind his back. He damped strings to imitate glottal stops, wavered a bottleneck

on them to produce vibrato, and put a lot of percussive bite into his playing. It was 1930 and House laid down tunes entitled "My Black Mama," "Preachin' the Blues," "Clarksdale Moan," "Mississippi County Farm Blues," and "Dry Spell Blues." The sleeve art for one of the records called him "Father of the Delta Blues."

> *Hey, black mama, what's the matter with you?*
> *Said it ain't satisfaction, don't care what I do.*
>
> (from "My Black Mama" by Son House)

It's not a hard move from these lyrics to "Satisfaction" by the Rolling Stones or Muddy Waters singing the line "There's another mule kicking in your stall . . ." from the song "Long Distance Call." House's raunchy lyrics were likely part of the long foreground to contemporary blues and rock music. But what the latter two don't possess is that pull toward spirituality and the tinge of guilt for not living a godly life—something that leavens House's lyrics, no matter how raw they might seem upon first hearing.

His sensibility is conflicted like that—steeped in regret and worry over "Judgment Day" even as he wails and shouts about a woman's lipstick and powder, a milk cow he needs to come home, racetracks and his wicked soul. If you listen to these 1930 Paramount recordings (*Son House and The Great Delta Blues Singers 1928–1930*, Document Records CD; *Legendary Sessions Delta Style*, Autogram Records LP), there's a grunt and groan to his singing and he plays his National with emphatic ostinatos, percussive slaps, and strumming that sometimes seems to shred across the tops of his strings. His worrisome blues is possessed of a consistently heavy, driving feeling. He wails and his voice slides up into falsetto as the passion in the tune rises and he shimmies his slide finger across the treble strings. I think Son House is saying something about earthly compulsion and his relinquishment of the promise of deliverance. In an interview from 1967, he said, "I preached for a long time in Louisiana. I be a Baptist preacher so I wouldn't have to work." But I don't fully believe him. There is an ecstatic quality to his blues, a joy pressurized by guilt and pain, but more than that, I hear a cos-

mology in his lyrics, a sense of life as trial and desperation, while feeling aware of an afterlife that follows it, an eternity that can be either retribution or deliverance.

> Oh, I'd'a had religion, Lord, this very day
> But the womens and whiskey, well, they would not let me pray.
>
> (from "Preachin' the Blues" by Son House)

I'm reminded of the octave (the first eight lines) of the sonnet "My own heart let me more have pity on" by Gerard Manley Hopkins, the anguished British poet and Jesuit priest of the later nineteenth century. House's lyrics have that same kind of devastating torment of heart, a similar arc in exploring the contours of a spiritual guilt:

> My own heart let me more have pity on; let
> Me live to my sad self hereafter kind,
> Charitable; not live this tormented mind
> With this tormented mind tormenting yet.
> I cast for comfort I can no more get
> By groping round my comfortless, than blind
> Eyes in their dark can day or thirst can find
> Thirst's all-in-all in all a world of wet.
>
> (from "My own heart let me more have pity on"
> by Gerard Manley Hopkins)

Hopkins hurts no more than Son House in his torment. And House's playing too, deceptively simple, has a tortured rhythmic quality, an emphatic syncopation at emotional climaxes, a brief hesitation and then a note played fortissimo or a cascade of them, as he squeezes down on his bottleneck, shaking it like the neck of a rag doll he's trying to strangle. The steely guitar notes plink and snap off from the metal top of the f-holed, sieve-domed single-cone resonator he played, throwing off brief sparks in the living air and giving witness, ephemeral in time, to an eternal pain in its player's heart, testifying to a short life of trouble.

After Alan Lomax recorded him for the Library of Congress in 1941, Son House moved to Rochester, New York, and, as far as his

music goes, fell silent for over two decades thereafter. He enjoyed a brief revival beginning in 1964, when he was rediscovered by a group of white blues devotees who persuaded him to play publicly again. He subsequently performed at various clubs, the Newport Folk Festival, colleges and universities, radio shows, and even the Montreux Jazz Festival and Carnegie Hall. In 1965, he re-recorded his catalogue for Columbia Records. Son House died in Detroit, Michigan, in 1988 and was buried there at the Mt. Hazel Cemetery, far from the train whistles of the Dockery Plantation, Coahoma County, and Parchman Farm in the Mississippi Delta, once his territories of travail and ecstatic ascension.

With Hirsch at the MLA

In the early 1980s, I met the poet Edward Hirsch, hilariously, at the annual conference of a professional association. I was stationed at the book fair in the ballroom of the Hilton Hotel in Los Angeles, invited there by my publisher, a university press, to share a book signing with my famed teacher, Charles Wright. Charles had published *Country Music: Selected Early Poems* (a book that later became co-winner of the National Book Award). I'd only just published my first "slender volume," issued simultaneously in hard- and softcover (a rarity nowadays). The publisher's strategy, of course, was that Charles would attract the throngs and I'd piggyback.

That day, for the first half hour or so, I think about two dozen folks had come by, singly and in pairs, buying Charles's book and getting him to sign their copies. None had asked for mine. None had even given me or my book a glance. This was the academic crowd—university professors and poets teaching at universities, men dressed in sport coats or blazers with dress shirts over casual slacks or jeans, women in professional skirt suits or slacks and coats over blouses in muted colors. Still a graduate student and trying to fit in, I wore a brown sport coat over twill trousers and a Sta-Prest polyester dress shirt I'd bought on discount. Charles wore starched jeans pressed with a sharp crease, a blue blazer, and a plaid L.L. Bean–type shirt. On his feet were cordovan leather loafers over white socks, Michael Jackson style. Aside from Charles, we weren't

a colorful crowd. And I was getting envious and frustrated. After another few browsers drifted by, not one looking my way, I told myself I'd get in the face of the next *motherfucker* who walked in.

This turned out to be a tall, athletic-looking guy in jeans, a light blue dress shirt, and a gray herringbone blazer. Early thirties, six feet something, wearing a pair of beat-up New Balance running shoes. He carried a small stack of books held against his hip with one of his large hands, and walked up to Charles, who was standing at the booth's entrance. I'd sequestered myself in a corner in front of a long table that displayed the press's array of academic publications. The man's hair was closely cropped on the sides, but dark brown and curly on the top of his head, sticking up a bit stylishly. He was blade-faced, like a hatchet, clean-shaven, with a prominent nose. I took him for an arrogant assistant professor at UCLA—casual, hip, rakishly rumpled, and eminently disdainful of any but the most intellectually elite. It was a type that dominated the convention, I thought. He immediately engaged Charles in an intense conversation, speaking rapidly, leaning forward, furrowing his brow, and I assumed, praising Wright's poems, demonstrating critical prowess, holding forth the *Selected Poems* for Wright to sign. He'd smoothly steered my teacher against the side of the booth, facing the entrance, screening anyone away from engaging them by turning his back and sealing the two of them off like a basketball big man setting a pick in the lane. I spotted the trick right away. I admired it.

I gave up my own position by the table, then strolled up to the two of them locked in conversation on the wing of the booth. The guy was still set in his screen, his back to me, but I tapped his shoulder, my finger feeling the soft padding under the wool fabric of his coat. He turned and scowled at me with irritation.

"What is it?" he said, impatient at the interruption.

"Hey, look," I said. "You come to the booth and you buy Charles's book but you don't buy mine. It's hurting my feelings."

"Oh, umm, what book is that?" he asked.

I showed him my paperback with its yellow cover.

"I *have* that book," he said, and quickly turned to face Charles again, showing me his back. I tapped the shoulder a second time.

"Yeah," I said, "but do you have the *hardcover?*" I gestured to the table behind where a couple dozen were still stacked.

When he turned, I saw that his face had broken from its scowl into a kind of innocence.

"Uhh, no I *dun't*," he said, a Chicago accent emerging in his speech.

"Well, today you get a twenty percent discount and a free signature from the author," I said.

"Okay," he said. "That sounds good."

He took out a brown leather wallet, battered and compressed, faced the attendant who'd been watching us the whole time, and paid. I opened up my book and turned to the title page, planning to inscribe it with flourishes. It would be my first sale.

"Whom shall I sign this to?" I asked.

"My name is Edward Hirsch," he said, standing over me, shoulders hunched, caught by my little gambit.

"Edward Hirsch!" I yelled. "You're fucking *Edward Hirsch*? I love your poems! I *love* them! I read your book standing up at a bookstore every day for a week I loved them so much! You wrote ' "Dance You Monster to My Soft Song!" ' You wrote 'Walking the Upper West Side with Lorca'! I fucking *love* your poems! They're so good, you piss me off!"

Hirsch's book had come out a year before mine from a prestigious New York press, and finding it in my local university bookstore, I'd admired its marine-blue dustcover with a white window that displayed the decorous gray etching of a tree. For nearly a week, I returned to read it every day until a store clerk said to me—"Are you gonna buy it or you gonna just wear the thing out?" Shamed, I bought it, though it cost more money—nearly twenty dollars—than I could afford on my grad student stipend. But the poems were astonishing—so beautifully structured, gifted in rhetoric and memorable images, with urban settings and many about working-class people in Chicago where he'd grown up. Best of all, the emotions ran a full range, from the somber and reflective to the ecstatic. And I could tell that the man knew his stuff. I recognized influences and borrowings from poets everywhere—Spain, Chile,

Philadelphia, Detroit, Paris, and San Francisco. I read echoes of the English Metaphysical poets, saw images like those of a raving Spanish Surrealist, heard shouts and murmurs in his supple poetic lines as though a bluesman were singing them, was consoled by prayers and elegies for the gone and tattered worlds of an immigrant past. This was a poet of soul with a deeply learned style, my own contemporary unmatched in skill and compassion.

His face cracked a big, wide smile, the deep furrow on his brow disappeared, and he grabbed my ears, pulling gently back and forth, saying, "O, I love you" as though he were a grandfather blessing a child.

Embarrassed but polite, our elder Charles Wright bowed away. Hirsch and I went off to lunch together, at a Chinese restaurant I knew close by, talking *poetry, poetry, poetry.* We shared stories of our apprenticeships in different books and cities, compared his formalist mentors and my stylishly louche ones. We talked about poets we loved, women we loved, and places we'd traveled. It turned out we'd both won the same fellowship straight out of college. He went to Eastern Europe and Russia on the trail of his ancestors and lost families. I went to Japan to do the same. After those long and parallel trails, we became brothers.

Seminar at Waimea Bay

One summer, after we'd traded visits during the academic year—I went to see him in Detroit, he came and stayed a few days with me in Columbia, Missouri—Hirsch invited me to join him in the islands where he'd been interviewing the poet W. S. Merwin for the *Paris Review.* Interview done, we spent a few days on Maui and then hopped over to O'ahu so I could show him where I'd grown up in Kahuku and Hau'ula along the North Shore.

On our way from Honolulu to our hotel near Kahuku, we kept flipping back and forth between a pop rock station he found and one that played "local" music—songs in Hawaiian mixed with English sometimes. Being back in Hawai'i was rare enough to me that, while I was there, I wanted to *feel* local, return to my roots, as it were, get into the *kani ka pila* spirit of the islands, so I wanted

Hawaiian guitars and singing. Eddie drove and wanted Motown—hits by Marvin Gaye, the Temptations, Supremes, and Four Tops—stuff I loved too, but we couldn't find an R&B station. Instead, he tuned to Top 40s, the strongest signal—blaring songs by Madonna, Peter Gabriel, Bon Jovi, Van Halen, and, worst of all, John Cougar Mellencamp. I could barely stand that, so in the lulls filled with DJ patter and ads, I kept flipping to a station on the far left of the AM band that featured old-school and Hawaiian Renaissance music—Sons of Hawaii mixed with Kalapana, Gabby Pahinui, and the Makaha Sons of Ni'ihau. All along the drive, we played dueling stations—"Talk to Me" by Stevie Nicks versus "Sweet Lei Moki-hana" by Hui Ohana, "Kiss" by Prince and the Revolution versus "Ku'u Home O Kahalu'u" by Olomana. When something by Stevie Wonder came on, we rejoiced and I sang along, trying to get Eddie to as well. But he didn't know the lyrics and couldn't carry a

tune, he said. When we had Hawaiian music on, I didn't know the lyrics either, but tried to hum and murmur stray syllables that I recalled from childhood, leaning into the curves as we drove, rolling down the passenger-side window of our rental car, and yelling out the few Hawaiian words and phrases I thought I knew to stands of sugarcane beside the road.

After not quite an hour of driving, we stopped at Waimea Bay, a beautiful scythe of a beach nestled between cliffs at the foot of a

deep, green canyon on Kamehameha Highway. Eddie was captivated by the scene—the rise of road to a short crest where the white
half-shell of sand opened up to our sight below and the expanse
of blue ocean ran to a horizon under the white knots of clouds
far away. Cars had parked alongside the *makai* (ocean) shoulder of
the highway all up to the entrance of the beach park, where he
insisted we stop and visit. To me, it was a tourist place that I usually
skipped whenever I came to see my grandmother in Hauʻula, but I
humored him. He was excited to be at the spot celebrated by sixties surf movies and songs by the Beach Boys. We turned into the
parking lot and, miraculously, found a space right away. Somebody
was backing out just as we rounded the first lane. People, mostly in
their teens and twenties, strolled back and forth to the cars from the
beach, hands full of rolled straw mats, gaudy towels, portable plastic
coolers, ball caps and pork-pie hats, and, occasionally, a bouncing
kite or a beach umbrella furled closed. It was like a broken parade
of varicolored paints, Paul Klee's watercolors and pencil drawings
of Tunisia, moving across the black asphalt lot, backgrounded by
beach sand and the huge stripe of blue that was the Pacific. Moss-
green monoliths of the Koʻolau rose up behind us and the deep vale
of Waimea retreated up a curving emerald river into cool shadows.

We got out of the car and felt the hot cuff of air from the beach
hit us like a wave and we peeled down to shorts, keeping our rubber
sandals on. Eddie pointed to the divers from the parapet of earth-
colored rock near shore. Men and boys were swimming out to its
makai and shorter end, climbing onto it, then traversing its craggy
top to the summit nearer the beach. They took turns diving into
the translucent aqua waters below, calm and deep enough not to be
risky.

"We're *not* gonna *do* that," I said.

As we walked over the beach toward the shore, I heard snippets
of songs coming from transistor radios and boom boxes along the
way. With the soft crash of waves and soft gusts of onshore winds
flapping through our loose clothes, it made a gentle cacophony of
sounds, a sinuous melody from Fleetwood Mac and the screech of a
gull intermingled with susurrant sighs from the sea.

Now, here I go again, I see
The crystal visions.

(from "Dreams" by Stevie Nicks)

We'd been talking about what had inspired us early on, as teenagers and young men pursuing this strange dream of creating poems from our own lives. The problem was how to do it in our own time, with the rough tools of our working-class backgrounds, with the daunting examples of the great poets of the past, with inklings only faintly felt, either as impossible ambition or the desolate miseries of our anxious loneliness. For Hirsch, it was daydreaming another world during the halftime of a varsity football game in high school. He played tight end and defensive line, getting run over by a powerful fullback on the opposing team during the entire first half.

"I couldn't stop the guy," he confessed to me. "He was like a forklift busting me off my feet every time they ran to my side. And, once they exposed me, they ran at me a lot."

"Hirsch!" his coach had yelled. He was across the locker room, dressed in sweats and a whistle. He picked up a metal trash can and banged it on the concrete floor. "Stop thinking about poetry!"

For me, it was at first all about Alina, who sat in front of me my junior year in a writing class. She was beautiful, loved poetry, and I tried to learn it to impress her, in my patter mixing poems by Rainer Maria Rilke and Dylan Thomas together in weird verbal bursts of teen angst and longing. I wanted to conjure the magic of a world she could love and found, in memories of my childhood in Hawai'i, an imaginary place as rich and heron-priested as the shores of Wales were for Thomas. I loved the girl but ended up loving the world of my conjuring almost as much. After Alina moved away, it was poetry that stayed. It was its feeling that guided me, a quiet rapture that had the mind spellbound and the breath steady. In that state, I traveled worlds, leaping from the grit of tire rubber and concrete by freeways in L.A. to the broken brains of seawashed, white corals flagged with green banners of *limu* on a country beach where I'd grown up.

E kau mai ana ka hali'a . . .
No sweet tubarose poina 'ole . . .

*

Memory comes to me in affection—
The sweet tuberose, so unforgettable.

(from "Pua Tubarose" by Kimo Kamana,

translation mine)

I told Ed that when I heard Hawaiian songs—the gentle ones like love ballads—it put me in the same mood as poetry did when I wrote it, a kind of concentrated affection bathed in an ocean of life, pitching waters that floated my heart and mind as though a buoyant sea were lifting under me to the rhythms of a slow dance.

"You're kidding," he said. "Don't you get pulled along by a form of some kind—a structure filling itself out, calling the words out of you?"

I shook my head. "No, not really," I said. "You mean like in a sonnet or strict quatrains? I can't work that way."

"Not necessarily," he said. He explained he was thinking of how the poem builds a hidden meaning throughout its course, the language saying something almost without you, inspired by the form of a poem that could be something like a sonnet, that Renaissance structure of fourteen lines in a scheme of rhymes. But it didn't have to be.

"It could be like the paradox in John Donne," he said. "The poem seems to go one way, arguing with itself or with God about religion, faith, and belief; then, at the end, the poet's twist of words comes in and the argument resolves itself by using language from sex, erotic love. It ambushes us—and maybe God too. And all along, this other kind of reasoning was running along in the background, like a tight end in the slot going in motion behind the line. You can't tell what he's gonna do." Hirsch laughed and started to quote Donne, the seventeenth-century profligate who became a cleric despite his carnal tendencies:

Batter my heart, three-person'd God, for you
As yet but knock, breathe, shine, and seek to mend;

That I may rise and stand, o'erthrow me, and bend
Your force to break, blow, burn, and make me new.

I was astonished Hirsch could quote from memory. It was like a little recital there on the beach. He delivered the poem like it was an alderman's speech—with a deep Chicago accent and its flattened *a*'s.

I, like an usurp'd town to another due,
Labor to admit you, but oh, to no end;
Reason, your viceroy in me, me should defend,
But is captiv'd, and proves weak or untrue.

It was so incongruous, this big Jewish guy from Chicago standing in shorts in front of me on the beach at Waimea, quoting from Donne's "Holy Sonnets," ignoring the scene around us. Beautiful women in bikinis, tube tops, and shorts lay on towels over the small pillows of sand, kids on skimboards and donut floats disported at the skirts of surf rushing our way, and divers leapt from the big rock and plunged through turquoise waters.

Yet dearly I love you, and would be lov'd fain,
But am betroth'd unto your enemy;
Divorce me, untie or break that knot again,
Take me to you, imprison me, for I,
Except you enthrall me, never shall be free,
Nor ever chaste, except you ravish me.

"At the end, it's that word *ravish* that kills you," Hirsch said, his open mouth like a cave capturing the wind. "You think the poet's asking God to take him over and banish sex from his life—strip him down—all those words about love unseated, breaking the knot of sex . . ."

A red kite soared over his shoulder as he spoke, rocketing like a skylark and then madly twirled, trapped by its limiting cord.

"But, in the end, Donne says he can't make it unless it's God who *ravishes* him! *God!* It's so weird and sicko and surprising."

The kite dove and disappeared behind him, and I felt the afternoon wind whip specks of sand against my bare legs.

Hirsch certainly had a point, but it was faint to me, as the poem sounded more like argument than from feeling. It felt so *procedural,* had the progress of a verbal experiment hatched in a clinical laboratory, albeit about the infection of sex colliding with an antiseptic faith. The poem and its paradox, a verbal trick, was foreign to me. I just wasn't the scholar Hirsch was, nor was I ever enamored of rhetorical traps in traditional verse. Though I'd studied it, poring long hours over the canon, I had followed mostly free verse, imagistic mentors, after all, and pursued something else.

"But what about the feeling?" I said. "Doesn't the poem you write yourself take you up and unravel all the stitchery—mundane concerns and how they fracture emotions? We get all busted up by the bullshit the world sends our way and poetry brings back a piece of feeling we hadn't paid enough attention to when we'd felt it? It's *emotion recollected in tranquility . . .*"

I cited William Wordsworth, the British Romantic poet who wrote those words in his famous Preface to the Second Edition of *Lyrical Ballads,* a collection he and Samuel Taylor Coleridge wrote featuring commoners and themselves in common, everyday scenes, written in what they felt was "common language," unadorned with high diction, linguistic filigrees, or sacrosanct emotions. I was, I thought, like them, arguing against strict poetic structure, outmoded forms, tricks of rhetorical sophistry. I wanted organic form that retrieved emotion, that rounded back to lost feelings and perceptions and, through the act of making a poem, perfected them. I was for *feeling* as the guiding principle, not form or manipulations of speech.

A piece of Hawaiian song floated over from a radio on the beach nearby. It was tuned to my local station. I can't remember now, but maybe it was one of those nostalgic ones I loved that puzzled Hirsch so much. There was some wondrous guitar work, picking and plucking and strumming to an adagio pace. The group played in unison with a heavily strummed downstroke on the beat, a quick *down-up-down* triple-strum syncopation, and then the lead singer swung his voice between baritone and a lilting falsetto that dropped

two octaves back to baritone again, all in Hawaiian I did not under-stand, except for the feeling. To me, it was simply beautiful—the tapestry of guitar work, the weaving voice of the singer, and the powerfully stately pace of the song itself. I wanted to tell Ed to listen to the music, but caught myself when it drifted away or the station got switched to pop-rock. Instead, I invoked Wordsworth again.

"Isn't it a *feeling* that we're chasing when we write?" I asked. "Don't you brood and meditate till you get the song of it inside yourself? Till it's strong and won't drift away? And then you write like it's dictation, listening to that song inside you, feeling the music and the soul rising within, 'trailing clouds of glory' like Words-worth said in 'Intimations of Immortality'?"

The Hawaiian music and the English poetry I loved represented a legacy I hoped would adopt me, a child taken away from his birthplace, risen up reveling in an aggressive urban culture that was the antithesis of the gentle songs of the islands, brutal opposite to the sweet airs of English poetry. They both represented aspirational worlds to me. Like the music I sang to myself in college to soothe my rage while I walked from the dorms to classes, from East Asian Religions in Pearsons Hall to the Gibson Dining Room: I let voices murmur and reverberate, inviting the arcane language of poetry into my heart, hoping to invoke memories and inspire a mood, lifting my lost spirit into a kind of loyalty, asking the wind sighing through cane to adopt me back, imploring generations of good, gray poets to cover my forehead with their colonial hands, myself *hanai* to land and culture, one of the lost asking for welcome.

The bright day was almost gone. Beachgoers took up their bags and kits, their bottles of sunscreen and leftover cans of soda. Squealing kids zoomed past us, chasing each other, tossing blue and pink flip-flops in the air, kicking up sand that, blown by the wind, dusted over our eyelids, shoulders, and chins. I said to Eddie we should go. He grinned and nodded, joking we were bad tourists, having skipped swimming and diving for the mutual seminar we'd conducted on the beach—"The Seminar at *Wye-ah-Meeyah,*" he said. He pronounced it like the Beach Boys did in "Surfin' USA."

We turned our backs to the big blue waters we'd never breached and trudged through thick sand back to our rental car. Next, we'd

go to Kahuku some miles away, my home village on a sugar plantation where I planned to take him to the beachside graveyard of my
ancestors.

Ancestral Voices

About ten years ago in 2010, I heard a cane worker's plaintive song
on a recording sent to me by Franklin Odo, a friend who is a scholar
of Japanese American history. He was researching what would later
become his book-length study *Voices from the Canefields: Folksongs
from Japanese Immigrant Workers in Hawai'i*. The song was called *horehorebushi*, meaning "song of *hole-hole*," a Hawaiian word referring
to the action of stripping cane leaves from the stalk with a machete.
Four generations ago, immigrant Japanese pronounced the Hawaiian as *horehore*, and its suffix, *bushi*, is from Japanese, meaning a tune,
an air, a knot of wind. Odo had found a source, an older man named
Harry Urata, who'd made it his calling to collect hundreds of songs
that were sung by these first Japanese immigrants who worked in
the Hawaiian canefields. Urata, like my maternal grandfather, was
born in Hawai'i but educated in Japan, returning to Honolulu for
high school just as war broke out in 1941. He'd made his living

as a music teacher, bandleader, and radio host, but late in life (as Alan Lomax had done for the blues and field hollers of the Black South), Urata found a folklorist's passion for going around with a tape recorder to retirement homes, asking residents to recall the songs they'd sung in the sugarfields. Urata's songs haunted me, fixed a feeling in my blood about that first generation of my own people, stooping to do hoe-work, chopping at the dirt, and swinging machetes to trim leaves from green stalks of cane, raising mournful voices that twisted in the hot air like fists of wind.

> *To Hawai'i, Hawai'i I came,*
> *Land of my dreams.*
> *But I grunt like an animal*
> *And my tears fall*
> *Amidst endless fields of cane.*

(from the Japanese, translation mine)

The tune was simple, sung almost haltingly, in a spiraling but descending line to a rhythm that I recognized from tunes my grandfather had sung during my childhood. It started at a slow, grieving pace that was almost funereal, then had a hitch, a kind of catch in the middle of its verses, where the melodic line would pirouette from its descent and climb way high in a pentatonic interval before cutting itself off abruptly and dropping to the next line of verse, quavering out a long vowel. It put a grip on your blood like a wraith was squeezing the tough muscles of your heart, then let go just before it stopped, so you felt a flush in your veins, a spot of blue terror, your soul suddenly clarified and made serious like the rain drilling down on a tin roof. It stopped just before you began to cry.

> *I grunt like a beast.*
> *I hack and I slash through canes.*
> *I trample them under my boots.*
> *But, in the evening, when crickets sing,*
> *I keep, just for them,*
> *An island of stalks I never cut, their leaves still whispering*
> *In winds that are soft and not cruel.*

(translation mine)

The songs told stories in just a few lines, most of them depicting lives of hard labor and disappointment, a nostalgia for Japan alongside invoking the items of their lives—wooden lunch pails, ditches where the water ran, knives and hot sun, the sticky pitch from the cane soaking their gloves, the needles of leaves getting under their skin. I thought of these songs as my own blues, a legacy of deep feeling even through the pains of hard labor. They came in a creole of Japanese mingled with words from Hawaiian sometimes, or sometimes in English, all pronounced in the Japanese way, sung to set tunes that the *gannenmono,* the first workers, brought over from their native chores of planting and harvesting rice, tea, and millet in Kumamoto, Fukuoka, and Okayama.

I imagined women in gingham dresses, straw hats, big aprons, and long gloves stripping the cane of leaves, *hore-hore,* flailing with hoes and machetes, their hats like baskets covering their faces, the thick work gloves drenched with sap and oil. In my mind, I saw a row of cane and a *hippari*-man, the pacesetter hired by the plantation, who rushed angrily down it, braying out a nag, a scolding song full of insult to the women:

> *Faster, go faster, you whores.*
> *Stop your grousing.*
> *You can't do honest work*
> *With your mouth!*
>
> (translation mine)

Then the women chanted back, throwing chaff and sticks over the tops of the cane at the *hippari*-man, trying to slow him down. And the insult they sang back to him, each from their own row of cane, was sly, direct, and not gentle:

> *Why should we keep up with you?*
> *You're a damn sellout.*
> *You're the one who gets paid*
> *For working your mouth!*
>
> (translation mine)

My mother's people came from southern Japan—in 1888, my grandfather always said, talking about his father's passage. The family came from Hiroshima, Fukuoka, and Wakayama. They were laborers and strike leaders on the sugar plantations in Hawai'i. They were a dancer, a boilermaker, a country storekeeper, and drivers of mules in the canefields. They were a kind of golden mystery to me for a long time. There was no brief of their passage. No record that I knew of that marked their stories down for me to inscribe myself within them as the living image of their ancestral shadows.

I knew only a few facts about Hideo Kubota, my maternal grandfather, and his early life as a shopkeeper on the North Shore of O'ahu. He was born in Hawai'i in 1899, then sent back to Japan at six to get an education in Hiroshima at a military school. He returned at sixteen, proficient in mathematics, martial arts, and calligraphy. He got a job as a stock boy at Tanaka Store near the Castle & Cooke sugar plantation in Kahuku. He rose to the position of clerk and wore dress shirts and bow ties over his khaki trousers and boots. He was given a store of his own in the village of Lā'ie nearby when he was twenty-three. That year, he married my grandmother,

Tsuruko Shigemitsu, who was seventeen years old. They lived in that village throughout most of their lives.

When I was younger, in my teens, hanging around the house in Gardena, that postwar Levittown of Los Angeles, doing chores or escaping homework, I'd snap on the television and Hop Sing would be there in living color, on *Bonanza,* doing his chop-suey English thing, catering to his bosses the ruling Cartwrights, making a damn fool of himself—and me too, I thought. Oriental minstrelsy, *chop-chop.* Or Peter Tong would come on in *Bachelor Father* with his houseboy act, garbling phone messages, *Ikallupusutay,* gooneying for the camera. This was the early sixties, and being Asian was a joke in America—a sidekicking, demeaning one. I saw it all. I was full of rage.

I wanted a story about my grandfather then. I'd ask my mother, or my aunt when she was visiting—they were both in a better mood whenever they got together, laughing and carrying on and talking pidgin and remembering the old days in Hawai'i. "What was Papa

like back then?" I'd ask, using the name they called him. "What were his routines? How did he know calligraphy? How come he kept so many Japanese books? What did they say? What were they about?"

"We don't remember," they would answer irritably. "It was a long time ago. Who cares about that stuff, anyway?"

It wasn't long after that I decided to dedicate myself to the study of art and literature. It would be as if I were an apprentice in some religious practice, laying down the foundation of learning in letters and values both spiritual and moral that I would draw upon in later days. My yearning was intense, I thought, and my devotion almost absolute. I read and I read and I read, only and all the time. I was away at college, of course, and I was overjoyed. I had escaped the noisy house of my upbringing in that Gardena neighborhood crowded with loneliness and working-class anger. I read fiction in the afternoons out on the porches and lawns of the campus and poetry in my room after supper in the dining halls. After midnight, with the company of a cup of wine, I practiced calligraphy. In the mornings, I studied languages and read as little as possible in my science texts—the explanations and charts therein befuddled me. I was for literature and that was it. On Botany field trips, I carried Shakespeare's comedies with me and read them on the bus as we swayed along small roadways through the California woodlands, spring lupines, purple brushes of salvia, and yellow buttercups burgeoning under the oaks and in the fields around us. Strolling the dusty campground at night, I recited the love lyrics of Thomas Campion and made eyes at the fires of my own learning.

It was, alas, a somewhat hermetic experience. I began to long for things: for companionship, of course, the true thing always elusive—*When to her lute Corinna sings, / Her voice revives the leaden strings*—but also, and in the most earnest way, for ancestry, for a sense of descent from noble things, not only from a people, as was being chronicled for me in the Mississippi novels of William Faulkner, but from a tradition of thought, of speech without desperation or the angry pollutions of human affairs. I who was so filial to the texts of my studies, so observant of the mores and principles both articulated and implied in Boethius and John Gower, was beginning to reflect on the contradiction that I, with my southern Japanese ancestors

arrived so recently, was in no direct way tied to these authors, however great, to that Western cultural tradition.

I had, quite simply, a profound rage for story, for a counter tale that justifies, in the powerful way that literatures do, my own presence in my own time in history. Obviously, I had realized that the literature I was studying could not account for that, that I was not being given "a national tale," a cultural identity that spoke to the convergence of global histories making me a fourth-generation American. Unlike the child in John Steinbeck's Salinas story "The Leader of the People," I could not be ushered into the sweet fiction that my grandfather had come West, leading the wagon trains along the Oregon Trail. And, with little direct knowledge, with almost nothing then available in the archive to study and learn about we peoples who came from Asia to America, I began to fabricate my own legend.

I invented a book. In secret. At first, I told no one, but I wrote that it was so in a diary of my own dreaming, as if it were a memory, though I knew I had not lived it, that the book did not exist. But I convinced myself it did. In my diary, I wrote that I found it when I was five or six, rummaging around in the basement garage of my grandfather's house on Kamehameha Highway on the island of Oʻahu near the town of Hauʻula. I had been exploring in the shelves alongside the polished green Chevy, careful to climb up on the floorboards, step on the seat, and push quickly with my bare foot from the dashboard and lowered passenger window on up so I could reach the high shelves crowded with things. My grandfather kept boxes full of spark plugs, rayon lure skirts, seashells, and beach-washed glass up there out of which he fashioned curios for the tourists. A crèche of toy hula girls in the polished half-shell of a coconut. A kind of Cornell box with *opihi* shells, chips of colored glass, spotted cowries, and a starfish.

There were a few books up there too, mildewing and coated with a fine, powdery black rot: high school yearbooks with teenage pictures of my mother and aunts in them, paperback adventure novels—Westerns mostly, and some with fake leather covers in red and green. Their titles were embossed in gold. *For Whom the Bell Tolls. The Great Gatsby. Anna Karenina. The Wisdom of the East.* But,

among these, the cache of my father's postwar subscription to a Reader's Digest Book Club, I told myself I found a storekeeper's ledger—a book made for accounts and balances, for inventories and expenditures, for profits and debts. Yet when I opened it, shaking off its dust, what I discovered instead was a journal—the diary of my learned, Neo-Confucian grandfather's—written in *sōsho,* a brush-written Japanese script like floating lotuses and reeds twisting in a swift stream, beautiful to apprehend but impossible for a child, or any modern, to decipher. I turned page after page, running my fingers down each neat, ribboning column, and I knew that this was sacred, a book of profound secrets.

The pages were yellowed on the edges and blackened near the spine where the stitching was, and they were soft, swollen with moisture from the nearby sea. The writing, composed of stylized ideograms and a linked syllabary, looked to me vaguely like the sutra scrolls the Buddhist priest chanted over at Kahuku temple, only finer, less like rows of black spiders and more like the surface swirls and eddies of Hau'ula Stream as it raced under the WPA Bridge. Another time I took it out, it looked like the banners draped over the sacrifice of *saké* barrels and *shoyu* on festival days, the headbands and flowing sleeves of the evening, fire-lit dancers on Bon-Odori, the Day of the Dead. The book crackled with esoteric energy.

I stole it. And I kept it with me throughout my days.

From that time in childhood when I snatched it from the garage shelf, through the move from Hawai'i to the Mainland, through Boy Scouts and juvenile gangs, football and girls, I told myself that I'd kept it. I called it the *kagami nikki,* a title I invented from what I knew from my rudimentary studies of Japanese literature. It meant "The Mirror Diary" and had the ring of medieval essay collections and Tokugawa travel diaries I loved so much. My vow was to become scholarly enough to read it one day. And, when I did, when I had trained myself properly and was ready, it would tell me, like the murmuring ghost of my own grandfather standing behind me in the bedroom's full-length mirror, the unshared secret of who I was and from whom I came.

For our homelands,
The far islands of the Rising Sun,
We try to soldier on,
Carrying the hoe on our sore shoulders
And not rifles,
Machetes and cane knives in our belts
Instead of short swords,
Hate brimming in our hearts
Rather than love.

(translation mine)

The *horehorebushi* Franklin Odo had sent me, though, came to me in middle life and was like an undiscovered blues legacy, mournful testimonies to the difficulties, not only of those first lives on the Hawaiian earth my ancestors had lived, but of their consciousness, how they felt about things, how much they yearned for a return home to Japan, then how determined they'd become that they would tough things out, scrape a living from out of the cane rows, the chaff and dust from the cane, smoke from the oil fires they lit to cure the stalks for harvest, all choking them. In their poetic language, the act of shouldering drooping bundles of cane from fields to rail cars was called *happaiko:* bearing the burden of the cane. When they took the quiet respite of a hand-rolled cigarette or ate their lunches from a pail by the side of a sluice where the air was cooler, they were surveilled by bosses who marked their breaks with whips and pocket watches.

In the brilliant light of dawn,
I wade through the fields.
My sharp machete gleams
To taste the blood of sugarcane.

*

I carry my lunch pail
From early in the morning,
Off to do hore-hore *work all day*
Just to put food on the table for another day.

*

I stop work for a break,
Light a cigarette for company.
Then the boss glares at me.

*

We drop our hoes,
We drop our knives.
We grab chopsticks
And sit by the sluice.

*

Lunchtime over, the boss whistles
Through his fingers
And the plantation cops crack their whips.

*

Who sings that sad song
That carries over the cane?
O, wind, take your cries
Back home to my mother.

*

I water the cane
And tears flow down my face.
I work without a break
Till the moon catches on cane leaves.

*

Cutting cane is death's work.
Thirty-five cents a day for three years . . .
At night, only my elbow for a pillow.

*

It was deep night when I waited for you,
Hidden amidst tall canes.
Then, the moon sailed like a silver boat
Through the leaves of coconut trees.

*

In this Kingdom of Hell
Lives are counted by the clock.
But when I come at night to you,
I cross the River of Heaven
To a Kingdom of Dreams.

*

In the worst of times,
I've at least a pair of straw sandals—
One foot for me,
The other my wife's.

*

Let me sleep if it rains.
Let me lie down.
Let the sky fill with clouds.
Let me drink till I drown.

(translations mine)

If you listen to a lot of them, the tunes are simple, angular, and sound a lot alike, except for odd, rhythmic accents now and then, great, emotional leaps of pentatonic octaves like a yodel or a coyote's cry. It's as though Thelonious Monk took a *saké* drinking song and twisted it into the brief measures of a dissonant blues. To each one there is an overwhelming tone of grieving, a sadness that rises from the gorge of a body up through the warbling throat that gives its muscular cry a tune carrying through the air so you can hear it, but not without feeling how it's wrenched from unrelenting travail and loneliness. These are songs of something beyond sadness, a combination of a chain gang's heaving woe and grievous lamentation.

I was interviewed for *Canefield Songs,* Odo's PBS special about them, and I refer to them as "the Buddhahead blues," clearly offending Jake Shimabukuro, the virtuoso *'ukulele* player from Honolulu, who was the narrator. He visibly recoils at the term *Buddhahead,* which, historically, derives from the Japanese-English creole term *buta*-head, or "pig-head," what the early laborers called themselves for their stubbornness and resolve. I don't think the young man realized that. He likely only heard a term that he mistook for a slur, something like *Chicano* may once have been for Mexican Americans before they adopted it as a term of honor. Like them, these first Japanese Americans, the Issei generation, the *gannenmono,* could not be broken, but would persevere, fostering descendants for six generations. When I call their songs "the Buddhahead blues," it is a term of tribute to their strength and endurance and an invocation of another

people's contribution to our common legacy as Americans. Their *horehorebushi* are songs that look forward to freedom and release.

> *With one willow trunk,*
> *A bachelor I came to Hawai'i.*
> *Sending for a bride by photo,*
> *The sting went away from cutting cane.*
> *Now I have children,*
> *Grandchildren too.*

(translation mine)

That the history is terrible rather than noble, that a varied literature does not exist, that my ancestors never wrote and no one bothered with them enough to transcribe their full lives into writing, provides me with the dark watermark of an absence that these songs and my own current dreams must fill. Over my adult life, I wrote poems of my own and books of my own, but the ghost of that old grandfather's diary—my imagined but necessary history, my invented pride—has slipped quietly away, without thought, as if it were a companion's hand I'd let go of after we'd swum out from shore to reef, and he was drifting away now, fairly quickly, caught by the swiftly receding tide and swept out to sea along with all my lost possessions and forgotten errantry. I've now at heart a canefield song instead, like the rising apparition of a moon that ascends a blank sky in the most brilliant light of day. In its verses, I can see cane workers arrayed around a locomotive, my great-grandparents trudging off their immigration boat in Honolulu Bay, a midday rainstorm drenching a village of thatched huts amidst an ocean of sugarcane. What words there are of theirs come to me in the real snatches of these few lyrics that have survived the canefields, *horehorebushi,* borne of unchronicled sufferings, sung by anonymous artists who imparted the trace of their labors to Harry Urata, who taped them, to Franklin Odo, who first translated them. Together they sing accompaniment to our early history in this country, the music and *meriyasu* for a walking tour through the broken rectangles of old plantation camps in the midst of stands of abandoned cane, along the spit of a sandy promontory, half-eaten by the sea,

studded with wooden grave markers, silvered by wind-wear and the sun, broken and rotting with more than a century of age.

Honky-Tonking in Nashville over Spring Break

In the spring of 2014, I was in Nashville to do a couple of events at Vanderbilt University and, first chance I got, headed off to the dives and honky-tonks of Music City's Lower Broadway. I'd been there ten years before, when the area was grittier and the clubs a lot seedier, but when I got out of my cab, I noticed things had gotten cleaned up and that the milling throngs were younger and hipper, more diverse, and more gleeful than I remembered. I headed for Robert's Western World, a nondescript place amidst the more glittery ones between Fourth and Fifth, smack in the heart of things. Its outside neon shill made the flamboyant claim that it was "Home of Sho-Bud, Honky Tonk Heaven, and Hillbilly Grill."

Robert's had a short queue and I made it through within minutes, finding a seat at a small, round table near the bandstand, along a side wall. I had to ask if I could sit, though, as there was an older couple on the bench seats across from me. It turned out they were seniors from Iowa, in Nashville expressly to enjoy the atmosphere and music, they said, as "there's nothing like this in Des Moines." They told me they came all the time and "just loved it." I bought them a round of Yuengling beers ("America's Oldest Brewery"), and we drank to the evening, settling into our seats under a gallery of signed photo-portraits of various stars who'd dropped by, as the Don Kelley Band, the second act of three that evening, took the stage.

It was a standard country-rock quartet—electric and acoustic guitars, stand-up bass, and drums—and the leader wore a Stetson hat pulled low over his eyebrows. He sang in a slight drawl I couldn't place and delivered a practiced patter between songs, goosing the set with a few canned jokes and shouts of "Thank you, hillbillies!" to flatter the crowd. The set consisted mainly of country standards I faintly recognized—George Strait, Johnny Cash, et al.—and just as I started feeling bored, despite the band's superior musicianship (*everyone* in Nashville plays well—they'd better), I heard rousing guitar chords strummed with authority, the *gut-thunk*

of a kick drum, and some familiar, cranking notes from the lead guitarist on his Telecaster. "Nineteen sixty-eight, my friends," yelled Don Kelley into the mike. It was "The Weight," that great and mournful ballad from the Band and their monumental *Music from Big Pink* album. Back when it came out, the song conjured up an antique ambience in shades of a melancholic narrative that, once I grasped it, became a crepuscular world I felt I could stand within to view the history of my own people, those contract laborers who had left me no sto-

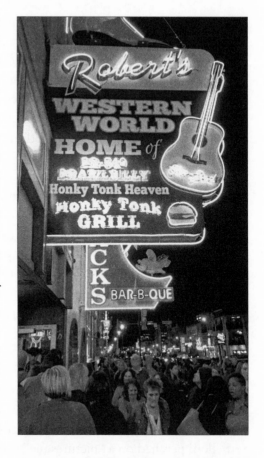

ries, only the relics of their shadows bobbing in between stalks of bending cane. The crowd around me applauded and shouted, and I felt a great momentum of feeling as the guys in the band took turns singing the bible-tinged verses. *This* was their music, not the country standards, it seemed to me, and they put their shoulders to the wheels of fire in the song.

The rest of the set just took it from there—inventive covers of a Waylon Jennings, a rousing Allman Brothers extended blues, then "Heaven" from Los Lonely Boys. I had a helluva good time and the place just rocked, the crowd swelling to hundreds packed into the long rectangle of the high-ceilinged bar. It was the beginning of spring break and there were lots of college students, keeping things lively. A few couples danced, Texas swing style, cramped up against the bandstand, and I heard blasts of a hog-caller's whistling cut

through the electric air. Around midnight, I decided it was time to go, said good night to the gentle couple from Des Moines, and started to pick my way through the throng.

I swear, as I left Robert's that night, I brushed by the prettiest girl in the bar. I confess I'd noticed her earlier when she came in and crossed in front of the stage—a slinky twenty-five or so, with ash-blond hair. She was saying something to me with drunken eyes.

"Can I kiss you?" she asked.

Well, I don't know what I said, but it must've been *Yes,* because she did kiss me—smack on the lips. I took a half step back, reeling a bit, taking in her look of boozy sincerity, then leaned toward her and asked, "Was this a goof or on a dare?"

She shook her head. "I was just at your reading last night and saw it was you over there." She pointed to where I was sitting. "You were *so* great."

Okay, I said to myself. *It's finally happened. Love for poetry!*

Bellagio

Some years ago, in late June 2005, I arranged to have a week in Florence, exploring its cultural splendors, and then a month's residency at the Rockefeller Foundation's Bellagio Center, the grand Villa Serbelloni perched on a "picturesque" promontory (both the classic Baedeker and contemporary Blue guides agree on this) overlooking the conjunction of lakes Como and Lecco in Bellagio. This was *the* Bellagio and not its Vegas mockery. The villa itself, a sumptuous, by-application-only hostelry for a crowd of international scholars and artists, consisted of a dozen suites with a private bath and study attached to each one. There were maids that gave us turn-down service (a brocaded duvet decorated my bed), a staff of chefs who prepared our meals (dinners lavish, lunches and breakfasts basic, always *buono*), waistcoated waiters who served us, a parlor with a grand piano and views of the lake, a listening room for recorded music, a useful library, and a computer room, plus office staff both brusque and polite. On the same grounds (they ran from a wood on the top of the cliff down to the lake, almost like a small Hawaiian *ahupuaʻa,* being a political and economic fiefdom derived from

the climatic subdivision of island land from mountains to seashore) was another structure—a former stable, called "The Maranese," just below the villa and its terraced flower gardens, which had been magnificently renovated with three suites as poshly appointed as the villa's, plus fireplaces. I was assigned to one of these.

I'd been there about a dozen years before and had finished a memoir, in poetic prose, about my long sojourn, during my thirties, back at the island of my birth and the exploration of my roots in Hawai'i. Bellagio is an inspiring place, with grand views of the lakes and sub-Alps from each balcony, and the knowledge that scores of

other writers have composed works of distinction there. Shall I just say that the scene from a famous wartime novel of a gallant Sikh and a Canadian nurse (who becomes his lover) hoisting and rappelling themselves via block-and-tackle alongside the wall of a bombed outbuilding so that they could view a magnificent medieval tapestry that hung there seemed to me likely based on the relict of a precious fresco still visible within the ancient, calcareous walls of the apse to an abandoned chapel on the grounds of the Villa Serbelloni? During that first stay, I read the soon-to-be-published novel in uncorrected, bound galley proofs given to me by my editor, himself a Sikh, though not a *dastār*-wearing one.

And what did I accomplish during this repeat posting at the most sublime of brick shithouses? Only the finish to my third book of poems, after a gap of some twenty years since my second, which had given me, such as it was, my early reputation. I worked happily in my studio there, reading my drafts with an afternoon cigar out under the grape arbor alongside the three-suite manse. My company was a doctor from Australia studying mental disorders, a professor of Arabic literature at Oxford, two political scientists, a dashingly handsome expert in international conflict resolution from Argentina (where conflicts are always resolute), the chair of Neuroscience at Harvard Medical School, an independent filmmaker who had made documentaries on both "Rainbow Man" (an American folk hero) and the Weather Underground (decidedly *not* heroic), a fiber artist from Spain whom I thought pushy, a fiery Russian journalist who gave great toasts and speeches, a former assistant attorney general under President Carter (who was modest yet debonair), an esteemed Polish ethnomusicologist from Stanford who would next bunk at Berenson's I Tatti for a sabbatical, and, like two turtledoves a-courting, a Columbia philosopher and his Caribbean lover. When I arrived in late June, the weather had been stormy and cool, with thundershowers over the lake and mountains at night. But soon the weather cleared and the afternoons got so hot that every day, after my writing was done, I would tramp downhill through the olive grove and a field of lavender to a cutout beach along the cliffside of the lake below, taking a cold dip among outcroppings and the naked roots of pines furling like tentacles emerging from slate-

gray piles of talus and broken boulders along the eroded shoreline. Afterward, on my return uphill, I'd wish someone would be there to shove me in the back, as though I were a stalled Smart car getting a push-start from behind. The hill was redolent with lavender, and looking back, I may have reached a kind of culmination there, surrounded by so much scenic grandeur and opulence of quarters among acknowledged intellects.

It was grand to take lunches on the terrace of the villa, sit with so many interesting minds and personalities. Trying to make me comfortable (as he knew I loved British Romanticism), the Oxford professor, a Bedouin from Alexandria, spoke of Coleridge and his creation of "Xanadu" in the poem "Kubla Khan," a distillation of colonial clichés regarding the Orient. But I gently turned our talk to Arabic poetry, of which he was the primary authority in English. The former assistant attorney general liked talking about the run-up to *Brown v. Board of Education* during the civil rights movement and what white wines went best with Italian fish. I suggested anything from Alto Adige. With the Stanford musicologist, I took long walks through the pathways of the villa's gardens and discussed postwar Polish poetry and its revival of the human spirit in the aftermath of the almost complete destruction of their civilization. Over dinner in the formal dining room, he let me ask a question about Mozart, and graciously spoke of how the composer had invented modernity through the concept of linear time in *The Magic Flute,* his fanciful opera. My old professor, who'd taught me how Sir Thomas Wyatt and the Earl of Surrey had invented the English lyric and its prosody through their translations of Petrarch's love sonnets, spent his afternoons amidst the olive grove overlooking the lake, drawing plein air landscapes in charcoal. When I stopped during one of my walks and remarked on his sketching, he confessed he was trying to imitate chiaroscuro technique in the Italianate landscapes of Jan Both, a seventeenth-century Dutch master.

As when I was in college, these conversations inspired me, and I wanted my poems to reflect as much learning and love of culture as I received from my company at afternoon drinks and around the dinner table. I envisioned my maternal grandfather, Hideo Kubota, imprisoned in the stockade of a Navajo trading post by the Justice

Department during World War II. In what I imagined might be his voice, anguished and seeking advice and solace, I wrote verse epistles to the poets of political incarceration. Kubota wrote to Miguel Hernández, dying of tuberculosis in a Fascist prison in Madrid. He wrote to the Chinese poets of Angel Island. He wrote to Tadeusz Różewicz, a Polish partisan who fought the Nazis, who returned from war to write a simple poetry that praised bread, a table, and the human spirit. I wrote dramatic monologues of a man like my father, traumatized by war, who comes back to the islands and paints, on the walls of a beach bathhouse, a fresco to commemorate his love of Hawai'i and all its varied peoples. In yet another poem, in my own voice, I recalled my resolve, at twenty, while staying at an aunt's seaside cottage on the North Shore of O'ahu, to swim out to the far reef of the lagoon before me. Never strong in the sea, I thought that if I could do it before the sun rose over the seam of the horizon, I'd become a poet. At Bellagio, it felt as though I'd come to fulfill a youthful promise to myself, that I'd reached that reef and received a vision of the flaring disc of the sun.

I had pre-planned an excursion to nearby Milan for the middle of my month-long stay at Bellagio. An officer at the American Consulate had helped me get a prize ticket to La Scala, the most famous opera house of them all. And I would visit Il Cenacolo Vinciano too—da Vinci's *The Last Supper,* housed in the refectory of the Convent of Santa Maria delle Grazie. I'd secured a good hotel (three stars) just a block away. And visits to the museums Pinacoteca di Brera and Pinacoteca Ambrosiana and Il Ristorante di Cracco Peck for a modest five-course, noon meal . . .

"Is *La Bohème* all right?" I remember the officer writing me.

Though I'd heard of it, I'd actually never seen it and didn't fully know its story, so I counted on the program synopsis, supertitles, and the music to carry me through to an appreciation of its musical character and narrative.

But how could this be? Well, I came to operas very late, as I was in my forties before I started listening to recordings from them—mostly CDs of select arias, or "compilations," as they are called. I'd read a review in the *WSJ* about a recording of castrati arias (no less) from the operas of Farinelli, himself a castrato of the late baroque who composed his own music. For some reason, this piqued my curiosity, although I missed it that the music arose, so to speak, from a severed root. In short, I flat *loved it,* playing the CD again and again on my car stereo wherever I went. I had been trying to buy a house and meeting with brokers to shop for a good mortgage and the stress of it all, with rates changing every day, with the various loan configurations—thirty-year fixed, 100 percent with mortgage insurance, seven-year ARM, balloon—dizzied my mind and made me feel as though I'd downed a bottleful of Ritalin or gotten sixteen shots of epi in my blood. The arias soothed and calmed me down. And I got such a crush on the mezzo who sang them, Viveca Genaux, a gorgeous Alaskan woman of indigenous background. I imagined her singing in an ice storm, the shadows of a baroque ensemble behind her, sleet slanting in almost white-out conditions, her oval face (made up with only a touch of tasteful, earth-friendly cosmetics) peeking out from the cinereous fur hood of a silver fox

parka. Was it a stereotype? Did I *objectify* her in order to heighten the sexuality?

Oh, my Frantz Fanon, maybe so, but I was *hooked,* jack. I started looking for more right away. I bought Cecilia Bartoli CDs, Angela Gheorghiu, Kiri Te Kanawa, Kathleen Battle, and—quickly my favorite—Renée Fleming, our All-American Good Housekeeping Seal of Approval baby-cheeks coloratura soprano from Churchville, New York. Whether they were mezzos, light, spinto, or the aforementioned coloraturas, I did not care, nor did I discern much of the differences in their registers or distinguish among the characters of their voices quite yet. I just loved and reveled in their artistry, what athletic leaps, spins, jab steps, and slam dunks they could do with their voices. And not just the sustained high notes (as from the tenor's role in *Il Trovatore* infamously parodied by the Marx Brothers in *A Night at the Opera*), but the delicate pianissimos, the ululating roulades and tremolos, the exalted fermata and liquid glissandi, rapid recitatives, and the ominous larghettos before the mad scenes. But I hadn't this vocabulary then, obviously, so I simply gloried in the exquisite ornamentations around a melodic line that I hadn't attended to in any of my listening before. I who had grown up around a diner jukebox in Hau'ula, filled with Elvis 45s and hula music, who had danced the boogaloo and funky chicken to inner-city rhythm and blues, who had worshipped at the shrine of sixties psychedelic blues-rock and gone, occasionally stoned, maybe a score of times to hear Blind Faith, CSN&Y, CCR, Quicksilver, the Grateful Dead, and the Mothers of Invention, had become a stone-cold *opera freak* if not an aficionado.

Or, should I say an *aria freak,* as, at the time, I'd only taken in two full-Monty, live opera performances—a student/faculty-issued *La Forza del Destino* while I was studying for my MFA and a decidedly *meh* performance of *Le Nozze di Figaro* at the Dorothy Chandler Pavilion in Los Angeles when I was visiting faculty at the Claremont Colleges just after the turn of the millennium. I was as innocent of operatic glory as was Mary of sex and man at the Annunziata, when the Archangel Gabriel came to deliver the news that from her virgin body a son would drop and he would be divinity on earth.

La Bohème *at* *La Scala*

Under pyramids of lamplight, I crossed the Piazza della Scala and its smoothed cobblestones that shone like glazed almonds. Whispering jets of water spouted from the orifices of marble cherubim in the square's modest fountain. The ink-black sky was frosted with a rim of urban auroras and the façade of the theater glowed under yellow floodlights that seemed to give its columns and porticoes a clown mask of comedic expressiveness, morphing into the downward-turning lines of a tragic frown as I approached the famous opera house. I'd tucked my ticket in the vest pocket of my black crepe blazer, and so went directly to the short entry queue of Italian men and women dressed in semiformal wear. The women, mostly middle-aged and some older, wore light evening dresses or cocktail gowns with scarves, sashes, and even a fur mini-cape on one or two, their necks, ears, wrists, and fingers only moderately bejeweled. It was easy to spot the tourists—mainly Americans dressed in the colorful polos and light khakis of summer sports clothing. And teenage girls in polyester sundresses, boys with their hands in their

pants pockets, downcast, as though being lectured to. The Italians moved inside quickly, as though boarding a subway train, while the tourist women jabbered and glanced backward, looking after their trailing spouses and kids. I slipped in behind an elegant, dark-haired Milanese woman, middle-aged and gym-trim, who wore an off-white, form-fitting lamé dress with a silver cloth clutch and silver heels, something Bulgari gladdening her tanned wrist. My own casual summer clothing wasn't completely out of keeping, thank god, but could not compare to the faultless attire of the natives, none of whom were impolite enough to give me a second look. The hubbub in the marble-floored lobby was all about conviviality and expectation of *il parlare d'amore*.

The brilliant but demanding *maestro* Riccardo Muti had been deposed as *direttore* only a few weeks before, I knew, and the orchestra would be led by the veteran choral master Bruno Casoni, a man in his seventies. But the program notes said that the set would be that of Franco Zeffirelli, famed for its lavish extravagance and attention to detail. "Overstuffed" was a popular descriptive shorthand for its sumptuousness, so I felt I was in for a treat.

The hall was much tinier than I expected, more a tower than an expansive auditorium. I'd not been in any European opera halls before, and La Scala's stunning intimacy was a revelation. I took my seat, mid-hall orchestra, about the sixteenth row, and marveled at how *close* I—and everyone else—was to the stage and musicians in the pit. From my vantage point, I could see about a third of the orchestra—the first and second violins, violas, and woodwinds. But what took my breath away was glancing upward. La Scala was like the interior of a red velvet wedding cake—six tiers of sashed boxes and buntinged balconies rising up above me, filling slowly with concertgoers ambling sideways like parakeets on golden rails.

Casoni walked in to the applause, took up his baton, and I could see him glance to his left at the concertmaster, not quite visible to my view, before he deftly dipped his baton to cue the music. His face was as expressionless as the music was thrilling. It opened with thundering fanfares from the horns, loud and blaring as circus music, and then sailed into sprightly strings and woodwinds for a

few bars before the singing started. The curtains lifted and the audience gasped and applauded lightly as the stage came into full view. The set was an expansive flat full of mismatched furniture—worn tables, easels, a desk, dining chairs, and a love seat upholstered in faded brocade. It emphasized bold angles and a skyline visible above the fancifully incomplete interior walls of the building. There was a kind of visual symphony of ochres, browns, and siennas dashed with bold strokes and banners of black that crossed the set like structural kelsons above the action. Two men in scarves and tatty overcoats shuffled before a wooden stove.

If you are—as I was then—ignorant of the story, *La Bohème* is Giaccomo Puccini's tragicomic opera based on Henri Murger's story collection *Scènes de la vie de bohème* (1851). The libretto, written in a *verismo* style by Luigi Illica and Giuseppe Giacosa, tells of the life of a small group of Parisian bohemians barely eking out a living in the Latin Quarter, banding together for companionship and love, self-celebrants of their youth, devotion to art, and each other. The male principals are romantics—a philosopher, a musician, a painter, and a poet. The two women are a demure seamstress and a coquettish singer. But aside from this brief synopsis and the character sketches in the program, the opera came as a completely new thrill and mystery about to be revealed.

And what is this *verismo*? It's characters with tuberculosis and others who simply struggle at life. Think Mimì, the seamstress in *La Bohème,* in fact, or Violetta, the courtesan in Verdi's *La Traviata,* both tubercular and cursed. *Verismo* is an early-twentieth-century style of plebeian opera (as opposed to operas regarding rulers and gods) that came into vogue just after the period when literary naturalism arose, the kinds of stories told by French greats Gustave Flaubert, Émile Zola, and Guy de Maupassant, and American Stephen Crane. They featured prostitutes, coal miners, and young toughs up from the streets, ingénues getting and losing their chances. In painting, I suppose this is analogous to the pre-Modernist, "ashcan" school of artists who depicted common urban street scenes with an eye for the grit as well as the beautiful.

So I followed the opening scenes of the play—somewhat

wooden, with lots of babbling and light amusement—as Colline, the baritone who was the philosopher, and Rodolfo, the tenor poet, tried to warm their cold flat. It was Christmas Eve in Paris and Rodolfo was lighting the woodstove with the manuscript pages of his own play, which he'd charitably donated to the cause. After another roomie, the painter, entered the scene and their stumble-bum landlord arrived to collect the rent, hijinks ensued as they evaded the payment through trickery and confusion. It was kind of a Marx Brothers gambit, the three bohemians dispatching the landlord, unpaid, just before their roommate the composer showed up with money from the sale of one of his compositions. They all decided to celebrate and repair to their favorite watering hole, the Café Momus nearby. But Rodolfo had a review to complete and he stayed behind, telling them he'd join in later. Up to this point, the opera was a blend of the bare mechanics of drama (the introductions of characters, the narrative setup, etc.) and some serviceable, even humdrum, comedic arias. Or so, in my arrogance, was my thought.

Well, then magic struck. Mimì knocks at the door as Rodolfo is trying to write. Rodolfo answers and Mimì tells him her candle has gone out and that she cannot see her way to her flat—will he help? He sees how beautiful she is and plots to keep her talking. She holds up her key and then, by accident, she drops it on the ratty carpet. Rodolfo sees his chance, blows his candle out, and plunges them in a darkness that is pretend (as the stage is lit, though they act as though it's not). The two of them get on their hands and knees to search for the key. This business goes on until, patting over the floor, Rodolfo's hands brush fortuitously against Mimì's. When she let out a startled gasp, I heard a single plush note from a bassoon, and then, almost simultaneously, but just a hair's breadth thereafter, a harp was delicately plucked. In the theatrical murk of Zeffirelli's stage garret, the poet took up the seamstress's hands, pronouncing them so cold, he wished to warm them with his own breath. They got to their feet together. Italian tenor Massimiliano Pisapia sang Rodolfo's part, and his aria, like the set, was also a little overstuffed, slowly taking over the entire stage and then the auditorium as well, swelling with both plummy and ringing notes as Rodolfo declared

his identity and soul's ambition, rising up past every tier, I swear, like a pilgrim moving upward through successive rings of heaven.

I stole a glance at the woman, thirtyish and American, sitting next to me and saw that a crawl of supertitles was making a stately traverse over the seatback LED screen in front of her. Its little etchings of light threw small banners of clouds that fell like cuneiform across her face. I looked back at my own screen and read Rodolfo's words praising poverty and a resolute devotion to art—the poet's apprenticeship. The moon, a spherical cutout of paper or plastic, floated slowly above the flat and the surrounding skyline in time with Puccini's sumptuous music. Its angular blaze, cast by stage lights, cut across the floor where the two lovers-to-be stood and held each other.

At this moment in the opera, the Italian lyrics are simple, humble, and magnificently to the point. Rodolfo says that people often ask, *What are you?* And he declares, feet planted firmly on the stage, looking straight into Mimì's dark eyes, that he is *un poeta*—a poet. He speaks his lines with chest-filling pride and sonorous, almost burnished notes (without accompaniment except for punctuations from the bassoon), poised between declamatory recitative and sung cantabile (the slow part of an aria).

He says people ask, *What do I do?* And he says, *I write. But how do you live?* people say. *I live!* Rodolfo declares, boldly, Pisapia's voice sounding heroic and emphatic. And as the tenor sang Rodolfo's lines, a harp's tender, isolate notes fitted themselves between lush swellings of the orchestra. The music seemed nothing but a grand lagoon upon which the burly Pisapia had cast himself, gliding in song like a gondola across a calm canal, his passionate voice sailing before him like the reflection of the moon as he leaned out from the bark of his body, trailing his hand, scudding the waters with the palm of romance and the fingers of poetry clasping and unclasping together. *I waste my rhymes like a rich man does his castles,* he sang, now fully into the aria. And I felt a shudder growing within me like the faintest starts of grief or arousal.

Rodolfo declared Mimì had stolen all his castling dreams from the air:

Talor dal mio forziere
ruban tutti i gioelli
due ladri, gli occhi belli.
V'entrar con voi pur ora,
ed i miei sogni usati
e i bei sogni miei,
tosto si dileguar!
Ma il furto non m'accora,
poiché, v'ha preso stanza
la dolce speranza!

(from *La Bohème*, libretto by
Luigi Illica and Giuseppe Giacosa)

These lines in a poetic Italian, combined with the soaring, gleaming notes of Rodolfo's aria sung over the softly pummeling toms of the timpani, barely audible under the singing, had a powerful effect on me. It all grabbed me up by the spine and shook me until all delights and details of the touristing day vanished like vapor, my emotions welling up like waves, cresting with each rousing note of the music. I swear to you I wept like a baby and couldn't stop.

When Rodolfo declared to Mimì he was a poet, who painted the ascent of the moon in words, I thought to myself, *Oh, that's just what we do!* It was momentous, and I've marked it since that time as though it were my own rebirth. All that had been dormant in poetry and passion before that night blurred into these songs of La Scala, reverberant in the hall and within my heart. But the scene was not yet over.

"Now that you know me, it's your turn to speak," Rodolfo sang, so plainly after the magnificence of his aria. It was like a little break, a brisk piece of prose after the poetry.

"Who are you? Will you tell me?" he said.

"My name is Mimì," the soprano answered.

It was the gorgeous Hei-Kyung Hong, a regular at the Met, I'd read—a slender Korean woman with porcelain skin and an oval face who sang the part with a voice so expressive, even her whispers were tuneful, bathed in pathos.

"I embroider silk with false lilies and replicas of roses."

She bowed her head and placed a hand over her heart, then looked up into Rodolfo's face. Her aria moved between the demure and sprightly to something more grandly expressive, as romantic as Rodolfo's own.

"I love all who speak of chimeric things with their fragile and poetic names," she sang, and continued:

> *Each day, alone in my whitewashed room,*
> *I make the same lunch and look upon the roofs of heaven.*
> *But, when the spring comes,*
> *April's first kiss of sun is always my own.*
> *A rose blooms, and, petal by petal, I watch its slow birth.*

(translation mine)

This is what Hei-Kyung Hong sang as Mimì, her words phrased modestly, as opposed to the bravura, even boastful aria of Rodolfo, but they emerged from the same lyric source—an ocean of utter romance and devotion to seeing the world *as it could be* rather than as it was. It was Quixote seeing a golden helmet where Sancho Panza saw a barber's bowl. Through it all, what was common to the newfound lovers was the praise of poverty and *il parlare d'amore*—the talk of love. I thought of my own apprenticeship, so many years long that my gentle father did not survive to witness its completion or my becoming what I'd once said to him I'd be.

O soave fanciulla! Roldolfo sang. "Oh, you lovely young lady!"

Ah, tu sol comandi, amor! Mimì sang in response. "Love, you alone command me!"

Their voices soared like gleaming comets over the hall, and the singers left, sauntering arm in arm through a door at the rear of the stage, fulfilling the conceit that they had walked from the dark apartment and down a narrow street of the Latin Quarter. Sifting lightly from above, stage snow began to fall across the illuminated face of the paper moon and their ecstatic voices faded through the wondrous facsimile of a wintry ether.

The setting, the music, and lyrics were too much. A sob shuddered through me with every high note, with each swelling pulse of music from the orchestra. I could not believe the glory of being

there, amidst such direct and yet canny declarations about the power of love to rescue us from poverty and sorrow. Then, like Dante on his pilgrimage, overcome with revelations, I swooned and fell into a shallow sleep in the golden arms of the opera house.

Mozart's Great Mass in C Minor

The warm breath of the orchestra's strings captivated me from the first strains of Mozart's Mass in C Minor, recorded on CD in 2006 by Le Concert d'Astrée and its Choir, directed by Louis Langrée. Then, when the full choir came in, singing soaring notes on "Kyrie," I felt a spark of awakening as though a comet had just silvered across a dark and sleepy summer sky above me. It was a brilliant fire, shaking me from the vagary of a drowsy love of music into something more alert to its every pulse and shard of sound. And when the percussionist thwacked his bass drum to punctuate the score's opening measures, it was like the head monk's *keisaku* stick that smacked my shoulders at meditation in Ryōanji when I sat there one evening in early summer. Sound was made pure, the choir's singing sailing through the air and into my body, the tomming drums accelerating my blood.

> *Lord, have mercy on us.*
> *Christ, have mercy.*
> *Lord, have mercy upon us.*

Then, soprano Natalie Dessay commenced her sinuous vocal over the decorous orchestral accompaniment, her voice furling like a banner of light, note to note, aflutter with feeling. She sang slowly only two phrases, *Christe eleison, Kyrie eleison,* again and again, raising her pitches in shimmering arcs as voices from the choir swelled and faded, until a moment when she briefly hovered on a quiet note and then dipped away from it like a hummingbird escaping the bell of a flower.

These were merely the opening few measures of the great, solemn mass of Mozart's, a composition he wrote, scholars say, in response to his discoveries of the masses of baroque masters Bach

and Handel, as a vehicle for his new wife, the soprano Constanze Weber. For me, it was a different order of *being,* listening to this music, even on my home stereo. Uplifting just doesn't cover it. Like hearing Rodolfo and Mimì at La Scala, it was life-changing, introducing me to a major human glory I could not get enough of. The singing broke all the little bones of the nasty dragon in my heart every time I heard soloists Véronique Gens or Dessay lift their gorgeous soprano voices. They turned his bitter fire into tears of honey.

Crass as I was, I'd tucked the CD into a wallet I toted around my very first audio show—the Consumer Electronics Show in Las Vegas in January 2007. I was going from room to room, making the Mozart mass my "demo" music, asking the poor vendors who'd welcomed me to their showrooms would they please play it? With that music, too esoteric for the usual audio show buffs, I not only chased away the small crowds of dealers and other audio vets that may have gathered, but I strained the patience of the vendors too, whose normally fabulous systems were not always set up to play such complex recordings. Though a trade show system might do wonderfully with acoustic jazz, solo piano, or electronic rock, none of these put the kind of strenuous demand on it that a grand mass does.

Nobody kicked me out, but I started getting the message when, with strained smiles, vendors would turn their attention away from me, forestalling any conversations. Unknowingly, I made myself fairly unpopular at that first trade show I attended.

But the Mass in C Minor was like cake for at least one system I ran into. It was in one of the last rooms I found, a large, enclosed meeting area in the Convention Center, in a suite of rooms just off the big hallway and its awful checkerboard carpeting of puke-yellow hoops over dark gray squares. When I entered, only the attendant was there, an importer of European audio gear who'd once been a jet pilot in one of the services. A courtly, blade-faced man, he asked if there was anything I'd like for him to play. He had lovely bossa nova music on, something by João Gilberto as I recall—pleasant, room-filling, and untaxing of an audio system.

The demo equipment he had was all new to me at the time—large piano-black Darth Vader–looking speakers made in Germany,

called Ascendo Zs; on a stand on the floor, a huge chrome silver space cruiser of an amplifier with a double-rowed flotilla of eight output tubes along each side; and then an audio rack behind the amp with a preamp and a disc player in it. The amp and preamp, a two-piece unit with separate power supply and control box, were manufactured by the firm Convergent Audio Technology—called CAT by audiophiles—and the player was an SACD/CD player made by Ayre. The latter two were American companies, so I assumed their gear was from the importer's own personal system. It was the gleaming black speakers he was featuring, then, and they sounded spectacular. There was a ribbon tweeter mounted on a movable "hood" of sorts (think Darth Vader's mask and helmet) that could be maneuvered forward and back over the main cabinet of each speaker, where there seemed only to be a midrange driver and a small bass port on the front baffle. But the vendor explained to me that the bass woofer was mounted inside the lower part of the cabinet, down-firing and providing slam and low-end frequencies to the full audio band.

When he put my recording of Mozart's Mass in C Minor through that system, it was a "parting of the gates" moment for me. The angels of the music came floating from those speakers and lofted through the sterile convention meeting room, raising the miserable drop-down ceiling as if it were replaced by an apse painted with clouds and fat, bulbous-cheeked seraphim. Clouds opened and rays of music rained down in sparkling sunshowers. *Finally,* I thought, *an audio system that got it right.* That sound stuck in my ears as a reference point for a very long time, and I measured every new piece I considered over the next few years according to its standard. And I churned through a score of purchases before I put together an audio system that satisfied me as much as that one demo did.

For over two hundred years, the Mass in C Minor has been a treasured part of the repertoire for orchestras and choirs everywhere. From Salzburg in Austria, where it was first performed in 1783, to Salinas, California, where the descendants of Steinbeck's farmers

and ranch hands step out of four-wheelers and pickups to file into a holiday concert (albeit with only one or two of its movements on the bill, alongside Handel's *Messiah* and Irving Berlin's "White Christmas"), Mozart's grand composition gets its due. It's even in the movies. "Kyrie," performed by the Academy of St. Martin in the Fields, directed by Sir Neville Marriner, is featured in the soundtrack of the Miloš Forman film *Amadeus,* a somewhat free-wheeling biopic of the composer's life.

Though the mass is indeed grandiose—well, sublime and inspiring even for nonbelievers—its music is never bombastic nor its moments easily anticipated. Mozart worked at the top of his game, not only to respond to the baroque models of the genre (Bach's Mass in B Minor and Handel's *Messiah*), but to push the composition into unexpected progressions, dissonances, and toward Italianate, operatic solos. While I'm merely a fanboy of this music, what I hear beyond its essential beauty is the dramatic and musical progression of its five major movements (*Kyrie, Gloria, Credo, Sanctus, Benedictus*), each possessed of different rhythms, tonal colors, and emotional messages. The mass presents stages on an emotional journey, a soul's ritual progress through different realms of being and devotion, an aesthetic and spiritual travelogue. I think it conforms to the medieval notion that life and the world are a chaos from which a god-fearing humanity wrests its resistance and refutation in a schedule of prayers, a breviary as a guide through the day, from matins to begin, through to vespers and compline at its culmination. The movements of the mass emerge from this cultural inheritance, and my response to it, though not directly from my upbringing, is nonetheless instinctual, as though Mozart were my own ancestor; the wonderment comes easy, like looking upon the cave paintings at Lascaux.

When I was twenty-one and lived in Shōkoku-ji in Kyōtō Sugasama, my sub-temple's priest, would rise in the morning before dawn and chant sutras, hammering on a wooden, fish-mouth temple bell as he sat before the golden and black lacquer shrine in the *bustu-ma,* the Buddha room. He'd start with the *hanya haramita,* the Heart Sutra, a fairly short but primary text he'd intone in a gruff, basso profundo voice and somewhat speedily, in a kind of hip-hop

rhythm, praising emptiness, declaring its equal is form, that matter and nothingness are the same. Then he'd get into a chant about the basic truth of the world, handed down by Buddhas and bodhisattvas directly to us—that we are our bodies with five senses looking into a jeweled mirror, seeing shadow and substance, and that we idiots are kept on the right path by the true master, a sage like a tiger with tattered ears. It was terrific poetry and I think I always got its messages scrambled up, not being fluent in the Japanese version of Sanskrit, nor (at least back then) blessed with an English text of what he was chanting. But it always *felt* like it flowed rightfully after the Heart Sutra, its rhythms jauntier, its message less philosophic and more Pauline, doctrinal. After that, Suga would give us a recitation of the lineage of temple masters, a kind of praise of forefa-

thers, not unlike the *mele ko'ihonua* that Mahealani Pai sang before me once, reciting the names of his Hawaiian ancestors back several generations, as we walked along a breakwater of worn lava stones at Kaloko-Honokohau on the northwest shore of the Big Island.

These all are ritual enactments, patterns of a this-worldly homage to other orders of being that, by those very devotions emerging out of us, are heightened and made one with us as we recite their liturgies. More than a ghost rises from our craven hearts once we lift voices in song, once we hear the choiring surround us, spiriting off care and penetrating disbelief and cynicism with a sudden spear of sacred passion. *Kyrie.*

Outro

Wandering Rocks (Slight Return), October 2020

I moved to a new house in 2019. I chose it because, among other attractive features (good neighborhood, fine construction, mature plantings that included two beautiful Japanese maples, a spacious backyard bordering on a copse of oak and alder trees), it had a day basement that I immediately saw could be divided into a study and a listening room. I envisioned a partial wall I'd install, using my LP and CD cabinets as room dividers, sectioning things off from each other. I'd have a study at the far end by the sliding glass doors (looking out on the back lawn and bushes of hydrangea and rhododendrons) and a long listening room on the other. Before I moved in, I had an electrician install a dedicated power line for my audio gear, separating it from all appliances and dimmer switches (anathema to the finest sound) and wiring to it a small fleet of audiophile-grade duplexes specially made by Japanese manufacturers and spaced strategically, at useful intervals, around three of the four walls of the basement.

I'd gotten the idea to buy a new house while I was in France in fall 2018 at a retreat in Ménerbes, having lunch with an artist who'd taken me along while she looked at some hillside homes in the village. She'd made money and was looking for a townhouse to buy, wanting my opinion. We'd looked at three homes that morning, admiring their compact tidiness, their three tiers of spaces built into a hillside, the feeling of a charmed, cantilevered existence within each. For choosing, I was of no help—I liked them all. But over cheese, fish, and bread with a little white wine, narcissist that I am, I'd turned to my own wants in terms of a house. I complained that the latest volcanic eruptions on the Big Island of Hawai'i had not dropped prices enough for me to buy a second home there. And besides, I wouldn't be able to retire soon but would have to rent it

out, hire an old neighbor to manage and care for it, visiting only summers. I was a bore. She poked at the filet of sole and said, "Why not just buy a nicer place in Eugene?"

And *voilà!* There it was. The most reasonable solution. So, when I got back to Oregon later that month, I called my old real estate agent and started looking. It took me just over a month to find what I wanted—a three-bedroom in the South Hills of Eugene, just over the lower saddle of Spencer's Butte and a ravine or two from my old house. In quite a bucolic setting, it would be my own Sabine farm—an Horatian place of refuge and retirement from which I might write contented, punctilious, and advisory epistles to the young and listen, in my leisurely repose, to sweet pipings from my stereo system. I made the offer, and by January, I had a new home and moved, packing over all my books, LPs, CDs, and audio gear with the help of former students and friends. The furniture and household goods I had pro movers handle.

Once I moved in, I set up the audio system along the short wall, the farthest from the study, and placed my speakers about seven feet apart, the five-shelf audio rack between them. My electronics and powered speakers easily accessed three of the duplexes I'd had installed. About twenty-five hundred LPs (jazz, rock, chamber, solo piano, orchestral, and opera) were behind my listening seat in huge IKEA cabinets about nine feet away from the frontal plane of the speakers. Three thousand CDs were in a row of cabinets behind the LPs on the study side. I hung my mother's framed embroidery for the ideogram "fortune" on the left wall and perched a black-and-white photograph of my father (at work over his test equipment) in the TV nook there that I'd left empty. On the right wall I hung a large photograph, taken by a friend, of water lilies (nymphaeas) floating on the green waters of a pond. I placed acoustic panels at reflection points on the front and side walls. I had bass traps in the left and right corners of the front wall. I blocked the basement window with panels too. I'd laid a Chinese carpet, indigo-blue with bursts of silver florets, on the wood floor that was over concrete. It was almost as wide and long as the listening area. Here was my first room completely dedicated to audio.

But something was terribly wrong. Compared to the listening

space in my old house (twelve feet by fourteen feet) the new room was twice as large (twelve feet by twenty-eight feet). Though its listening area was almost exactly the same size, I hadn't accounted for the added space of the study on the other side of my LPs and CDs. Even though they'd divided the room into two areas, the study a quarter less than the size of the stereo room, the cabinets didn't reach the ceiling. Acoustically, this meant the entire volume of the basement was the actual space that needed to be energized by speakers—speakers matched to the much smaller volume of my old listening room. Consequently, the sound in my new, dedicated room was fainter. The midrange—violas and cellos, Bryn Terfel and Taj Mahal singing, the lower register of Coltrane's tenor sax, and Barry Tuckwell's French horn—all got swallowed in the larger space. The five-inch midrange drivers on my old speakers, once wondrous to me and bountiful in their sound, were overmatched by the size of the new room. So were the one-inch dome tweeters. They just could not drive enough air in a space twice as large as before. It was like the moment in *Jaws* when Roy Scheider, after tossing slugs of chum into the water, bloodying it, sees the snout of the great white rise out of the sea to feed, baring its teeth. He says to Robert Shaw, the shark hunter, "You're gonna need a bigger boat." I needed bigger speakers.

I called my friend George Radulesk in Portland. He suggested I get a pair of Ascendo System Ms, made in Germany. He'd had a pair for a few years and said their midranges were eight inches in diameter—three more than the speakers I had. And, like the Ascendo Zs I'd heard at my first Consumer Electronics Show in Las Vegas, the System Ms had a separate cabinet, housing ribbon tweeters over an inch wide and four and a half inches long—more than three times the area of a one-inch dome tweet. Plus there was a huge bass woofer (eleven inches) and a cabinet that gave the speaker all the slam and tonal foundation I'd need for a large room like mine.

"They move a lot of air," George said. "It's what you need."

At CES over a decade earlier, I'd heard Ascendo System Zs, a smaller version of the Ms. I remembered how glorious they sounded, especially on the grand mass music that I loved—Beethoven's *Missa Solemnis,* Bach's Mass in B Minor, and Mozart's Requiem and Mass

in C Minor. When other show demos shrunk the music, the Ascendos gave it glory. And I'd heard the Ascendo Ms at George's home in Portland. I had to get a pair of my own.

But the Ascendo Ms were old speakers. It had been over fifteen years since they were introduced, and finding a pair turned out to be intricately difficult. In his own search, George had said he'd found only a handful of pairs had made it over to the United States. I put want ads out on two sales websites for used audio gear. Right away, I got a solicitous response from an audio dealer in Mumbai who wanted nearly thirty thousand dollars, plus shipping. *Uh-uh. No thanks.* I scoured a site called *Hifi Shark* that was a compendium listing of used items for sale internationally. There was a pair in Corsica, the pair I'd nixed in Mumbai, and a pair in Istanbul all up for sale at varying prices. The pair in Corsica turned out to be a scam—the seller had posted photos that were clearly of two different sets of speakers. And when I wrote him, he'd no good answers for me. But the owner in Istanbul was for real, had the original wooden shipping crates, and had already dropped the price several times from twenty thousand euros to twelve thousand when I wrote him. It seemed fair. The trick would be getting them shipped.

I dickered with him for about a month, not about price but mostly about shipping and a method of payment. He wanted the full amount in a wire transfer before shipping the speakers. That was a big problem. Trustworthy and real as he seemed to be, by email and phone and via his Facebook page (he was an inveterate runner of 10K races), I was hesitant about handing over so much cash up front without a guarantee. So he volunteered a demo if a friend could visit the dealership in Istanbul where he'd deposited the speakers. They'd not been for sale there, he said, but "only for keeping."

I got in touch with a Turkish rug merchant in Los Angeles who had sold me two beautiful carpets and did regular business in Istanbul, but he had no one he could recommend. I wrote a poet friend from Tennessee who'd had a Fulbright fellowship to Turkey

in years past. After more than twenty years since he was there, he had no contacts left. But a former graduate student of mine, now a tenured professor of Turkish Studies at a major university in the South, had a brother-in-law in Istanbul whom I might ask to do me the favor, he said. The brother-in-law was himself a professor at Sophia University there. I wrote him, and he was more than willing to help, saying he drove near the dealership every weekend on his way across the city to visit his parents. He'd take a look at the speakers, but said that he knew nothing about audio and could not explain to me what the speakers sounded like. I told him he only had to verify they existed, that they were undamaged, and that there were shipping crates on hand. He agreed.

But then the coronavirus pandemic intervened and changed all our plans. The kind Sophia professor no longer felt comfortable visiting others, especially as he had elderly parents he saw regularly, and who could blame him? My seller thought we should wait as so much was uncertain. Plus, he said he was very stressed and occupied every night, trying to build an algorithm to buy and sell stocks via computers. He was a high roller.

Months went by and we just held fire. Then, in late September, my friend George wrote me a cryptic email saying, "Don't do anything yet about the Ascendos." He did not explain why. The next day he wrote again, saying, "I might have a pair for you."

It quickly evolved that his son Kyle, a man with a franchise for Pepperidge Farm, had a pair of the same Ascendo M speakers that George had, piano-black, only in much better condition. He was willing to sell them if I could make a deal fast, as he wanted to try a different pair of speakers himself. And though he had their original wooden crates, I wouldn't have to bother with shipping. He'd drive them down, crated, in his own liftgate truck and, with his father, would set them up for me. *Serendipity.* I paid his asking. I wrote my seller in Istanbul and backed out as gracefully as I could. Miraculously, the man was completely understanding.

A week later, George and Kyle, both masked, pulled up in the long, liftgate truck and parked out on the street in front of my house. George leapt out of the cab, walked the slight incline down my drive, and stood in front of my garage door. Then he waved

his arms, pointed fingers, waved some more, and guided Kyle as he backed the truck down, angling into the drive, then straightening out and parking. George leapt up on the lip of the folded liftgate and unlocked the truck's roll-up door. His son emerged and worked the controller to the gate's pneumatic pumps. This unfolded it, and they began shoving and unloading each of the huge speaker crates onto my driveway and then, with a pallet dolly, moving them into my garage. One by one, they opened the crates and unpacked each speaker.

Inside each crate, under a plastic sheet, there was a heavy chrome frame that arched above the two large sections of each speaker. This was a kind of gantry that suspended both the tweeter tower and the bass cabinet above the floor, decoupling each of them not only from each other, but from the ground as well. All resonances would be guided to the frame, which dissipated them to the floor beneath. It was an intricate job to assemble them, taking both father and son working in coordination, lifting each heavy tweeter module onto a kind of calibrated, horizontal pole assembly that reached out from the vertical part of the frame and mated with a smaller tube that stuck out from the back of the tweeter box. Then, they lifted the huge bass cabinet and fitted that, via a tiny prong with a ball at the end of it, onto the frame that had a matching socket joint. The Ascendo speakers were an intricately coordinated suite of parts. And those parts were exceedingly heavy (made of MDF and bitumen)— the bass module over eighty pounds and the tweeter tower around seventy. The chrome frame weighed over a hundred (filled with forty-five pounds of sand). Each speaker assembly, including the elaborate frame, weighed two hundred and sixty-five pounds—just about a hundred pounds more per side than my old speakers. And they seemed huge in my room, rising fully five feet tall and measuring almost sixteen inches wide and just over two feet deep. They commanded the space at the front of my room.

After I took some triumphant photographs, George and Kyle (both over six feet tall) standing, briefly unmasked, each beside one of the gleaming speakers, we all wondered what they might sound like. Momentarily, George and I worried they'd overwhelm the room, they seemed so huge, ominous as monuments of obsidian.

We hooked the speakers quickly to my system, now a Zanden tube stereo amp, a matching Zanden pre, and the Zanden variable EQ phono. My digital source was an Esoteric SACD player. It was an all-Japanese system of electronics. I chose a CD of George Benson's *Breezin'* from 1976, smooth jazz with funk beats and shimmering, virtuosic electric guitar. Impatiently, I advanced the tracks to his version of Leon Russell's "This Masquerade," a kind of mordant love song in Russell's original, but a thing that became a spooky combination of erotic melancholy and almost operatic triumphalism in Benson's reworking. In an instant, the song sailed out from the Ascendo's drivers. Every molecule in the space seemed charged with the music, sound waves pressurizing the room. The speakers rendered Benson's complex singing style—a hybrid of jazz, pop, and personal swagger—with a wealth of detail and sensitivity, reacting to every shift of interpretive timbre, every octave interval, every swoop and dive of his voice. Benson's vocal, full of numerous midnote shifts, climbing glissandos, and the signature scatting that doubled his speedy flat-pick runs on electric guitar, all came through, wave after wave, invisible coils of sound seemingly without flaw. George and Kyle had set the time alignment of the Ascendo's tweeters perfectly, using a tape measure to position them to fire just outside where my ears were as I sat in my listening chair. And the electric bass had real thump, the woofers charging the whole room with Phil Upchurch's decorous but insistent ostinato. The three of us sat captivated, the soulful music and Benson's bravura vocals pressing us back in our seats. I felt myself wanting to sing along, surrounded by a perfect sound.

System: Ascendo System M speakers; Zanden 8100 stereo amp; Zanden 3100 linestage preamp; Zanden 120 phono stage; TW-Acustic Raven Two turntable with Ortofon RS-309D twelve-inch tonearm and Miyajima Zero mono MC cartridge and TW-Acustic 10.5 Raven tonearm with Zyx Ultimate 4D MC cartridge; Audience SX phono cable; Ypsilon MC-10 step-up transformer; Esoteric K-05x SACD player; Bluesound 2i streamer/DAC; fo.Q Modrate HEM-25 Pure Note Insulators; Synergistic Research Galileo universal speaker cells with Foundation jumpers; Audience frontRow speaker cables, interconnects (RCA and XLR), and

*powerChords; Audience Adept Response aR6-TSSOX power conditioner
with Audience fR powerChord; Oyaide R1 and Furutech GTX-D
duplexes; Furutech Alpha CB-10 OCC 10-gauge electrical cable (in-wall,
panel to outlets); Zanden AT-1 acoustic tubes and AP-1 acoustic panels;
Acoustic Science Corporation acoustic panels; GiK 4A Alpha Pro Series
Bass Trap Diffuser/Absorber acoustic panels; Box Furniture five-shelf rack.*

Shōkei: *Accession (Under the Stars)*

*Shall goodwill ever be secure?
I gaze up at the long river of stars . . .*
  (Li Po, eighth century)

My adult quest has been to retrieve a history I intuited as lost
through the negligence of culture and our lack of care for it as
descendants of laborers brought over from southern Japan in the late
nineteenth century. My thought since college days was to honor it
through words—the practice of poetry and remembrance through
an honorable application of literary attention. I wrote elegies in
mourning for the bodies and ashes of ancestors washed away from
their burial plots in a sandy outcropping ravaged by a tidal wave in
1946. I wrote paeans of praise for our old enactments and rituals in
homage to the lost. I studied, I chanted, I bowed my head in rites
strange to me until, performing them day after day in a temple in
Japan, they were as much second nature to me as the dropback cari-
ocas I practiced as a third-string linebacker in high school. I came
away a strange mongrel of a devotee, wearing jeans and a T-shirt as
I made my genuflections, my long hair like tresses of a willow tree
tangling in my hands as I swept them, palms up, beside my ears.

When the priest chanted, I might have heard his voice intone in rhythms less familiar to me than the blues were. When I practiced my rows of ideograms, scripting them vertically down a page, I saw a floating fly line being paid out in S-curves over the glassy surface of a spring creek. When I remembered my father, dead now nearly forty years, I heard the baritone of a race announcer's stretch call at Santa Anita. And I heard luscious music wrap around me like a languorous current freshened and spilling in from the open sea.

For most of the time I was young I had not put much stock in any tokens of our former lives, or sought to preserve any relics or heirlooms as I moved through my twenties, thirties, and beyond. But as I made acquaintance with others from other worlds and noted their casual preservations of legacies, I admired a cameo and its fine silhouette, a tortoiseshell comb, mother-of-pearl pendants, a gold pocket watch, a grandfather clock, gaudy necklaces, massive oak dining tables or a handsome secretary made of bird's-eye maple. They represented a kind of continuity of things and even values that were absent in my own family, who may never have had thoughts to celebrate and honor their own kind in these casual ways, who considered the past a log of the forgettable rather than anything to preserve. We'd not any wish to keep hold of the *bango* tags the plantation issued to assign numbers to each laborer in the family. None of the wooden bowls or bento boxes they used for midday meals in the canefields came down through the generations. No silk kimono or embroidered *obi* laced with gold thread were preserved, and I doubt anyone in our line ever owned such fine things, except perhaps a beauteous aunt from Molokai, or a grandmother who danced in teahouses and then ran away, abandoning her children, to escape a cruel husband, my paternal grandfather, who beat her. There were no intimate icons on our mantelpiece except a trio of disposable curios—a brass Buddha, a purple plaster bull from Tijuana, a Japanese doll that was a replica of a *maiko* in fine kimono, one arm dipping a brocaded sleeve in a dance pose, her head tilted so the *kanzashi* hair ornaments made of mother-of-pearl would catch the yellow light from the three-headed, aluminum pole lamp standing rigid in the corner of our living room. In time, my own first few books replaced these, sealed in Glad bags

and propped like decorative Blue Willow plates on wooden stands. To my parents, my books that had reclaimed the forgotten, that upheld our ancestral past, were not unlike the gaud of trophies that my father had won for lunch hour Ping-Pong competitions at work.

But, as I grew older, I collected a few small things. My maternal grandmother gave me my grandfather's wooden *soroban,* the abacus he used to total customer purchases at the counter of his plantation store. On its back was incised the Japanese ideograms and Roman characters for *S. Hata Shōten,* the wholesaler who supplied general stores throughout the islands. My sons collected dark brown potsherds, a nail, and broken ceramic light-bulb sockets from the grounds of Manzanar, the War Authority concentration camp near the Eastern Sierras in California when I took them there as pre-teens. A friend who taught at a college near Flagstaff, Arizona, gifted me with a Hopi rattle and a Navajo clay pipe to commemorate my grandfather's incarceration near there in the stockade of the trading post on the Navajo Nation during three years of World War II. I keep on my bed a Hawaiian quilt my grandmother made for my mother. And I've a silver *saké* cup the Japanese emperor sent to acknowledge her birthday when my grandmother reached her centennial year. Of my father's, I have his Sony radio, his gambling hat, and, luckily, at least one piece of his precious stereo.

The celebrated Beat poet Gary Snyder has among his early poems one that addresses the regret over lost but trivial things—a lover's comb he'd kept after the breakup, a gold earring he'd once worn as a grad student in Berkeley, maybe a paperback book. I can't recall exactly, and searching my library and his books, I couldn't spot where my memory of it came from. But its point was to practice the Buddhist tenet of nonattachment. *Get over it,* the poet seemed to be saying to himself. And get on with the work of living, of being in the present, of entering the Heraclitean stream of continual change and ephemerality. Who needs mementos?

It's true that I've my own swag of trivial losses: a fancy umbrella I left in a taxi once one rainy night in Kyōtō over forty years ago, a paperback of Vincent van Gogh's letters to his brother Theo that I left behind in the seatback of a plane, a Mont Blanc rollerball pen that likely fell out of my pocket in the Honolulu airport, and,

most recently, a reporter's notebook filled with scribblings on the spectacular images of the nymphaeas floating on the ponds, the bankside willow trees, and the wisteria-covered Japanese bridge in Monet's garden at Giverny. That was a sore loss I'd incurred, carelessly placing it on the trunk-lid of my car one afternoon in late fall, as I hurried to run errands and drove off. It's true that each small shard of pain faded with time, though I still toss at night sometimes over the absence of those notes. But Snyder and Buddhism are right that one can't mourn grains of sand that slip down a mountainside as you climb. You can't constantly be looking back downhill like Orpheus. You've got to keep headed to the empyrean.

Still, there are a trove of things that are of a different order than lost trinkets. Family heirlooms are real treasures; often they are cultural artifacts that inspire the continuity of human care—my grandmother's humble garnet wedding ring, a Leica my father brought back from his time as a guard at Nuremburg. And then there are the great paintings of the Italian Quattrocento, Etruscan pottery, the cave paintings at Lascaux, the Great Buddha at Kamakura, and the magnificent ruins of Angkor Wat. An Italian friend, who is a great translator of American poetry, a scholar of Native American literature, and the mayor of a hilltown at the foot of Mount Subasio in Umbria that overlooks a vast Virgilian landscape, once proclaimed to me, disparaging the nouvelle cuisine praised by an American memoir set in his *terra natia,* "How can one improve on the *sen-chure-rees* of Italian cooking, I ask? It is the exact *flavor* that is handed down from mother to daughter, from one cook to another—the absolute *taste* of not just former times, but respect for the culture of generations *through* time! It is *proven!* Why desecrate what has been preserved?"

I make a Japanese *nishime,* a pork stew with vegetables, the way my grandmother taught me. I make *chikin hekka* like my grandfather did, chopping straight through the bones of a fresh carcass with a heavy cleaver so that the marrow can invade the soup when I simmer it. I was attracted to the black, Lucite-covered faceplate and gold push buttons of a Norwegian amplifier because it possessed an aura reminiscent of my father's '59 Plymouth Fury. The effort to remember, taste what they once made for me, honor what was

the compilation of their earnestness, the gradual build of sincerity and sensibility over time is my pursuit. I wanted in my life to build a humble shrine of items that would acknowledge the past, signify for me both travail and splendor—my grandfather's resolve and stoicism, my grandmother's cooking, my father's devotion to the pleasure of hearing music come from a couple of mail-order amps he built himself. They are ashes. But from the work and the play of effortful lives I wish to maintain what was the shine of their presence on earth.

In April 2018, about six months after my mother died, I was cleaning out the garage in my parents' house, going over what they'd left behind. All of it was still in the same boxes that my brother and I had packed in 1984 when my father died. My mother wanted everything associated with my father moved out of the house, and one of her sisters who came to console her, an orderly woman, decided to get rid of all the excess as well, since my mother was a hoarder throughout her life. We packed up silverware, cookware, blankets and quilts, old clothes, Japanese *kakebuton* (duvet-like quilts for use as blankets), Japanese enameled trays, a few of my father's textbooks from trade school, and all his electronic gear we didn't dump or sell. There were assorted vacuum tubes in coffee cans, a soldering gun, an oscilloscope, a voltmeter, and tools for working on TVs and audio circuitry. Now, more than thirty years later, when I started to clear things out, everything was as we'd left it, a thin layer of grit covering the topmost boxes. It seemed no one had touched the great pile my brother and I had shoved alongside what had been my father's workbench in the north end of the garage. I went through it all and found books from my high school and college years, my brother's weights and lifting bench, photography equipment that had outfitted my home darkroom, and, enshrouded in a crust of oily dust, my father's old University Mini-Flex speaker. I'd thought it long ago lost or discarded. But he'd kept it, shoved deep on a metal shelf alongside his tools. I set it aside and, after my cherry-picking was over, placed it in a U-Haul box I included in

the bunch of things I shipped home with fly-by-night movers to Eugene, where I lived.

The speaker was one of a pair that had created that magical sound throughout our living room back when I was twelve, my father playing his big band tunes and Hawaiian LPs, sometimes after he'd tinkered with his amp or preamp by changing tubes or modifying the circuit somehow. As an electronics technician, he loved studying circuit charts, calculating voltages, gains, resistance, impedance, and all those electronic properties that are mostly all still Greek to me. Like calculating odds at poker or the racetrack, he'd often do this all in his head after glancing at a sheet showing the circuit of an amp. Sometimes he'd scribble with a pen—he had an elegant hand—in the margins or right over the values printed on the sheet, figuring some adjustment he could make. In today's terms, he'd be called a "modder," but, to me, he was a wizard, poring silently over his work, shifting the after-dinner toothpick from one corner of his mouth to the other, alternately leaning into and leaning back from his work, stretching his arms out from his sleeveless cotton undershirt, rocking on his haunches (he sat cross-legged on a *zabuton* placed under him on the carpeted floor).

When I came upon the speaker in my mother's garage, I knew it was something I wanted to remember him by, even though it was just a single from the original pair. I had so few things left of his life—a steel box of his pens and cuff links, his wedding ring, and a couple of old electronics textbooks and manuals from his studies at L.A. Trade-Tech. I'd always thought he'd never got to finish off his life as he deserved, dying as he did some six years short of retirement. He lived stoically, silently, often in his own world as his poor hearing cut him off from most everyone except those patient enough to slow their hurry and adjust to the ways he could communicate. That was through the music he loved, through quiet activities like studying his racing forms. When I was twelve, I listened with him. When I was twenty and home from college, we'd sit side by side on lawn chairs in the tiny backyard and I'd read my books of poetry while he calculated odds and read the handicap sheets.

The other day I took to rummaging in the garage of my new home in Oregon. It's an astonishing heap of disordered boxes, bicy-

cles, camping gear, fishing tackle, a portable air conditioner, left-over cans of house paint, metal and wood file cabinets, and even a shipping pallet or two (relics from my years-long speaker quest). Though the house I've moved to is bigger in living space, it's much smaller in storage capacity than my old one about two miles away, and I was forced to cram all the excess into my pretty new place and its two-car garage. I filled the latter to bursting, and though I made a mental map of where a few crucial things were stashed, I'd already lost track of where I placed my dad's speaker, having shifted boxes numerous times. An electronic keyboard bridged over boxes of Christmas ornaments. Empty boxes for audio gear got stacked atop a built-in desk I'd ripped out from where I installed a bay window for my study. Rattan chairs (from a former dining set) teetered over boxes of manuscripts and files. I told myself I'd have a garage sale the summer after I moved in, but I traveled too much to organize things and my midden pile of unused possessions just kept growing.

Among these was that speaker in the U-Haul box I'd hastily marked almost two years before. I just knew it was *somewhere,* and to find it, I'd have to attack the garage in plausible sectors like an archaeologist mapping a site, trying to speculate how prior inhab-itants might've discarded things—bones, broken tools, rope and string, and the blood of their lives alongside their defecation. *Going through my own shit,* I told myself. *What was I thinking when I stored that old Tangkula ceiling fan next to the Sharp A/C?* I spent one afternoon plowing through boxes of files from my first teaching job at USC. Among them, I came across a box of audio cables I'd been missing—two Wireworld power cords, Audience and Cardas digital and USB cables, AudioQuest Cheetah silver interconnects—a small trove of treasures that would enhance my own meticulous tinkering with my audio system. But not my father's speaker. The next afternoon, I attacked the pile of empty audio-gear boxes I keep even after unpacking items, wondering if I'd put the boxed speaker under them. After about an hour of shoving things around, disassembling a small pyramid of cardboard boxes and accidentally dousing myself with Styrofoam peanuts—nope. *Nada.* Finally, twist-shoving and sliding aside a pair of tidy, five-foot-high wooden crates that once cradled my wonderful German speakers, I found behind them the

big accordion caddy of my father's old audio and TV tubes that I'd had moved up from Gardena. And with that, I found the U-Haul box that housed my father's Mini-Flex speaker. The box, contrary to my memory that I'd packed hastily, was carefully inner-lined with fine green Bubble Wrap I'd bought at the gritty U-Haul on Carson Boulevard in Torrance. The speaker it housed was my grail, an heirloom of another time.

I remembered how Charles Wright kept on a shelf near his writing desk a shoe his father, a clubfoot, wore as he surveyed sites likely for the mighty dams of the Tennessee Valley Authority. Below it on their own shelf were his own painted toy soldiers made of lead. In his living room where a coffee table might be was the handsome cedar chest hinged with blackened steel his parents once owned. He preserved things.

What I'd taken away from the house on Dalton Avenue in Gardena was more friable and modest. There was a photo collage my father once put together of old two-inch-square black-and-whites and sepias. Primary school color photos of my brother and me floated among them. There was my young mother wearing, uncharacteristically, a ribbon in her hair and standing against a low lava-rock wall; my brother as an infant staring out from a bundle of blankets, the mop of his hair like the curl of an 'okina (a reverse, single quote) atop his head; Mrs. Fukamachi (the housekeeper who raised my father) clutching her purse, standing next to her bespectacled brother (who looked like a slim version of the young Emperor Hirohito) on a ledge near the Blowhole at Koko Head on the island of O'ahu; my potbellied father in middle age, wearing a yellow ball cap and baseball tee, watering plants in the backyard; and the last photo of all of us together, sometime late in the seventies, my slim, six-foot-tall brother in cycling clothes, my chubby mother in a puffy blouse and pedal pushers, my father again in a blue-sleeved baseball tee, a camera dangling on a strap around his neck, smiling as though a horse had just come in, and me in three-quarters profile, wearing shorts and a V-neck soccer shirt, my own belly just starting to thicken.

I'd packed in cardboard art crates about a dozen framed *bunka* my mother had embroidered—scenes of Swiss postcard landscapes

of mountains, a river, and a village; a *geisha* standing before the Golden Pavilion in Kyōtō; gold bamboo stalks and leaves on red cloth; the intricately articulated purple and white petaling of irises on a gold background; white cranes with black-fletched underwings flying toward the snow-crested cone of Mount Fuji. I'd packed boxes of family photos and one single 35mm contact print of Alina, my beautiful muse during my junior year in high school. She was surrounded by tall grasses and wore a light, woven, likely acrylic sweater-top, her tanned face ashine with early morning sunlight. And I'd packed a woodblock diptych of nineteenth-century Japanese courtesans in brocaded kimono that I'd sent from Kyōtō to my parents as a gift upon leaving Japan after my year living in temples and wandering the countryside. When I bought them, on leftover fellowship money, I'd felt they'd later return to me as my own legacy, albeit a falsified one that represented more the breach of continuities than their maintenance. And so they were. But what was true—a real legacy—were my father's things: his Weller soldering gun, his caddy of vacuum tubes, and his prized Mini-Flex speaker.

Once I'd found it in the garage, I placed the speaker on my desk behind my laptop. I wanted to absorb its presence again, draw some of whatever it had left back into me, maybe feel my father's spirit drift from its aged drivers and wiring into the air in front of me like some benign version of Hamlet's ghost, demanding not that I avenge, but simply remember. I liked the way it looked there, just peeking from behind the frame of the MacBook's view-screen, its top edge of dark walnut almost parallel to the white line at the bottom of my bay window, which looked out over a wood and wire fence to the garden lot next door. Each morning a Steller's jay had been visiting me, alighting on the top of the fence for a moment, turreting its black-crested head and beading its eye my way, then fluttering its wings and sailing off over the green lawn next door. A squirrel came too, hopping along the planking that made the top edge of the fence, hummocking its back like a furry, miniature porpoise. One day, the jay alit and the squirrel chased it stage right,

both going out of my view. Then, an instant later, the jay was chasing the squirrel toward stage left, a hovering flurry of blue feathers coptering above the brown flight of fur. It was a sequence like a kid's cartoon and I laughed as the sun came out and lit all the leaves on the trees outside a glorious yellow-green of spring.

Then, it came to me. I should just hook the speaker up. It might still work. Its innards were all intact, I decided, but the back, where the hookups were, looked dusty and maybe frozen in place, the screwheads rusty-brown with oxidation and age. So I got a flathead screwdriver out of the toolbox I stored in the cabinet over the washer and dryer and notched up one of the heads. It moved easily, but it was still rusty. I took a tiny tube of DeoxIT from my small, metal-mesh box of audio tools and squeezed a drop onto each screw and its contacts. From my closetful of cables, I dug out a pair of speaker wires with the kind of tiny spades that fit old gear, from when screwheads connected with wires instead of fancier binding posts. I rifled through my CD collection, carefully categorized and alphabetized in wooden cases on my study wall, and picked out a few titles—"old-time" Hawaiian music I thought my father might've liked. And I traipsed upstairs to the audio system—pig that I am, I've two such—in my family room to give things a try.

The amp upstairs is a humongous Viva Solista weighing nearly eighty pounds, a piece of bespoke audio electronics made by my friend Amedeo Schembri in Vicenza, Italy. It's a contemporary design but uses old 845 vacuum tubes of the kind radio transmitters once employed. The output is only 28 watts per channel—within spec for the power handling printed on the patch of paper on the back of the Mini-Flex. I didn't want to blow it out by using an amp that was too powerful. When I got upstairs with the speaker, I gave a quick glance around for a likely spot to place it and decided it would fit well on top of the REL subwoofer I had stashed in the corner behind the left channel speaker, a Reference 3A L'Integrale—itself fairly efficient, so it could work with the 845 amplifier. To make it fit onto the REL, I had to stand the box of the Mini-Flex vertically, so that it looked slightly odd, yet a bit hidden as well, just as it had been in my father's house all those years. I hooked the Mini-Flex to a single run of speaker cables, snugging their tiny spades below

the screwheads I'd loosened, and tightened them up. I was getting excited now. I told myself to slow down. I disconnected the speaker cables I'd installed on the amp's left channel and slid the banana connections of the wires attached to the Mini-Flex into its taps. I fired up the system, letting the tubes warm a bit, and then I realized I could do something special.

I'd brought up that handful of CDs from my collection, but they were motley—titles like *Blue Hawaii* and *Hawaii Calls,* compilations from fifties radio and TV shows, and it was not all that likely they'd have tunes my father liked, I thought. But I'd bought a wireless music streamer recently, with a built-in DAC (digital-to-analog converter). This was a device that could wirelessly stream any music available on iTunes—or any other streaming service, for that matter. My iPhone functioned as a remote controller, once I'd installed special software. I'd long been a holdout from such things, skeptical of their sound quality and resistant to the way they encouraged hobbyists to listen—randomly, playlist-ly, in a manner totally uncurated by human taste but by an algorithm. I was above that. But I'd broken down when I found their prices had dropped and that there was a fairly inexpensive device that I'd had no trouble trying out. It served mostly as a radio while I cooked dinner or had casual company, playing Simon & Garfunkel's *Greatest Hits,* Hawaiian slack key guitar music, and jazz for easy listening. Its catalogue was immense—everything under the Amazon sun. But did it have my father's music?

It did. Using the search window in the iPhone's app, I typed "Arthur Lyman," and a magical column of icons, colorful thumbnails of the vintage album covers, appeared. I saw *Taboo, Yellow Bird, Hawaiian Sunset,* and *Polynesia,* each exotic one clickable on my smartphone. I picked *Taboo,* that album with the 1959 fountain eruption of Kīlau'ea-Iki on its cover. The tubes glowed magenta-red against the dark murk of the window behind my amp. It was night and the house was cold, quiet except for the wash and small, intermittent *thunks* of my pulse in my ears.

The speaker played at first faintly, as I imagine my father's best hearing might've been all those years ago. Then, about a minute in, the Mini-Flex opened up into its character, gaining a bit more

volume, stirring to life in the corner. I heard a plaintive flute, then Lyman's shimmering vibraphone, and the screech of birdcalls, lightly explosive, made by men in the band to announce the exotic otherness and dated claims of authenticity in their music. What would fake Hawaiian music be without *birdcalls,* their whoops and hootings, their macaw-like swoops of faux-primitive vocalizations? The Mini-Flex played them; then came the airy horn of conch shell and Lyman's vibes again, sweet and delicate, not quite ashimmer as I remembered in my father's house, from his precious old system. When Ellington's "Caravan" came on, the third track in, the sound got bolder and Lyman's vibes went buttery smooth, their sinuosities almost palpable in air. I goosed the volume for more sound. Then I switched to *Yellow Bird,* another Lyman album, and heard the vibes take the melody of the title tune, with a valiha-sounding harp chiming above them. I thought to play Artie Shaw next—his tune "Moonglow" with its bellyful of mellow, swinging clarinet— a player and a bandleader my father loved for having heard his band live in Honolulu once or twice. I'd found the old monophonic album among my father's records and brought it back with me in a suitcase, storing it for a rainy day. And now, controlled by my iPhone, streaming through the new wireless device to the DAC, and amplified by the Viva Solista into sound from my father's bookshelf speaker, I heard Shaw's expressive solo—stylishly bold, sophisticated, and swinging. The horn chorus was solid and strong and the brass section played with mutes, accentuating the delicate shimmer of metal in the bells of their horns. But the sound was still not as rich as my memory of what the speaker once made. It was attenuated, faint, a music carried to me from a distance, like the sound of a group of schoolchildren upstream as I stood at the mouth of a coastal river, their voices drifting to me in a smooth and pleasant babble, like the sound of water running over the slick stones at my feet. I wept. I stifled a sob.

I come from an affectless, unsentimental people, their emotions battered by three generations of brutal life on the plantations and conditioned to its harsh disappointments, resigned to limitations of birth and station. When I was a child, I cannot remember kind words, let alone testaments of devotion between us, and the

embraces we made were public and ceremonial, triggered by the company of others, mostly *kānaka 'ōiwi* friends, neighbors, and those married into the family. When I screamed for joy as a child, I was hushed if not struck. When I wept for the beauty I saw in the landscapes and seascapes that surrounded us, I was mocked by cousins and uncles. When I witnessed shameful acts or creditable achievements, the tacit messages were always to ignore them. What was important was to persist and move on, and any singular moment of beauty was the least sibyl of our universe. But my father's music overwhelmed this impoverished inheritance like a wave enfolding its barrier reef, barreling softly in glassy curls, foam, and folds of lace.

It overcame what was taboo.

Acknowledgments

Some of the material in this book appeared, in prior and much different versions, in *The Georgia Review,* the *Los Angeles Review of Books, The Massachusetts Review, Plume, SoundStage! Global, SoundStage! Traveler,* and *SoundStage! Ultra.* My thanks to the editors of these publications.

A book like this has many sponsors and confidantes over the years it takes to live and write it. Peter Morrison was my first mentor in the audio pursuit. At its beginning, Liz Darhansoff gave me confidence in the idea for the project and persuaded Daniel Frank, Edward Kastenmeier, and Deborah Garrison to take it on. The intricate task of coordinating serial rights was handled masterfully by Michele Mortimer. Edward Hirsch, T. R. Hummer, David Mura, Russell Tomlin, Kevin Shuler, and George Radulesk were with me throughout the journey.

My thanks also to Lance Patigian, Jules Coleman, Albert Von Schweikert, Kara Chaffee, Dan Meinwald, Amedeo Schembri, Jonathan Halpern, Mehran Farahmand, Karen Tei Yamashita, Nicholas Christopher, Sandra Zane, John McDonald, Eric Pheils, Jeff Catalano, Bob Clarke, Margot Haliday Knight, Jeffrey Stolet, Lynn Freed, Alec Stone Sweet, Bart Scott, William Taylor, Lori A. Wood, Frank Abe, and Brandy Nālani McDougall for their friendship and advice. My special thanks to Kevin Hayes for technical advice about vacuum tube mechanics and the operation of tube equipment.

Charlie Kittleson, editor of *Vacuum Tube Valley,* was first to ask me to write about audio. Dan Davis introduced me to Marc Mickelson, who invited me to write for *SoundStage!* Then, through many years, Jeff Fritz and Doug Schneider kept me on as a contributor at *Soundstage! Ultra.* Richard Lehnert informed my writing there.

I wrote various sections of this book at several artists retreats: Ameri-

can Academy in Rome, BAU residencies at the Camargo Foundation, Djerassi Artists Residencies, La Macina di San Cresci, La Maison Dora Maar, Lucas Artists Residencies at Villa Montalvo, MacDowell Colony, and Virginia Center for the Creative Arts. I wish to thank them for their support of this project.

Finally, I wish to express gratitude to Edward Hirsch, T. R. Hummer, and George Radulesk for reading the entire manuscript and making crucial suggestions. My editor, Deborah Garrison, together with copy editor Holly Webber and production editor Kathleen Cook, assayed every word of the final draft and made the book better. They each helped me through their sensitive criticism and sharp comments.

Notes on Sources

75   The Penguins, "Earth Angel (Will You Be Mine)," by Jesse Belvin, Gaynel Hodge, and Curtis Williams, Dootone Records, 1954.

94–95   Rainier Maria Rilke, *Letters to a Young Poet,* trans. M. D. Herter Norton (New York: W. W. Norton & Company, 2004; 1934), 16–17.

98   "Cathy, I'm lost . . ." is a line from Paul Simon's song "America" on Simon & Garfunkel's *Bookends,* Columbia Records KCS9529, 1968.

130   Bob Nobuyuki Hongo, *Hey, Pineapple!* (Japan: The Hokuseido Press, 1958).

142–143, 147   From the website for Vintage Tube Services: http://vintage tubeservices.com.

146   All issues of *Vacuum Tube Valley* are available online: https://web.archive .org/web/20130604042043/http://www.jumpjet.info/Pioneering-Wireless/eMaga zines/VTV/vtv.htm.

200   D. T. N. Williamson, "Design for a High-quality Amplifier," *Wireless World,* April 1947, pp. 118–121.

203   Victor Brociner, "Speaker Size and Performance in Small Cabinets," *Audio,* March 1970, pp. 20–23, 69, and 79.

204–206   Steve Birchall, "Acoustic Suspension," *Journal of the Audio Engineering Society* 41, November 1993, pp. 970–971.

293–294   Vitruvius, *On Architecture, Volume I: Book I, Architectural Principles,* trans. Frank Granger, Loeb Classical Library 251 (Cambridge, MA: Harvard University Press, 1931).

295–296   Vitruvius, *On Architecture, Volume I: Book V, Public Buildings: Theatres (and Music), Baths, Harbors,* trans. Frank Granger, Loeb Classical Library 251 (Cambridge, MA: Harvard University Press, 1931).

296–297   Tom Chao, "Mystery of Greek Amphitheater's Amazing Sound Finally Solved," *Live Science,* April 5, 2007.

297   Nico F. Declercq, Joris Degrieck, Rudy Briers, and Oswald Leroy, "A full simulation of the Quetzal echo at the Mayan pyramid of Kukulkan at Chichen Itza in Mexico," *Journal of the Acoustical Society of America* 113, no. 4 (2003): 2189.

301   Virgil, *Georgics, Book III,* trans. A. S. Kline, Poetry in Translation, https:// www.poetryintranslation.com/PITBR/Latin/VirgilGeorgicsIII.php.

301   George Seferis, "Delphi," trans. C. Capri-Karka, *The Charioteer: An Annual Review of Modern Greek Culture,* no. 35 (1993–1994).

302–303   Edward Hirsch, *A Poet's Glossary* (New York: Mariner Books, 2017), 168.

307–309   Several albums by Michael Levy are available on iTunes, but the most pertinent is *An Ancient Lyre:* https://music.apple.com/us/album/an-ancient-lyre/338 195640. The most pertinent YouTube video is "The Oldest Known Melody (Hurrian Hymn no. 6—c. 1400 B.C.)": https://www.youtube.com/watch?v=QpxN2VX PMLc.

312   "The Cause of Sound and Music," *Scientific American* 13, 38, 302 (May 1858).

313   Paula Findlen, ed., *Athanasius Kircher: The Man Who Knew Everything* (New York: Routledge, 2004), 329.

313   *Scientific American* Supplement, No. 483, April 4, 1885.

316   Anthony Grafton is quoted in Sarai Kasik, *The Esoteric Codex: Christian Kabbalah* (Lulu.com, 2015), 96.

316   Hector Berlioz, *Voyage musical en Allemagne et en Italie* (Paris: Jules Labitte, 1844).

316–317   Thomas Hankins and Robert Silverman, "The Aeolian Harp and the Romantic Quest of Nature," in *Instruments and the Imagination* (Princeton, NJ: Princeton University Press, 2016). I learned much about the foreground to Coleridge's fascination with the harp among the Scottish poets.

317   Robert Bloomfield, "Aeolus," from *The Remains of Robert Bloomfield* (1824).

320   "They flee from me . . ." is from a 1557 poem by Sir Thomas Wyatt entitled "The louer sheweth how he is forsaken of such as he somtime enioyed," from *Tottel's Miscellany: Songs and Sonnets of Henry Howard, Earl of Surrey, Sir Thomas Wyatt and Others,* ed. Amanda Holton and Tom MacFaul (New York: Penguin Classics, 2014).

321   "I hear you singing in the wire. . . ." is a line from Glen Campbell's song "Wichita Lineman," written by Jimmy Webb, Capitol Records 2302, 1968.

335   "This Is the Way I Do," from *Hank Williams: Lost Highway,* libretto by Randal Myler and Mark Harelik (New York: Dramatists Play Service, Inc., 2005).

339   Etheridge Knight, "The Idea of Ancestry," in *The Essential Etheridge Knight* (Pittsburgh: University of Pittsburgh Press, 1986), 12–13.

341   Japanese text is from Franklin Odo, *Voices from the Canefields: Folksongs from Japanese Immigrant Workers in Hawai'i* (Cambridge, MA: Oxford University Press, 2013), 156.

353   Charles Cros's poems "To Mademoiselle Nelsy de S" and "Au Café" are available in the original French on PoemHunter.com: https://www.poemhunter.com /charles-cros/poems/.

359   Carl Haber's presentation at the Schenectady Museum of Innovation and Science can be viewed on YouTube: "The Earliest Voice Recording—Heard 1st Time Since 1878—Schenectady GE," October 27, 2017, https://youtu.be/NL9-PrmeG7g.

359–361   All six Volta Labs recordings can be found on YouTube, courtesy of the National Museum of American History: "Volta Labs Recordings, 1880–1885" playlist, https://www.youtube.com/playlist?list=PL6F59F72775B4EA64.

369   Enrico Caruso's arias can be heard on *Enrico Caruso: His First Recordings,* Everest, 2008.

371   Bessie Smith's version of "St. Louis Blues," written by W. H. Handy, can be found on *Bessie Smith: Queen of the Blues, Vol. 1,* TSP Records, JSP929C, 2006, 2012.

379   Django Reinhardt's version of "St. Louis Blues," written by W. C. Handy, can be found on *Django Reinhardt: Swingin' with Django, Vol. 4, 1937,* Naxos Jazz Legends, 2004.

381   Billie Holiday's version of "St. Louis Blues," written by W. C. Handy, can

be found on CD six in the ten-CD box set *Lady Day: The Complete Billie Holiday on Columbia 1933–1944,* Legacy Recordings, Sony Music 86979 30362, 2001.

394–395 The Empire 398G is advertised on page 141 of the *Allied Electronics for Everyone,* Catalog 220 (Chicago: Allied Radio Corp., 1963).

407 John Keats, "Ode to a Nightingale," *Lamia, Isabella, The Eve of St Agnes, and Other Poems* (1820).

438 My interpretation of Charles Wright's aesthetic comes from Stuart Friebert and David Young, "Charles Wright at Oberlin, *Field* 17 (Fall 1977), pp. 46–85, later reprinted as "At Oberlin College" in Charles Wright, *Halflife: Improvisations and Interviews, 1977–87* (Ann Arbor: University of Michigan Press, 1988), 59–88.

442 "The silvery alphabet of the sea" is taken from Charles Wright, "The Other Side of the River," *The Other Side of the River* (New York: Random House, 1984), 25.

443 Charles Wright, "California Dreaming," *The Other Side of the River* (New York: Random House, 1984), 70–73.

447 Son House, "My Black Mama," from *Raw Delta Blues,* Not Now Music NOT2LP138, 2011.

451 The Edward Hirsch book referenced here is *For the Sleepwalkers* (New York: Alfred A. Knopf, 1981).

460 All my translations of *horehorebushi* in this chapter are based on Romaji (Japanese language rendered in English lettering) versions included in the appendix of Odo's magisterial book *Voices from the Canefields: Folksongs from Japanese Immigrant Workers in Hawai'i* (Cambridge, MA: Oxford University Press, 2013).

465 Thomas Campion, "When to Her Lute Corinna Sings," from *A Book of Ayres* (1601).

476 The famous wartime novel is, of course, Michael Ondaatje's *The English Patient* (New York: Alfred A. Knopf, 1992).

477–478 The poems I wrote at Bellagio eventually became *Coral Road: Poems* (New York: Alfred A. Knopf, 2011).

479 The CD mentioned in the *Wall Street Journal* is Vivica, Genaux, *Arias for Farinelli,* Harmonia Mundi, HMC 90778, 2002.

483–487 The scenes described are from Act I of *La Bohème,* librettists Giuseppe Giacosa and Luigi Iliac.

Illustration Credits

437 Photograph by Holly Wright. Used by permission.

481 Wikimedia Commons

495 "Dust to Dust" by Alan Lau © 2021. Courtesy ArtXchange Gallery and Hongo Collection.

504 Calligraphy by Johana Yoda. Courtesy of Hongo Collection.

Garrett Hongo was born in Volcano, Hawai'i, and grew up on the North Shore of O'ahu and in Los Angeles. He is the author of three poetry collections, including *Coral Road* and *The River of Heaven,* for which he was a finalist for the Pulitzer Prize for Poetry; *The Mirror Diary: Selected Essays; Volcano: A Memoir of Hawai'i;* and two anthologies. He has been the recipient of several awards, including fellowships from the NEA and the Guggenheim Foundation. A regular contributor to *SoundStage! Ultra,* Hongo lives in Eugene, Oregon, and teaches at the University of Oregon, where he is Distinguished Professor in the College of Arts and Sciences.

A NOTE ON THE TYPE

This book was set in a version of the well-known Monotype face Bembo.
This letter was cut for the celebrated Venetian printer Aldus Manutius by
Francesco Griffo, and first used in Pietro Cardinal Bembo's *De Aetna* of 1495.

The companion italic is an adaptation of the chancery script type designed by
the calligrapher and printer Lodovico degli Arrighi.

Typeset by North Market Street Graphics, Lancaster, Pennsylvania
Printed and bound by Berryville Graphics Berryville, Virginia
Designed by Maria Carella